Nanotechnology for Wastewater Treatment

Edited by

Anjaneyulu Bendi
Innovation and Translational Research Hub (iTRH) &
Department of Chemistry
Presidency University
Bangalore, Karnataka
India

Nanotechnology for Wastewater Treatment

Editor: Anjaneyulu Bendi

ISBN (Online): 978-981-5322-82-8

ISBN (Print): 978-981-5322-83-5

ISBN (Paperback): 978-981-5322-84-2

© 2025, Bentham Books imprint.

Published by Bentham Science Publishers Pte. Ltd. Singapore. All Rights Reserved.

First published in 2025.

need for a court order if at any point you breach any terms of this License Agreement. In no event will any delay or failure by Bentham Science Publishers in enforcing your compliance with this License Agreement constitute a waiver of any of its rights.

3. You acknowledge that you have read this License Agreement, and agree to be bound by its terms and conditions. To the extent that any other terms and conditions presented on any website of Bentham Science Publishers conflict with, or are inconsistent with, the terms and conditions set out in this License Agreement, you acknowledge that the terms and conditions set out in this License Agreement shall prevail.

Bentham Science Publishers Pte. Ltd.
80 Robinson Road #02-00
Singapore 068898
Singapore
Email: subscriptions@benthamscience.net

BENTHAM SCIENCE

CONTENTS

FOREWORD

I am pleased to write the foreword for the book *"Nanotechnology for Wastewater Treatment"* edited by *Dr. Anjaneyulu Bendi*. This book marks a significant advancement in environmental science, focusing on the innovative application of nanotechnology to address critical wastewater treatment issues. As global environmental challenges intensify, especially concerning water pollution, the need for advanced technologies to provide practical solutions has never been more urgent. Nanotechnology emerges as a transformative force in this arena, promising to revolutionize the treatment and purification of wastewater through its cutting-edge approaches.

The chapters in the book are well-written to represent the most recent findings and advancements in nanomaterials and their applications. The book comprehensively examines numerous nanotechnological techniques to enhance wastewater treatment effectiveness and sustainability. Every chapter provides insightful analysis of practical elements, laying the groundwork for more in-depth study and creative thinking. The collective expertise of the authors and the editorial team has resulted in a work that is both enlightening and inspiring.

I am confident that *"Nanotechnology for Wastewater Treatment"* will be an invaluable resource for researchers, practitioners, and policymakers, fostering further exploration and contributing to a more sustainable future for our global water resources.

S. Ganapathy Venkatasubramanian
Department of Environmental Science & Law Expert Member
State Environment Impact Assessment Authority
Tamil Nadu
Ministry of Environment, Forest and Climate Change
Government of India

PREFACE

Is nanotechnology beneficial for wastewater treatment?

Nanotechnology offers significant benefits for wastewater treatment by providing innovative solutions to remove contaminants more effectively than traditional methods. Nanomaterials can adsorb, break down, and filter out pollutants at the molecular level due to their unique properties. Water treated and made safer by these substances can effectively target and eradicate microorganisms, heavy metals, and organic contaminants. Furthermore, nanotechnology can minimize operating costs, consume less energy, and improve the efficiency of current treatment methods. Wastewater treatment can be made more sustainable and effective by utilizing the enhanced properties of nanomaterials, thereby tackling critical environmental issues and enhancing public health.

Modern experimental research on various nanomaterials used in wastewater treatment is highlighted in this book. **Chapter 1** outlines the recent advancements in nanotechnology for wastewater treatment, highlighting the innovative approaches and techniques that have emerged. It explores how nanomaterials are being utilized to improve the efficiency and effectiveness of removing contaminants from wastewater. **Chapter 2** discusses the precise applications of zero-valent metal nanoparticles in wastewater treatment with particular emphasis on removing heavy metals and dye degradation, which poses a significant risk to human health and the environment.

Metal oxide nanoparticles play an essential role in wastewater treatment because they improve the removal of pollutants via catalytic, adsorption, and degradation processes. Using them in wastewater treatment is an essential advancement toward practical and sustainable water purification techniques, and hence, the significance of metal oxide nanoparticles has been thoroughly discussed in **Chapter 3**, and **Chapter 4** offers a thorough summary of the latest developments, difficulties, and potential uses of carbon nanotubes and their composites in wastewater treatment.

Chapter 5 highlights the use of graphene-based nanoparticles in wastewater treatment, focusing on their unique properties and effectiveness. This chapter explores their role in enhancing pollutant removal and advancing sustainable water purification methods, and **Chapter 6** focuses on the use of carbon-based nanomaterials, including fullerenes, carbon quantum dots, and nanodiamonds for dye degradation and heavy metal removal from wastewater.

In **Chapter 7**, the authors focus on exploring the chemistry of metal-oxide-based nanocomposites and their relevance for the eradication of heavy metal ions and organic dyes from polluted water. **Chapter 8** examines the properties of various nanosorbents and the sorption methods employed for the removal of pollutants from wastewater.

Information on upcoming aspects of the nanofiltration process is provided in **Chapter 9**, which is anticipated to address present and future problems with water treatment and open the door to cleaner and safer water supplies globally, and **Chapter 10** aims to shed light on the extensive research on the creation and use of both natural and manufactured zeolites in the effective removal of diverse pollutants from wastewater.

In conclusion, this book aims to provide a comprehensive overview of the innovative research efforts in the design and synthesis of novel nanomaterials for wastewater treatment. I hope that the content will captivate the interest of aspiring researchers worldwide and inspire them to continue exploring novel nanomaterials for treating various types of wastewater. Additionally, I hope that this book will serve as a valuable resource for researchers, academics, and industry professionals with a strong interest in this field.

I would like to express my profound admiration and gratitude for the dedicated efforts of all the authors who have tirelessly prepared the content of this book. Additionally, I extend my heartfelt thanks to the publisher and its staff for their efficient management of the project at every stage. Finally, I wish to acknowledge my family members for their unwavering support throughout this entire journey.

Anjaneyulu Bendi

Innovation and Translational Research Hub (iTRH) &
Department of Chemistry
Presidency University
Bangalore, Karnataka
India

List of Contributors

Allu Udayasri Department of Chemistry, Aditya Institute of Technology and Management, Tekkali-532201, Andhra Pradesh, India

Anirudh Singh Bhathiwal Department of Chemistry, MMV, Banaras Hindu University, Varanasi, UP, India

Anjaneyulu Bendi Innovation and Translational Research Hub (iTRH) & Department of Chemistry, Presidency University, Bangalore, Karnataka, India

Aditi Tiwari Intertek India, Gurugram, Haryana-122001, India

Akanshya Mishra School of Applied Sciences (Chemistry), KIIT Deemed to be University, Bhubaneswar- 751024, India

Arpit Sand Department of Sciences, School of Sciences, Manav Rachna University, Faridabad, Haryana-121003, India

A. Jayamani Department of Sciences, School of Sciences, Manav Rachna University, Faridabad, Haryana, India

Chinmay Mittal Department of Chemistry, SGT University, Gurugram-122001, Haryana, India

Chanchal Vashisth Department of Chemistry, Kurukshetra University, Kurukshetra-136119, Haryana, India

G. B. Dharma Rao Department of Chemistry, Kommuri Pratap Reddy Institute of Technology, Hyderabad-500088, Telangana, India

Harsha Devnani Department of Sciences, School of Sciences, Manav Rachna University, Faridabad, Haryana, India

Jaya Tuteja School of Basic Sciences, Galgotias University, Greater Noida, India

Jasaswini Tripathy School of Applied Sciences (Chemistry), KIIT Deemed to be University, Bhubaneswar- 751024, India

Lakhwinder Singh IIMT University, Meerut-250001, Uttar Pradesh, India

M. Radha Sirija Department of Chemistry, Vardhaman College of Engineering, Hyderabad-501218, India

Minakshi Department of Chemistry, Indira Gandhi University, Meerpur, Rewari -122502, Haryana, India

Neera Raghav Department of Chemistry, Kurukshetra University, Kurukshetra-136119, Haryana, India

Prabhjot Kaur Department of Chemistry, Kurukshetra University, Kurukshetra-136119, Haryana, India

Ravi Kumar Rana Department of Chemistry, Baba Masthnath University, Rohtak-124001, Haryana, India

Rashmi Pundeer Department of Chemistry, Indira Gandhi University, Meerpur, Rewari -122502, Haryana, India

Rajni Department of Chemistry, Faculty of Basic and Applied Sciences, SGT University, Gurugram-122505, Haryana, India

Roopa Rani	Department of Sciences, School of Sciences, Manav Rachna University, Faridabad, Haryana-121003, India
Shubhika Goyal	Department of Chemistry, SGT University, Gurugram-122001, Haryana, India
Taruna	Department of Chemistry, Faculty of Basic and Applied Sciences, SGT University, Gurugram-122505, Haryana, India
Versha	Department of Chemistry, Baba Masthnath University, Rohtak-124001, Haryana, India
Vishaka Chauhan	Department of Chemistry, SGT University, Gurugram-122001, Haryana, India

CHAPTER 1

Role of Nanotechnology in Wastewater Treatment

Versha[1] and **Ravi Kumar Rana**[1,*]

[1] *Department of Chemistry, Baba Masthnath University, Rohtak-124001, Haryana, India*

Abstract: Earth's water is a valuable and finite resource, continually recycled through the water cycle. Wastewater is defined as water that has been contaminated with organic and inorganic elements, heavy metals, pathogens, or other toxins that have changed its chemical, physical, or biological properties and made it dangerous for the ecosystem. It is impossible to overestimate the importance of technology innovation in enabling integrated water management. With the safe use of non-traditional water sources, nanotechnology has the potential to significantly increase the efficiency of water and wastewater treatment while also increasing water supply. Several types of nanomaterials are utilized for wastewater treatment, depending on the kind of contaminants and the required level of treatment efficiency. Novel and emerging nanomaterials are emerging together with the field of nanomaterial development. This chapter outlines the recent advancements in nanotechnology for wastewater treatment.

Keywords: Nanotechnology, Nanomaterial, Photocatalysis, Reverse osmosis, Wastewater treatment, Zeolites.

INTRODUCTION

While several materials are required for life's survival and development, water is the most fundamental. Earth is called a "blue planet" because water surrounds more than 70 percentage of its facets. About 97.5 percentage of water is thought to be saline, with the remaining 2.5% being considered pure water, of which 68.9% is found in the form of glaciers, ice, and permanent snow [1 - 7]. In addition, only 0.3% of pure water is readily accessible, with ground water making up 30.8% of the total [8 - 12]. Scientists are still looking into many new innovations to improve inexpensive ways of water filtration [13, 14]. Recently, the field of developing nanotechnology has offered the possibility of removing contaminants from water at a reduced cost, with a high removal effectiveness of impurities and the capacity to be reused [15]. Because of their distinctive active surface area, nanotechnologies can be very significant. Nanomaterials, on the other hand, can be used for extensive purposes, including catalytic membranes,

* **Corresponding author Ravi Kumar Rana:** Department of Chemistry, Baba Masthnath University, Rohtak-124001, Haryana, India; E-mail: hit.ravirana@gmail.com

Anjaneyulu Bendi (Ed.)

nano sorbents, bioactive nanoparticles, and metal nanoparticles like titanium oxides, iron, and silver [16]. Nanomaterials usually have a size that varies from 1-100nm. Because of their tiny size, they have significantly less atoms than bulk materials, which gives them a very wide range of properties. Because of their tiny size, nanomaterials have an extraordinarily large ratio of surface area in relation to volume, leading to increased surface-relying properties. It has been put in place that due to their dominant physicochemical characteristics, nanoparticles are effectively penetrable for a diversity of water adulterants [17 - 22].

Measurements, including the capacity for adsorption, elimination proportion, disintegration rate, and the rate of reaction, are used to gauge how well nanomaterials eliminate contaminants. Important measurements of performance consist of:

The Elimination Effectiveness

The proportion of treated samples with a lower pollutant concentration.

The Regeneration and the Possibility of Reuse

Cyclic testing evaluates how well a nanomaterial continues to function. The following are several applications.

A Catalytic Activity

The rate at which pollutants degrade under particular circumstances (such as UV exposure) is frequently used to gauge the effectiveness of photocatalytic devices.

The Adsorption Capacity

It is the quantity of the pollutant consumed, which is usually expressed in milligrams per gram of nanomaterial. These data are also impacted by surroundings like pH level, temperature, and contaminant content [23].

Technological developments in nanomaterials have revolutionized water treatment methods, making it possible to safely use unconventional water sources like polluted water, brine water, and saltwater. By enhancing the filtration process, contaminant deterioration, bacterial oversight, and tracking, these materials improve purification and provide more effective, economical, and environmentally friendly options. A summary of their efforts is provided below:

Enhanced Pollutants Elimination for Reusing Wastewater

The elimination of toxic metals, organic contaminants, and medication leftovers is enhanced by the improved process of oxidation and catalytic qualities of nanomaterials.

Control of pathogens and the infection: Excellent antibacterial qualities of nanomaterials provide secure drinking water by removing microorganisms from unconventional sources.

Better Brackish Water and Seawater Desalination

By enhancing the membrane efficiency, decreasing energy usage, and minimizing dirt, nanoparticles improve salinity.

Thin-film Nanocomposite (TFN)

The permeability and rejection of salts are increased when nanoparticles, such as carbon nanotubes and graphene oxide, are embedded in membranes.

Photothermal Nanomaterials

By successfully converting daylight into energy that evaporates liquid, materials such as carbon-based nanocomposites allow solar desalination.

Nanocoating that Prevents Fouling

The longevity of membranes is prolonged by anti-fouling nanocoating, which is composed of titanium dioxide (TiO_2) or silver nanoparticles and limits the growth of bacteria and scalability.

Energy-Saving Water Treatment Methods

Desalination and filtration are made environmentally friendly by the use of nanomaterials in the creation of low-energy methods of treatment. By promoting the electro-sorption of salts, carbon nanomaterials improve the effectiveness of CDI and provide a more environmentally friendly substitute for the process of reverse osmosis. Uninterrupted desalination using sunlight is made possible by photothermal nanomaterials, which speed up the evaporation of water. By improving the heating conductivity and decreasing pore wetness in membranes, nanoparticles increase the energy efficiency of distillation [24].

In order to highlight the possible use of these approaches to address various issues faced by the current wastewater treatment technologies, this chapter primarily examines some recent breakthroughs and utilization of nanotechnology in tainted water treatment. The latest findings on a range of nanomaterials, including zero-valent metal-nanoparticles, metal-oxides nanoparticles, carbon-nanotubes, graphene-based nanoparticles, carbonaceous nanoparticles, nanocomposites, nano absorbents, nano filters, and zeolites, that are employed in these methods are also discussed.

APPLICATION OF NANOTECHNOLOGY IN WASTEWATER TREATMENT

Conventional Methods for Waste Water Treatment and their Limitations

The process of treating wastewater involves eliminating contaminants or impurities using chemical or physical techniques before releasing the mixture back into the environment. Wastewater treatment attempts to eliminate adulterated and toxins less than the uppermost permitted limit while also recovering micronutrients and water to reduce risks to living souls and environmental salubrity [25, 26]. Primary, secondary, or tertiary treatment are the three basic phases of the contemporary wastewater treatment system. The degree of pollution removal and the methods used to remove pollutants determine how many steps are involved. The goal of primary treatment is to rid wastewater of both organic and inorganic particles. During the 2nd method of purification, fine dangling particles, dispersion solids and diffused organics are eliminated *via* assimilation, biotransformation, and volatilization into the residue. The goal of 3rd more complex treatment is to improve the grade of water released into natural waters by employing a diversity of physical, biological, and chemical-based techniques. The level of water purity might range from ninety-five percent to ninety-eight percent, depending on the techniques employed.

Coagulation

In order to remove turbidity particles and naturally occurring organic matter from wastewater, two crucial physicochemical processes are used: flocculation and coagulation. Masaoki Kimura *et al.* [27] state that the most popular coagulants are iron salts and hydrolytic aluminum. For $Al(OH)_3$, a pH of 4.5 is ideal, while for $Fe(OH)_3$, a pH of 8 is ideal.

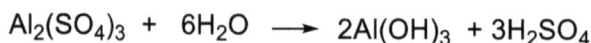

$$Al_2(SO_4)_3 + 6H_2O \longrightarrow 2Al(OH)_3 + 3H_2SO_4$$

$$FeCl_3 + 6H_2O \longrightarrow Fe(OH)_3 + 3HCl$$

The creation of significant amounts of chemical sludge is the primary drawback of these procedures. Moreover, coagulants based on aluminum also increase the concentration of left over aluminum in contamination-free water. Such left-over aluminum is related to various difficulties, including rising turbidity, less disinfection effectiveness, reduced hydraulic capacity, and major side effects that include Alzheimer's disease. The remaining aluminum levels can be governed by lowering the potential of hydrogen value to moderately acidic levels. Sadly, this method is not popularly employed since it adds to the process's cost and demands a pH increase after treatment in order to avoid corrosion in the distribution systems for water. Novel flocculants, such as polyaluminum chloride (PACl), which are organic polymers derived from nature or artificial sources, to replace traditional aluminous coagulants have drawn the curiosity of numerous advanced research in past years, as proposed by Weiying Xu *et al.* [28].

Flocculation

The process by which flocs link to one another and then clump together into larger agglomerates is known as flocculation. This is caused by polymers binding the particles together. Filtration or flotation can subsequently be used to get rid of these agglomerates. Fenglian Fu and Qi Wang *et al.* [29] proposed that there are several materials that can be utilized as flocculants, including polyacrylamide (PAM), polyferricsulfate (PFS), PAC, and PAV.

A few flocculants, including Konjac-graft-poly (acrylamide)-co-sodium xanthate and mercaptoacetyl chitosan, have been demonstrated to be significant in removing ions of heavy metals from wastewater, even with the help of some turbidity, according to Qing Chang *et al.* [30] and Jiacai Duan *et al.* [31].

The characteristics of particles vary; hence, a universal flocculent cannot be used. Consequently, flocculent can be categorized into multiple groups:

- Amphoteric: They are made up of proteins with both cationic and anionic groups.
- Cationic: They include groups $-NH_2$ and $=NH$. Due to their lower cost compared to cationic species, anionic flocculants constitute a large portion of synthetic flocculants available today.
- Anionic: These comprise groups $-COOH-SO_3H$.
- Non-ionic: These comprise COOH and $-OH$ (Natural polymers include alginates, gums, glues, and starch);
- Large amounts of sludge are frequently formed when inorganic flocculants are used; in contrast, natural polymers are more efficient and biodegradable.

Adsorption

Adsorption is acknowledged as a practical and cost-effective technique toremove contaminants from wastewater. According to Mojdeh Owald *et al.* [32], the number of molecules present on a receptive surface gives the procedure its unique character. Adsorption has important benefits that include economical, wide accessibility and financial success, design and operation adaptability, and reversibility of the process, especially when viewed from the perspectives of the environment and economy. One of the most popular sorbents is activated carbon, which is typically used to remove organic pollutants due to wastewater's large surface area and huge micropore and mesopore volumes. Based on how it is made, four types of activated carbon may be identified: powdered carbon, granular activated carbon, cloth-activated carbon, and fibre-activated carbon. Each and every type has a distinct use. Much research has looked into the cancellation of organic pollutants and transition metals utilizing activated carbon that is produced starting with a variety of basic sources. It was discovered that altering or treating the carbon before application is necessary to boost the effectiveness of metal ion removal.

According to W.S. Wan Ngah and M.A.K.M. Hanafiah [33], activated carbon is not a proper sorbent for sophisticated wastewater processing procedures because of its high cost. Consequently, there exists plenty of opportunity for the production of inexpensive, easily accessible, and economically cost-effective sorbents derived from natural materials or, we can say, particularly from waste products of commercial or farming operations. The heavy metals from wastewater are eliminated by a variety of inexpensive sorbents, including rose hulls, sawdust, weeds, alfalfa biomass, and chitosan.

Precipitation

In the course of precipitation processes, chemicals and transition metal ions mix to make insoluble precipitates that are then removed from the treated water by sedimentation or filtration. Charemtanyarak *et al.* [34] stated that metals could be taken out of wastewater *via* sulfide precipitation or hydroxide precipitation at high pH levels. Radioactive elements, phosphorus compounds, and metal ions are often removed by precipitation.

Hydroxide treatment is the most often used precipitation method because of its automated pH control, affordability, and relative simplicity. Precipitants include substances like calcium hydroxide and sodium hydroxide. The process of removing heavy metals by precipitation of chemicals can be shown using the following equation.

$$M^{2+} + 2(OH^-) \longrightarrow M(OH)_2$$

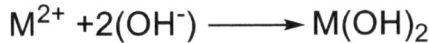

A huge volume of very low-density silt is formed during hydroxide precipitation, which poses challenges for disposal and dewatering. This is the principal drawback of the phenomenon.

It has been shown that sulfur dioxide precipitation is a viable substitute for hydroxide precipitation. The primary benefits are the potential for selective metal recovery and strong metal removal even at reduced pH. Additionally, compared to similar metal hydroxide waste, metal sulfide slurry has a superior performance in drying and thickening. As stated by Robert W. Peters and Linda Shem, the technique has limitations since it produces harmful hydrogen sulfide vapors and sulfide colloidal residues. Precipitation and coagulation are sometimes used together [35].

Ion Exchange

Ion exchange is among the processes that are most often utilized popularly to extract heavy metals *via* waste water. Berta Galan *et al.* [36] state that the primary advantages of exchanging ions technique are its ability to recover metal, its increased selectivity, and its reduced sludge quantities. The basic principle of the process, according to Kurniawan *et al.* [37], is the chemically identical inter-change of ions among the solid resin phases and liquid electrolytic solution, all in the absence of compromising the resin's structural stability. Groups of the most popular cation exchangers are as follows:

- Anionites containing amino groups that are weakly basic.
- Strong basic anionites with NH_2 groups.
- Slightly acidic (resins) containing carboxylic acid groups;
- Extremely acidic (resins) containing sulfonic acid groups.

Many varieties of commonly used ion-exchange resins are both weak and powerful in adsorption procedures. For instance, in Lewatit MP-64, chromium was eliminated using an anion exchange resin of the weakly basic macroporous type. Juang *et al.* [38] demonstrated the efficacy of using Amberlite IR120 resin to effectively absorb nickel, manganese, and cobalt from complex synthetic solutions. Natural minerals, in particular, zeolites, are extensively employed to extract transition metals fromaqueous solutions in addition to manufactured resins due to their economical and easy availability. Natural minerals, particularly zeolites, are widely available and reasonably priced; hence, they are frequently utilized in addition to manufactured resins for the removal of transition metals from aqueous solutions.

The traditional procedures of coagulation, precipitation, and adsorption are employed to reduce the elevated concentrations of different chemical compounds and metal ions to levels that are permissible by law. When a pollutant is present in low concentrations, membrane technology works better.

Membrane Filtration

Since membrane filtration may be used to remove contaminants from various sources, it has attracted a lot of attention lately. An industrial process currently in operation may see cost and energy usage reductions through the application of membrane technology. H. Bessbousse *et al.* (2008) [39] identified three membrane processes that are currently in use: reverse osmosis, nanofiltration, and ultrafiltration.

Reverse osmosis (RO): Water moves through a membrane during the pressure-driven membrane process known as reverse osmosis, but the polluting metal ions are held in place. When it comes to eliminating metal ions from inorganic solutions, RO is more effective. Moreover, the technique works well over a broad pH range of 3 to 11 and a pressure range of 4.5 to 15. In order to push the water to go through the semi-permeable membranes in RO, high-pressure pumps are also needed. This results in the reject stream holding between 95% and 100% of the dissolved salts. The amount of salt in the water directly relates to the pressure that is needed. The advantages of this technology are its outstanding efficiency and inexpensive cost.

Ultra Filtration (UF): Ultrafiltration is a method of removing suspended particles, transition metals, and macro-molecules from solution. It uses a pore-equipped porous membrane with diameters varying from five to twenty nanometres and a molecular weight of separating substances ranging from 1000-100,000 Da. High-mass molecules are retained by gradient concentration or pressure, whereas low-mass molecules flow through the membrane. The purification of protein solutions is the primary use of this technology in the industry. Ultrafiltration membrane operation can be enhanced by using a complexing polymer that is soluble in water and interacts with the cations, which should be taken off and create large-sized complexes. The major advantages of ultrafiltration methods are the prevention of chemical usage and the outstanding standard of the finished product with ninety to hundred percent of pathogen removal. In spite of this, the method is constrained by the membrane's expensive cost.

Electrochemical Treatment

To fulfill the requirements of numerous businesses, including the purification of various kinds of wastewater, other physical-chemical technologies are in competition with electrochemical approaches. The primary benefit of using electrochemical techniques for metal recovery treatments is their ability to utilize clean electrons as reagents. The method's basic idea is to plate out metal ions on a cathode surface and then recover them in their elemental form. The low environmental effect, ease of use and handling, and absence of hazardous or poisonous residues are some of the many advantages of electrochemical processes (electrocoagulation, electro flotation, electrooxidation, and electrodeposition). Ignasi Sirés and Enric Brillas [40] proposed that despite the many benefits of electrochemical technologies, their high electricity consumption and large capital investment requirements make them unsuitable for widespread use.

NANOTECHNOLOGIES FOR WASTEWATER TREATMENT

Many different forms of nanomaterials have been characterized as having possible applications in wastewater treatment. These include carbon-based nanomaterials, metal nanoparticles, polymeric nanoparticles (NPs), zeolite, self-assembled monolayers on mesoporous supports (SAMMS), biopolymers, and many more [41]. Nanoscience is the study of materials at the nanoscale that exhibit exceptional properties, functions, and phenomena due to their small size. Nanotechnology is the practical application of this science. The basis of nanotechnology is the use of atoms and molecules to make materials, structures, components, electronics, and systems at the nanoscale. The development of new instruments and methods made possible by nanotechnology in recent years, particularly in the subject of filtration of water, presents a viable new option for the more economical and efficient treatment of waste water. Photocatalysis and nanofiltration are two of the many possible water treatment techniques and tools that nanotechnology has brought forth [42].

Nanofiltration

In wastewater treatment and water purification, filtration is the most popular and crucial process, which uses a membrane or filter medium to separate the liquid and solid components. Different particle sizes and types are filtered out using various membrane-based filtration techniques. Because of its special charged-based repulsion ability and huge rate of penetration, the membrane separation process driven by pressure, known as nanofiltration, is rapidly growing in wastewater treatment and purification of water in industries. Reverse osmosis (RO) processes demand higher pressure requirements (20–100 ATM), but NF requires less; hence, it is becoming more and more popular as a lower-energy technology [43]. When it comes to treating wastewater, NF is still relatively new, but a number of industries, including textile, pharmaceutical, dairy, and petrochemical, are showing great interest in this technology [44]. Due to its distinct filtration mechanism and the wide range of membrane types that are available, NF can effectively filter out nearly all inorganic and organic pollutants, including a number of dangerous bacteria, from wastewater [45]. The majority of NF membranes are composed of synthetic polymers because they are simple to manufacture, have great flexibility, and are low in cost. Nevertheless, due to their reduced chemical resistance and tendency to foul quickly, polymeric membranes have restricted durability. In contrast, membranes composed of inorganic ceramics are less flexible and have a higher manufacturing cost. However, they also have a longer lifespan and great chemical and thermal resistance. As a result of their ability to be made flexible and synthetically inexpensively, freshly discovered nanomaterials may be crucial in the development of NF membranes. Below is a discussion of some of the most promising nanomaterials and how they are being used in wastewater treatment procedures.

Zero-Valent Metal Nanoparticles

Nano zero-valent iron: Nano zero-valent iron is a revolutionary nano-material that is being developed to eradicate various organic and inorganic impunity from water. For nanoremediation, iron nanoparticles are a very helpful component. Fe (II) and Fe (III) are reduced using borohydride to produce iron at the nanoscale. The diameter of the zero-valent iron at the nanoscale particles ranges from 10 to 100 nm. Its structure is a conventional core shell. The mixed valent oxide shell, which includes Fe (II) and Fe (III) iron, is created when the metallic iron oxidizes, whereas the core is mainly composed of zero-valent or metallic iron. Due to its huge surface area, greater number of reactive sites compared to micronized particles, and dual adsorption and reduction capabilities, nanoscale zerovalent iron is typically chosen for nanoremediation [46]. The most extensively studied environmental nanotechnology application to date has been the cleaning of groundwater using nanoscale zerovalent iron (nZVI). A wide range of common pollutants, including trihalomethanes, chlorinated ethenes, chlorinated benzenes, various polychlorinated hydrocarbons, herbicides, and dyes, have been shown to be effectively destroyed by nanoscale metallic iron [47]. Zerovalent iron corrodes in the environment, which is the source of the reaction:

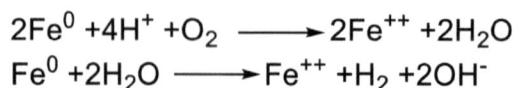

$$2Fe^0 + 4H^+ + O_2 \longrightarrow 2Fe^{++} + 2H_2O$$
$$Fe^0 + 2H_2O \longrightarrow Fe^{++} + H_2 + 2OH^-$$

Inorganic anions such as chromate, perchlorate, selenate, arsenate, arenite, and nitrate have also been found to be reduced by nZVI in addition to organic pollutants. nZVI has a much higher sorption capacity, and its response speed is many times faster than that of regular granular iron. Metals that are dissolved in solution, such as Pb and Ni, can also be removed with ZVI. Either zerovalent or lower oxidation states are achieved for the metals. Sodium borohydride is typically employed as the main reductant in the preparation of nZVI. For instance, a 1:1 volume ratio solution of $FeCl_3.6H_2O$ (0.05 M) is mixed with 0.2 M $NaBH_4$. The following process occurs when the borohydride reduces ferric iron [48]:

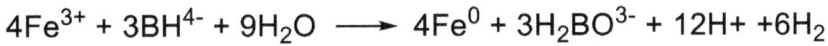

$$4Fe^{3+} + 3BH^{4-} + 9H_2O \longrightarrow 4Fe^0 + 3H_2BO^{3-} + 12H+ +6H_2$$

The permeable reactive barrier (PRB) technology is a new approach to groundwater remediation that uses reactive materials to cure contaminated groundwater *in situ* physically, chemically, or biologically. Reactive barriers made of granular ZVI have been utilized for many years to remediate both inorganic and organic pollutants in groundwater at various locations throughout the world [49]. With this technology, nZVI has become a more appealing candidate in recent years. Without using water excavation, the contaminants are treated by placing the reactive materials in subterranean trenches downstream of the polluted zone and forcing the water to flow through them. In comparison to alternative techniques, this affordable removal methodology generally results in less environmental damage.

Metal Oxides Nanoparticles

Another reasonably priced option for creating NF membranes is metal oxides. Furthermore, under the influence of light, the majority of metal oxides exhibit photocatalytic activity, which assists in the deterioration of various organic and inorganic impurities, such as many harmful bacteria present in water, turning the membranes into reactive materials as opposed to merely physical barriers [50].

Among the numerous ions and metals present in the wastewater taken out of oil refineries are Ca^{2+} and Cu^{2+}. By magnetizing and carboxylting GO to get rid of these metals and ions, Lei He *et al.* developed recyclable nano-adsorbents established on $Fe_3O_4/GO-COOH$. The nano-adsorbents absorb 78.4 percent & 51 percent of the Ca^{2+} and Cu^{2+} after sixty minutes of treatment, respectively. The nano adsorbent maintains good rates of recovery (82.1 percent for Ca^{2+} ions and 91.8 percent for Cu^2ions) and elimination percentages (72.3% for Ca^{2+} and 49.33% for Cu^{2+}) after 5 adsorption-desorption cycles [51].

Marcos E. Peralta *et al.* synthesized magnetic nano adsorbents utilizing silica and assessed their efficacy in removing organic pollutants, including aliphatic hydrocarbons (AHCs), polyaromatic hydrocarbons (PAHCs), and contaminants of emergent concern (CECs). A combination shell made of silica and 3-(trimethoxysilyl)propyl-octadecyldimethyl-ammonium chloride was employed to create magnetic iron oxide nanoparticles. This shell served as a structure-directing agent. Trimethoxyphenylsilane was utilized to further alter the magnetic mesostructured silica nanoparticles (MMSSNPs-ph) that were created in order to construct nano adsorbents functionalized with phenyl. Following the evaluation and characterization of both materials for batch sorption tests using both single and mixed contaminants, the investigation showed that MMSSNPs-ph is more effective in adsorbing AHCs and PAHCs. The mesostructured silica scaffolds' phenyl and 3-TPODAC moieties were discovered to be crucial for achieving high PAHC absorption from aqueous conditions. When MMSSNPs-ph were evaluated against CECs such as carbamazepine, diclofenac, and ibuprofen, they displayed significant uptakes of both medications even though MMSSNPs exhibited better CEC adsorption capacities [52].

A membrane with TiO_2 nanowire mesh was developed by Xiwang Zhang *et al.* and used to filter wastewater for total organic carbon and humic acid. Under ultraviolet light irradiation, the researcher declared almost complete elimination of humic acid as well as greater than ninety percent removal of TOC with the help of combining the filtering and photocatalytic properties of the TiO_2-based membrane. The NF membranes also show a notable organic dye retention rate when combined with TiO_2 and γ -alumina [53].

Carbon Nanotubes

Because of their large targeted surface area, excellent chemical and thermal stability, and capacity for surface functionalization, another material that is getting a great deal of attention as an improved adsorbent is one-dimensional carbon nanotubes. In particular, the reactivity of CNTs can be precisely controlled. These CNTs' superior performance as adsorbents over other adsorbents has been attributed to their highly porous structure, tuneable surface chemistry, high specific surface area, low mass density, strong interaction with the contaminants, and chemically inert nature. These attributes render them highly suitable for application in the treatment of wastewater [54, 55].

The rising population and increasing rates of illness have resulted in a significant increase in the need for pharmaceutical items. Since antibiotics are among the medicinal products that are most commonly used, their release into the natural ecosystem has also grown. When people or other living things consume these products, major health problems arise. I. Kariim *et al.* synthesized MWCNTs using wood sawdust-derived activated carbon and loaded it with nickel-ferrite to facilitate the sorption of levofloxacin and metronidazole from pharmaceutical effluent. The surface area of Ni-Fe-supported activated carbon CNTs and pure activated carbon CNTs was 650.45 m2/g and 840.38 m2/g, respectively, based on a surface morphological analysis. The adsorption process results showed that the produced MWCNTs had a good adsorption capacity for both levofloxacin & metronidazole [56].

When Chao-Yin Kuo *et al.* used CNTs to study the adsorption of organic dyes from water, they found that the adsorption of colors on the CNTs' surface was caused by a physisorption process. Carbon nanotubes and activated carbon are suitable for use in water treatment procedures due to their high adsorption rates and capacities, along with their chemical and thermal stability. However, because of their small sizes, it is challenging to completely separate the powdered activated carbon and CNTs from the water. In order to solve this problem, it was discovered that integrating magnetized nanoparticles with activated carbon and CNTs was very successful since magnetic separation techniques make it simple to remove these nanosized composite absorbents from the aquatic phase [57].

Graphene-based Nanoparticles

Graphene is a 2-dim material consisting of hexagonally arranged sp2 hybridized carbon atoms. At room temperature, it displays ambipolar electric field effects, quantum hall effects, and classical thermal conductivity. Its high surface area and porosity also make it a great option for adsorbing various gases, including carbon dioxide, methane, and hydrogen. Because of its huge delocalized π-electron systems, high surface area, improved active sites, and better chemical stability, it is suitable for use as an adsorbent for wastewater treatment. The number and stacking of layers can be changed to modify the properties of this substance [58]. The single monomolecular layer of graphite known as graphene oxide (GO) contains groups that may hold oxygen, including epoxide, carbonyl, carboxyl, and hydroxyl groups. These days, GO and magnetic particles have drawn a lot of interest as an adsorbent for waste-water treatment due to their straightforward construction, reduced sensitivity, affordability, and convenience of usage with regard to harmful contaminants [59].

According to Hang Chen *et al.*, malachite green dye adsorption in wastewater was intended to be achieved by creating an aerogel of GO/aminated lignin (GO/AL-AG). The adhesion capacity was examined at various aerogel dosages, pH levels, contact times, and reaction temperatures. Experimental results showed that, under ideal conditions, pre-pared GO/AL-AG had an ultimate adsorption ability and efficacy of 113.5 mg/g and 91.72%. The combination of the amine of aminated lignin and the carboxyl group on the surface of GO considerably improved the capacity for adsorption of GO/AL-AG when compared to other AG in the experiment. After five adsorption-desorption cycles, the GO/AL-AG's adsorption effectiveness was almost 90% [60].

Alvin Lim Teik Zheng and colleagues created a GO hydrogel coated with silver nanoparticles (Ag NPs), which they then integrated with a porphyrin complex in order to remediate wastewater by adsorbing dyes. When the capacity for adsorption of the hydrogels was evaluated in relation to different porphyrin complexes, it was found that the hydrogel modified with tetraphenylporphyrin had the highest adsorption capacity (130.37 mg/g) for MB [61].

Sirajudheen *et al.* created a hydro composite comprising GO surrounded by a biopolymer, chitosan (CS) (GO/CS-HCP), to effectively remove organic pollutants found in wastewater. Congo red dyes in wastewater had the maximum capacity for adsorption of 43.06 mg/g for the manufactured GO/CS-HCP, which was followed by acid red 1 at 41.32 mg/g and reactive red 2 at 40.03 mg/g. At pH 2, there was enhanced adsorption, and as pH rose, the removal effectiveness dropped. In the 0.1 M NaOH solution, the prepared GO/CS-HCP demonstrated perfect desorption with a capacity for regeneration of over 65% [62].

Carbonaceous Nanoparticles

The absorption of many organic and inorganic pollutants' in water has been widely accomplished by the use of carbon nanoparticles. Excellent chemical, mechanical, and electrical qualities are typically exhibited by carbonaceous nanomaterials. They combine exceptional physical-chemical properties at the nanoscale with the distinctive features of sp2 hybridized carbon bonds. They have garnered a lot of interest in a variety of applications, such as medication delivery, environmental cleanup, supercapacitors, environmental transformation, and catalysis. The rapidly advancing field of nanotechnology means that carbonaceous nanomaterials will play ever-more-important roles in future living. Because of its great resilience to attrition losses, high thermal stability, low cost, and strong adsorption capacity, the most common type of activated carbon that is widely used among these nanomaterials is carbon. Both odorous pollutants and different organic impurities were eliminated by utilizing granular activated carbon to remove water [63].

Asenjo *et al.* investigated industrial wastewater's benzene and toluene adsorption on activated carbon and found that both had a great adsorption capacity of approximately 400–500 milligrams/gram for benzene and approximately 700 mg/g for toluene. Activated carbon has also been shown to be helpful for the removal of heavy metal ions such as Hg(II), Ni(II), Co(II), Cd(II), Cu(II), Pb(II), Cr(III), and Cr(VI) [64].

Nanocomposites

Nanocomposites are solid multi-phase materials with a number of physically distinct components, one or more dimensions less than 100 nm, and a visible interface among them. This description might only apply to materials that are at the reinforcing stage for structural purposes, such as fibers or components that are supported by a matrix or binder phase. Since these materials can mix different qualities to meet specific needs, they have a wide range of uses. Nanocomposites represent an entirely new class of materials, serving as substitutes for traditionally filled polymers. There are nanoscale inorganic fillers in this recently developed family of materials (in at least one dimension). The polymer properties exhibit significant enhancements when juxtaposed with their bulk-size counterparts due to their elevated surface area, elevated percentage of atoms on the surface, and diverse linking interactions [65].

According to Tsu-Wei Chou *et al.*, some composite materials have demonstrated strength differences of up to 1000 times when compared to their respective material constituents. This is because of improved properties such as high heat resistance, chemical resistance, and tensile strength [66]. The most popular class of multifunctional polymer systems throughout the last 10 years has been polymer composite forms. These classes of materials have drawn a lot of interest from researchers, marketers, and manufacturers all over the world. Composites are already widely employed in many different industries, including manufacturing, transportation, electronics, and consumer goods. They have shown an amazing weight, strength, and stiffness mixture that is difficult to achieve separately in the component parts. Due to their nanometre dispersions, nanocomposites have significantly better characteristics than pure polymers and traditional composites.

Types of Nanocomposites

Ceramic Matrix Nanocomposites

Composites in the ceramic matrix are made of ceramic fibers placed into ceramic matrices. A matrix and fibers, especially carbon fibers, are the components of all ceramic materials. Often, metal is used as the secondary component in ceramics, which accounts for almost all of the volume. These oxides include borides, nitrides, and silicates. It is necessary to finely disperse the tribological, resistance, corrosion, and other protective qualities of both components in order to provide additional optical, magnetic, and electrical properties [67].

Metal Matrix Nanocomposites

Another name for metal matrix nanocomposites is composites of reinforced metal matrix. Both continuous and non-continuous materials can be used to describe these composite structures. A popular type of nanocomposites is the carbon nanotube-metal matrix composite. Dawid Janas and Barbara Liszka synthesized a novel material called CNT-MMC, which is being developed to take advantage of the exceptional tensile strength and electrical conductivity of carbon nanotubes [68]. The commercially feasible production of synthetic technology will guarantee the homogeneous dispersion of nanotubes within a metal matrix. As a result, there is great interfacial adhesion between the carbon nanotubes and the metal matrix, which is required for a CNT-MMC to have the best properties in those regions.

Polymer Matrix Nanocomposite

When applied to a polymer matrix appropriately, using the properties of a nanoscale filler will also greatly boost the efficiency of nanoparticles. This technique works particularly well for producing high-performance composites when the filler at the nanoscale has properties that are noticeably different or better than those of the matrix and the filler dispersion is carried out uniformly. In all nanocomposites, phase separation is guided by thermodynamics to maintain distribution homogeneity [69].

Nano Sorbents

The process of a substance, called sorbent, adhering to another substance, called substrate, through chemical or physical interactions is known as sorption. Sorbents are frequently employed as separation media in the purification and treatment of contaminated water in order to remove both organic and inorganic impurities. Three steps typically comprise the sorption process of contaminants in water on the sorbent surface: (i) the pollutant is carried through the water and onto the sorbent surface; (ii) adsorption occurs at the sorbent surface; and (iii) the pollutant is borne inside the sorbent. Because of their nanoscale pores, nano sorbents aid in the sorption of impurities. In order to regenerate nano sorbents, the absorbed pollutants must be removed. For instance, it has been shown that organic dyes and heavy metal ions found in polluted water can be efficiently absorbed by self-assembled 3D flower-like iron oxide nanostructures [70].

Wide-ranging characteristics of nano sorbents, such as their strong sorption capacity, make them more effective and suitable for treating water. Despite the fact that these nano sorbents are somewhat rare in the market, scientists and experts are putting a lot of effort into producing them commercially and in larger quantities. Additionally, polymeric nano sorbents and metal/metal oxide were known to exist. The composite of various materials, such as carbon/ag/polyaniline, carbon/TiO$_2$, *etc.*, is extremely important for minimizing the toxicity effect during the wastewater treatment process [71].

Carbon Nano Sorbents

Depending on how they are synthesized, carbonaceous materials like carbon nanotubes (CNTs) with a cylindrical form can appear as single-walled or multiwalled nanotubes. CNTs have quantifiable adsorption sites and sustained surfaces due to their vast surface area. The hydrophobic surface of carbon nanotubes (CNTs) must be stabilized to prevent aggregation, which would lower the quantity of surface-active sites. As a result, it is an effective material for adsorbing contaminants. Analogously, dendrimers and other polymeric nano adsorbents work well effectively to eliminate organic pollutants and heavy metals from wastewater [72, 73].

Biosorbents

It is impossible to completely eliminate certain organic contaminants from a water body due to their extremely low concentrations, which are typically in the range of picograms or nanograms per liter of water [74]. Biosorbents, usually made from biological or agricultural resources, have been shown to be promising for the effective removal of such contaminants. Biosorbents provide many advantages over conventional absorbents, including low cost, high efficiency, minimal biological and agricultural sludge, no additional nutrient demand, and regenerative properties. Liu *et al.* created a DNA matrix made of salmon milt DNA hydrogel beads, which was effectively utilized for the targeted absorption of dioxin derivatives [75]. All that has to be done to get the DNA beads to regenerate after being adsorbent of dioxins is to rinse them with hexane. There have been several studies on the use of triolein-embedded biosorbents to eliminate organic pollutants from water. Heavy metal ions have been removed from water using biosorbents. Additionally, chitosan-based sorbents have demonstrated encouraging outcomes with a very effective metal ion adsorption capacity; the metal ion adsorption is accomplished by chelating on the chitosan's amino acid groups [76]. Xueyan Guo *et al.*149 looked into the sorption capability of heavy metals and created a biosorbent from black liquor, a waste product from the paper industry. According to reports, the biosorbent's sorption affinity for different heavy metal ions is ranked as follows: Pb(II) > Cu(II) > Cd(II) > Zn(II) > Ni(II) [77]. Heavy metal removal from water has also been accomplished by the use of biosorbents made from various agricultural wastes and materials [78].

Nanofilters

The procedure for eliminating or lowering the amount of harmful biological and chemical contaminants, suspended particles, and other particulate matter from tainted or dirty water to produce potable water that is safe to drink and use in pharmaceuticals and medical applications is known as water filtration [79]. The most important advancement in membrane technology, which has garnered a lot of attention recently, is nanofiltration membranes. As the name suggests, NF membranes have a molecular weight cut-off (MWCO) for uncharged particles in the nanoscale region. NF membranes are a relatively new technology that is the method of choice for purifying drinking water and wastewater. NF is a membrane process powered by pressure that is situated in between RO and UF (ultrafiltration). Its pore size ranges from 0.5 to 2.0 nm, and its pressure ranges from 5 to 20 bars. Its characteristics include low rejection of monovalent ions, significant rejection of divalent or higher-valent ions, high flux, low energy consumption, and high rejection in comparison to RO [80]. This is a relatively new advancement in membrane technology, with the ability to operate in an aqueous or non-aqueous environment. Due to its distinct filtration method and range of membrane options, one of the most significant and widely applied techniques in the wastewater treatment sector is NF. NF works well to remove almost all of the inorganic and organic contaminants from wastewater, including a significant number of dangerous microorganisms. NF membranes are inexpensive, simple to make, and incredibly flexible. The two most widely utilized forms of NF membranes are ceramic and polymeric membranes. Due to their poor chemical resistance and high fouling rate, polymeric membranes have a limited lifespan. Conversely, ceramic membranes have greater thermal, chemical, and mechanical stabilities [81]. Mostafavi *et al.* created an NF based on CNTs to extract the MS2 virus from water. A member of the Enterobacteriaceae family of bacterial viruses, the MS2 virus infects Escherichia coli and other bacteria. The porosity and surface shape of the NF were assessed, and its effectiveness in eliminating the MS2 virus was subsequently assessed. The high removal efficiency was shown in the result at 8–11 pressure bars [82].

An NF membrane for water filtration on a mesoporous substrate was developed by Yi Han *et al.* The synthetic filtration membrane, which measured between 22 and 53 nm in thickness, showed effective retention of the organic dye that was in the water. Ionic compounds, however, showed a moderate retention.

Submicrometric-thick NF membranes, with the exception of water, were created by Nair *et al.* and showed remarkably good impermeability to all vapors, gases, and liquids [83].

Photocatalysis

There are two Greek words in the term "photocatalysis." "Photo" denotes "light," and "catalysis" refers to any material that modifies a chemical reaction's pace without getting entangled in it. Thus, a light-induced reaction that is catalyzed and accelerated by light can be characterized as photocatalysis [84]. In other words, photocatalysis is the procedure of absorbing light (photons) *via* a solid catalyst to initiate a chemical reaction. Photocatalysis, the *in-situ* creation of extremely potent chemical oxidants with the aid of a catalyst, Fenton's reagent, hydrogen peroxide (H_2O_2), UV light, or ozone (O_3), is one of the advanced oxidation processes (AOPs). The generated hydroxyl radicals possess sufficient strength to oxidize robust organic molecules [85]. AOPs are well-known methods for removing CECs from waste water. The scientific community has studied photocatalysis extensively for the treatment of wastewater because of its capacity to degrade a wide range of organic materials, estrogens, dyes, organic acids, pesticides, crude oil, microbes (which includes viruses and chlorine-resistant microbes), and some inorganic molecules like nitrous oxides. Metals (like mercury) in wastewater can also be eliminated by combining them with precipitation or filtering [86]. Researchers are increasingly interested in using nanomaterials as photocatalysts because of their unique quantum effects and superior surface, mechanical, chemical, electrical, magnetic, and optical features that cause them to behave differently from bulk materials [87].

The mechanism behind photocatalysis, a surface phenomenon, consists of the following five fundamental steps, as shown in Fig. (**1**) [88]:

- The processes of reactant/pollutant diffusion onto the surface of the photocatalyst, reactant/pollutant adsorption on the surface.
- Reactant/pollutant adsorbed on the surface.
- Product desorption from the surface.
- Product removal/diffusion from the interface.

Typical nanostructured semiconductor photocatalysts include zinc sulfide (ZnS), tungsten trioxide (WO_3), zirconium dioxide (ZrO_2), TiO_2, Fe_2O_3, ZnO, and copper sulfide (CdS) [89].

Photocatalysis has been transformed by nanomaterials, which increase the process's efficiency and specificity. Compared to bulk materials, their improved interaction with light and reactants is made possible by their large surface region, adjustable characteristics, and the effects of quantum mechanics. The main ways that nanomaterials affect photocatalysis are provided below.

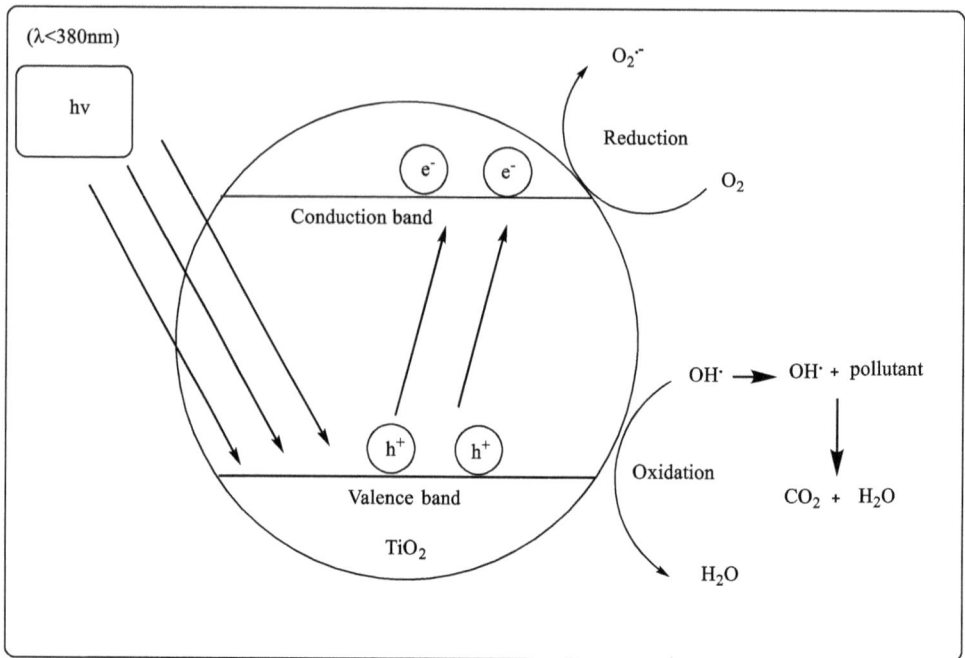

Fig. (1). General mechanism of the photocatalysis.

- Activated Sites and Increased Area of Surface: In comparison to their mass equivalents, nanostructures such as tiny particles, nanorods, and nanosheets offer a larger accessible area of surface. This raises the possibility of a reaction by allowing more molecules of reactant to be absorbed on the catalyst's surface.
- Better Band Gap Engineering and Absorption of Light: By altering their size or composition, nanomaterials permit adjustment of the bandgap potential.
- Modification of the Surface for Selective Catalytic Process.
- Enhancements in Reliability and Longevity.
- Effective Transport and Separation of Charges [90].

The process of altering a substance's electrical band arrangement to improve its photonic and electrical characteristics is known as "band gap engineering." It is essential for enhancing electrochemical productivity, separation of charges, and absorption of light in photocatalysis. Function in maximizing the activity of photocatalysis.

- Adjusting absorption of light: By reducing the band gap, photocatalysts (like TiO_2) can take in greater amounts of light, boosting solar efficiency.
- Fitting band edges together: Modifying the locations of the valence and

conduction bands guarantees that they satisfy the redox capacity specifications for processes such as the splitting of water.

- Cutting Down on Recombination: By encouraging the separation of charges, the heterojunctions or additives reduce the recombination of electrons and holes.
- Utilizing nano structuring: By creating size-tunable band gaps, tiny quantum dots and 2-dimensional substances improve absorbance and transfer of charge.

Band gap engineering enhances the photocatalytic efficiency by controlling these factors, which makes solar-driven chemical reactions and restoration of the environment more effective [91].

Zeolites

Zeolites are three-dimensional, crystalline, microporous materials having distinct void and conduit structures and sizes. Oxygen, silicon, and aluminum molecules, in their usual configurations, are easily accessible through pores with well-defined dimensions [92]. Zeolites are naturally occurring silicate minerals. Additionally, they can be produced artificially as bio-zeolite, magnetically altered zeolite, *etc* [93]. Nowadays, zeolitic nanoparticles as adsorbents have gained a lot of attention in ecological fields due to their environmental friendliness, stability in water, high surface area, low production cost, and selectivity [94]. Numerous investigations conducted to far have attested to their remarkable ability to extract metal cations from wastewater [95]. Zeolite has been shown to have several commercial investigation applications; however, its adsorbent characteristic has received the greatest attention due to its ability for regeneration and repurpose, making it an excellent choice for the treatment of wastewater. Yafei Zhao *et al.* synthesized cubic NaA zeolite, a microporous crystalline aluminosilicate zeolite consisting of Na_2O/Al_2O_3. Ammonium ions are recovered from naturally occurring halloysite minerals utilizing nanotubular structures in order to be adsorbed from wastewater. The produced NaA zeolite was found to have a maximum capacity for adsorption of 44.3 mg/g for NH_4^+ ions. The designed adsorbent system was reusable and proved a possible use for removing NH^{4+} contaminants from wastewater [96]. Samarghandi *et al.* created manganese oxide nanoparticles by using NZ adsorbents that were coated to remove CEX from a water-based solution. When the pH is seven, almost all CEX elimination capacity that was assessed was 28% for NZ and 89% for CNZ [97]. Zeolite and zeolite nanocomposite (Zeolite-Fe_3O_4@NC) were produced by Emmanuel Nyankson *et al.* to aid in the adsorption of organic molecules, such as MB, from solution. The synthesized Zeolite-Fe_3O_4@NC MB was adsorbed from the solution by using of kinetic, equilibrium, and UV-visible isotherm models. At 25 ˚C, Zeo-Fe_3O_4@NC's greatest adsorption ability and efficacy were 2.57 mg/g and 97.5%, respectively. Following regeneration, the highest adsorption efficiency at a pH of 7 was discovered to be 82.6% [98] Gugushe *et al.* produced a nanocomposite of MMWCNTs further coated with zeolite (Fe_3O_4-MWCNTs/Zeolite) for the purpose of adsorbing lead and thallium (TI) in complicated environmental samples. The Fe_3O_4-MMWCNTs/Zeolite that was synthesized was employed as an adsorbent in the magnetic solid phase extraction method with ultrasonic assistance for Pb and TI. The maximum adsorption capacity of the mixture was discovered to be 37.8 and 44.5 mgg^{-1} [99].

CONCLUSION

Current methods of treating wastewater are effective at managing organic and inorganic waste in water. However, these approaches are energy-demanding and cost-inefficient due to their incomplete water purification capabilities and the challenge of reusing the by-products. Nanotechnology will have a significant impact on the wastewater treatment industry. The goal of nanotechnology is to make existing methods better by making processes more efficient and making nanomaterials more reusable, which lowers the cost of running the plant or processes. This chapter holds great value for researchers seeking to innovate and devise fresh approaches for wastewater treatment by using nanotechnology.

LIST OF ABBREVIATIONS

AG	Aero Gel
AHCs	Aliphatic Hydrocarbons
AOPs	Advanced Oxidation Processes
AR1	Acid Red 1
CECs	Contaminants of Emergent Concern
CMCs	Composites in the Ceramic Mixture
CNT-MMC	Carbon Nanotube-Metal Matrix Composite
CNTs	Carbon Nanotubes
CR	CongoRed
CS	ChitoSan
GAC	Granular Activated Carbon
GO	Graphene Oxide
MB	Methylene Blue
MG	Malachite Green
MWCO	Molecular Weight Cut Off
NC	Nano Composites
NF	Nano Filtration
NP	Nano Particle

NZVI	Nano Zero-valent Iron
PAHCs	Polyaromatic Hydrocarbons
PRB	Permeable Reactive Barrier
RO	Reverse Osmosis
RR2	Reactive Red 2
UF	Ultra Filtration

ACKNOWLEDGEMENTS

The authors would like to express their sincere thanks to the management of Baba Masthnath University, Rohtak, Haryana, India, for providing the facilities to write and submit the book chapter for publication.

REFERENCES

[1] Narain-Ford, D.M.; Bartholomeus, R.P.; Dekker, S.C.; vanWezel, A.P. Natural purification through soils: risks and opportunities of sewage effluent reuse in sub-surface irrigation. In: *Reviews of Environmental Contamination and Toxicology*; Springer: Berlin/Heidelberg, Germany, **2020**; 250, pp. 85-117.

[2] Pham-Duc, P.; Nguyen-Viet, H.; Hattendorf, J.; Cam, P.D.; Zurbrügg, C.; Zinsstag, J.; Odermatt, P. Diarrhoeal diseases among adult population in an agricultural community Hanam province, Vietnam, with high wastewater and excreta re-use. *BMC Public Health,* **2014**, *14*(1), 978.
[http://dx.doi.org/10.1186/1471-2458-14-978] [PMID: 25239151]

[3] Yang, J.; Jia, R.; Gao, Y.; Wang, W.; Cao, P. The reliability evaluation of reclaimed water reused in power plant project. *IOP Conf. Ser. Earth Environ. Sci.,* **2017**, *100*, 012189.
[http://dx.doi.org/10.1088/1755-1315/100/1/012189]

[4] Nagar, A.; Pradeep, T. Clean water through nanotechnology: Needs, gaps, and fulfillment. *ACS Nano. American Chemical Society,* **2020**, 6420-6435.
[http://dx.doi.org/10.1021/acsnano.9b01730]

[5] Sato, T.; Qadir, M.; Yamamoto, S.; Endo, T.; Zahoor, A. Global, regional, and country level need for data on wastewater generation, treatment, and use. *Agric. Water Manage.,* **2013**, *130*, 1-13.
[http://dx.doi.org/10.1016/j.agwat.2013.08.007]

[6] Dickin, S.K.; Schuster-Wallace, C.J.; Qadir, M.; Pizzacalla, K. A review of health risks and pathways for exposure to wastewater Use in Agriculture. *Environ. Health Perspect.,* **2016**, *124*(7), 900-909.
[http://dx.doi.org/10.1289/ehp.1509995] [PMID: 26824464]

[7] Duan, B.; Zhang, W.; Zheng, H.; Wu, C.; Zhang, Q.; Bu, Y. Comparison of health risk assessments of heavy metals and as in sewage sludge from wastewater treatment plants (WWTPs) for adults and children in the urban district of Taiyuan, China. *Int. J. Environ. Res. Public Health,* **2017**, *14*(10), 1194.
[http://dx.doi.org/10.3390/ijerph14101194] [PMID: 28991185]

[8] Definition of Freshwater Resources. Available from: https://web.archive.org/web/20160411064155/http:/ /webworld.unesco.org/water/ihp/publications/waterway/webpc/definition.html

[9] Legnoverde, M.S.; Simonetti, S.; Basaldella, E.I. Influence of pH on cephalexin adsorption onto SBA-15 mesoporous silica: Theoretical and experimental study. *Appl. Surf. Sci.,* **2014**, *300*, 37-42.
[http://dx.doi.org/10.1016/j.apsusc.2014.01.198]

[10]   Kong, S.; Wang, Y.; Zhan, H.; Liu, M.; Liang, L.; Hu, Q. Competitive adsorption of humic acid and arsenate on nanoscale iron-manganese binary oxide-loaded zeolite in groundwater. *J Geo-chem Explor,* **2014**, *144*, 220-225.
[http://dx.doi.org/10.1016/j.gexplo.2014.02.005]

[11]   Ajith, M.P.; Rajamani, P. Nanotechnology for water purification-current trends and challenges *J Nanotechnol Nanomaterials.,* **2021**, *2*(3) Available from: https://www.scientificarchives.com/journal/journal-of-nanotechnology-and-nanomaterials

[12]   Qu, X.; Brame, J.; Li, Q.; Alvarez, P.J.J. Nanotechnology for a safe and sustainable water supply: enabling integrated water treatment and reuse. *Acc. Chem. Res.,* **2013**, *46*(3), 834-843.
[http://dx.doi.org/10.1021/ar300029v] [PMID: 22738389]

[13]   Umar, K.; Haque, M.M.; Mir, N.A.; Muneer, M.; Farooqi, I.H. Titanium dioxide-mediated photocatalysed mineralization of two selected organic pollutants in aqueous suspensions. *J. Adv. Oxid. Technol.,* **2013**, *16*(2)
[http://dx.doi.org/10.1515/jaots-2013-0205]

[14]   Umar, K.; Ibrahim, M.N.M.; Ahmad, A.; Rafatullah, M. Synthesis of Mn-doped TiO2 by novel route and photocatalytic mineralization/intermediate studies of organic pollutants. *Res. Chem. Intermed.,* **2019**, *45*(5), 2927-2945.
[http://dx.doi.org/10.1007/s11164-019-03771-x]

[15]   Baruah, S.; Najam Khan, M.; Dutta, J. Perspectives and applications of nanotechnology in water treatment. In: *Environmental Chemistry Letters*; Springer Verlag, **2016**; pp. 1-14.
[http://dx.doi.org/10.1007/s10311-015-0542-2]

[16]   Pérez, H.; Quintero García, O.J.; Amezcua-Allieri, M.A.; Rodríguez Vázquez, R. Nanotechnology as an efficient and effective alternative for wastewater treatment: an overview. *Water Sci. Technol.,* **2023**, *87*(12), 2971-3001.
[http://dx.doi.org/10.2166/wst.2023.179] [PMID: 37387425]

[17]   Lu, H.; Wang, J.; Stoller, M.; Wang, T.; Bao, Y.; Hao, H. *An overview of nanomaterials for water and wastewater treatment. Advances in materials science and engineering*; Hindawi Publishing Corporation, **2016**.
[http://dx.doi.org/10.1155/2016/4964828]

[18]   Das, R.; Ali, M.E.; Hamid, S.B.A.; Ramakrishna, S.; Chowdhury, Z.Z. Carbon nanotube membranes for water purification: A bright future in water desalination. *Desalination,* **2014**, *336*, 97-109.
[http://dx.doi.org/10.1016/j.desal.2013.12.026]

[19]   Kumar, S.; T, S.; Kumar, B.; Baruah, A.; Shanker, V. Synthesis of magnetically separable and recyclable g-C$_3$N$_4$–Fe$_3$O$_4$ Hybrid nanocomposites with enhanced photocatalytic performance under visible-light irradiation. *J. Phys. Chem. C,* **2013**, *117*(49), 26135-26143.
[http://dx.doi.org/10.1021/jp409651g]

[20]   Sharma, M.; Das, D.; Baruah, A.; Jain, A.; Ganguli, A.K. Design of porous silica supported tantalum oxide hollow spheres showing enhanced photocatalytic activity. *Langmuir,* **2014**, *30*(11), 3199-3208.
[http://dx.doi.org/10.1021/la500167a] [PMID: 24588721]

[21]   Epelle, E. I.; Okoye, P. U.; Roddy, S.; Gunes, B.; Okolie, J. A. Advances in the applications of nanomaterials for wastewater treatment. *Environments,* **2022**.
[http://dx.doi.org/10.3390/environments9110141]

[22]   Saleem, H.; Zaidi, S. J. Developments in the application of nanomaterials for water treatment and their impact on the environment. *Nanomaterials.,* **2020**, 1-39.
[http://dx.doi.org/10.3390/nano10091764]

[23]   Vela, N.; Calín, M.; Yáñez-Gascón, M. J.; Garrido, I.; Pérez-Lucas, G.; Fenoll, J.; Navarro, S. Photocatalytic oxidation of six pesticides listed as endocrine disruptor chemicals from wastewater using two different tio2 samples at pilot plant scale under sunlight irradiation. *J Photochem Photobiol*

A Chem **2018**, 353, 271–278.
[http://dx.doi.org/10.1016/j.jphotochem.2017.11.040]

[24]   Nishu, ; Kumar, S. Smart and innovative nanotechnology applications for water purification. *Hybrid Advances,* **2023**, *3*, 100044.
[http://dx.doi.org/10.1016/j.hybadv.2023.100044]

[25]   Aboelfetoh, E.F.; Zain Elabedien, M.E.; Ebeid, E.Z.M. Effective treatment of industrial wastewater applying SBA-15 mesoporous silica modified with graphene oxide and hematite nanoparticles. *J. Environ. Chem. Eng.,* **2021**, *9*(1), 104817.
[http://dx.doi.org/10.1016/j.jece.2020.104817]

[26]   Sher, F.; Hanif, K.; Iqbal, S.Z.; Imran, M. Implications of advanced wastewater treatment: Electrocoagulation and electroflocculation of effluent discharged from a wastewater treatment plant. *J. Water Process Eng.,* **2020**, *33*, 101101.
[http://dx.doi.org/10.1016/j.jwpe.2019.101101]

[27]   Kimura, M.; Matsui, Y.; Kondo, K.; Ishikawa, T. B.; Matsushita, T. *The manuscript published in water research minimizing residual aluminum concentration in treated water by tailoring properties of polyaluminum coagulants.* 2013, 2075-2084.

[28]   Xu, W.; Gao, B.; Wang, Y.; Yue, Q.; Ren, H. Effect of second coagulant addition on coagulation efficiency, floc properties and residual Al for humic acid treatment by Al13 polymer and polyaluminum chloride (PACl). *J. Hazard. Mater.,* **2012**, *215-216*, 129-137.
[http://dx.doi.org/10.1016/j.jhazmat.2012.02.051] [PMID: 22410719]

[29]   Fu, F.; Wang, Q. Removal of Heavy Metal Ions from Wastewaters: A Review. *Journal of Environmental Management.,* **2011**, , 407-418.
[http://dx.doi.org/10.1016/j.jenvman.2010.11.011]

[30]   Chang, Q.; Zhang, M.; Wang, J. Removal of Cu2+ and turbidity from wastewater by mercaptoacetyl chitosan. *J. Hazard. Mater.,* **2009**, *169*(1-3), 621-625.
[http://dx.doi.org/10.1016/j.jhazmat.2009.03.144] [PMID: 19414213]

[31]   Duan, J.; Lu, Q.; Chen, R.; Duan, Y.; Wang, L.; Gao, L.; Pan, S. Synthesis of a novel flocculant on the basis of crosslinked Konjac glucomannan-graft-polyacrylamide-co-sodium xanthate and its application in removal of Cu2+ ion. *Carbohydr. Polym.,* **2010**, *80*(2), 436-441.
[http://dx.doi.org/10.1016/j.carbpol.2009.11.046]

[32]   Owlad, M.; Aroua, M.K.; Daud, W.A.W.; Baroutian, S. Removal of hexavalent chromium-contaminated water and wastewater: A review. *Water Air Soil Pollut.,* **2009**, *200*(1-4), 59-77.
[http://dx.doi.org/10.1007/s11270-008-9893-7]

[33]   Wan Ngah, W.S.; Hanafiah, M.A.K.M. Removal of heavy metal ions from wastewater by chemically modified plant wastes as adsorbents: A review. *Bioresour. Technol.,* **2008**, *99*(10), 3935-3948.
[http://dx.doi.org/10.1016/j.biortech.2007.06.011] [PMID: 17681755]

[34]   Charerntanyarak, L. Heavy metals removal by chemical coagulation and precipitation. In: *Water Science and Technology*; Elsevier Science Ltd, **1999**; 39, pp. 135-138.
[http://dx.doi.org/10.2166/wst.1999.0642]

[35]   Peters, R. W.; Shem, L. Separation of heavy metals: removal from industrial wastewaters and contaminated soil. 1993, 3-64.

[36]   Galán, B.; Castañeda, D.; Ortiz, I. Removal and recovery of Cr(VI) from polluted ground waters: A comparative study of ion-exchange technologies. *Water Res.,* **2005**, *39*(18), 4317-4324.
[http://dx.doi.org/10.1016/j.watres.2005.08.015] [PMID: 16221483]

[37]   Kurniawan, T.A.; Chan, G.Y.S.; Lo, W.H.; Babel, S. Physico–chemical treatment techniques for wastewater laden with heavy metals. *Chem. Eng. J.,* **2006**, *118*(1-2), 83-98.
[http://dx.doi.org/10.1016/j.cej.2006.01.015]

[38]   Liu, F.; Zhang, G.; Meng, Q.; Zhang, H. Performance of nanofiltration and reverse osmosis

membranes in metal effluent treatment. **2008**, *16*.

[39] Bessbousse, H.; Rhlalou, T.; Verchère, J.F.; Lebrun, L. Removal of heavy metal ions from aqueous solutions by filtration with a novel complexing membrane containing poly(ethyleneimine) in a poly(vinyl alcohol) matrix. *J. Membr. Sci.,* **2008**, *307*(2), 249-259.
[http://dx.doi.org/10.1016/j.memsci.2007.09.027]

[40] Sirés, I.; Brillas, E. *Remediation of water pollution caused by pharmaceutical residues based on electrochemical separation and degradation technologies: A review. Environment International*; Elsevier Ltd, **2012**, pp. 212-229.
[http://dx.doi.org/10.1016/j.envint.2011.07.012]

[41] Baruah, A.; Chaudhary, V.; Malik, R.; Tomer, V.K. Nanotechnology based solutions for wastewater treatment. In: *Nanotechnology in Water and Wastewater Treatment: Theory and Applications*; Elsevier, **2019**; pp. 337-368.
[http://dx.doi.org/10.1016/B978-0-12-813902-8.00017-4]

[42] Baruah, S.; Pal, S.K.; Dutta, J. Nanostructured zinc oxide for water treatment. **2012**, *2*.

[43] Bellona, C.; Drewes, J.E.; Oelker, G.; Luna, J.; Filteau, G.; Amy, G. Comparing nanofiltration and reverse osmosis for drinking water augmentation. *J. Am. Water Works Assoc.,* **2008**, *100*(9), 102-116.
[http://dx.doi.org/10.1002/j.1551-8833.2008.tb09724.x]

[44] Yangali-Quintanilla, V.; Maeng, S.K.; Fujioka, T.; Kennedy, M.; Amy, G. Proposing nanofiltration as acceptable barrier for organic contaminants in water reuse. *J. Membr. Sci.,* **2010**, *362*(1-2), 334-345.
[http://dx.doi.org/10.1016/j.memsci.2010.06.058]

[45] Thanuttamavong, M.; Yamamotob, K. *Desalination rejection characteristics of organic and inorganic pollutants by ultra low-pressure nanofiltration of surface water for drinking water treatment*; 2002; Vol. 145 www.elsevier.com/locate/desal

[46] Sharma, V.; Sharma, A. Nanotechnology : An emerging future trend in wastewater treatment with its innovative products and processes. *Int. J. Enhanc. Res. Sci. Technol. Eng.,* **2012**, 1.

[47] Zhang, W-X. *Nanoscale iron particles for environmental remediation: An overview*; , **2003**.

[48] Singh, R.; Misra, V.; Singh, R.P. Remediation of Cr(VI) contaminated soil by zero-valent iron nanoparticles (NZVI) entrapped in calcium alginate beads.**2011**.
[http://dx.doi.org/10.13140/2.1.3293.6969]

[49] Nowack, B.; Bucheli, T.D. Occurrence, behavior and effects of nanoparticles in the environment. *Environ. Pollut.,* **2007**, *150*(1), 5-22.
[http://dx.doi.org/10.1016/j.envpol.2007.06.006] [PMID: 17658673]

[50] Chan, S.H.S.; Yeong Wu, T.; Juan, J.C.; Teh, C.Y. Recent developments of metal oxide semiconductors as photocatalysts in advanced oxidation processes (AOPs) for treatment of dye wastewater. *J. Chem. Technol. Biotechnol.,* **2011**, *86*(9), 1130-1158.
[http://dx.doi.org/10.1002/jctb.2636]

[51] He, L.; Wang, L.; Zhu, H.; Wang, Z.; Zhang, L.; Yang, L.; Dai, Y.; Mo, H.; Zhang, J.; Shen, J. A reusable Fe_3O_4/GO-COOH nanoadsorbent for Ca^{2+} and Cu^{2+} removal from oilfield wastewater. *Chem. Eng. Res. Des.,* **2021**, *166*, 248-258.
[http://dx.doi.org/10.1016/j.cherd.2020.12.019]

[52] Peralta, M.E.; Mártire, D.O.; Moreno, M.S.; Parolo, M.E.; Carlos, L. Versatile nanoadsorbents based on magnetic mesostructured silica nanoparticles with tailored surface properties for organic pollutants removal. *J. Environ. Chem. Eng.,* **2021**, *9*(1), 104841.
[http://dx.doi.org/10.1016/j.jece.2020.104841]

[53] Zhang, X.; Du, A.J.; Lee, P.; Sun, D.D.; Leckie, J.O. TiO_2 nanowire membrane for concurrent filtration and photocatalytic oxidation of humic acid in water. *J. Membr. Sci.,* **2008**, *313*(1-2), 44-51.
[http://dx.doi.org/10.1016/j.memsci.2007.12.045]

[54] Moradi, O.; Fakhri, A.; Adami, S.; Adami, S. Isotherm, thermodynamic, kinetics, and adsorption mechanism studies of Ethidium bromide by single-walled carbon nanotube and carboxylate group functionalized single-walled carbon nanotube. *J. Colloid Interface Sci.,* **2013**, *395*(1), 224-229.
[http://dx.doi.org/10.1016/j.jcis.2012.11.013] [PMID: 23261335]

[55] Wilson, M.E.; Rukh, M.G.; Ashraf, M.A. The role of nanotechnology, based on carbon nanotubes in water and wastewater treatment. *Desalination Water Treat.,* **2021**, *242*, 12-21.
[http://dx.doi.org/10.5004/dwt.2021.27568]

[56] Kariim, I.; Abdulkareem, A.S.; Abubakre, O.K. Development and characterization of MWCNTs from activated carbon as adsorbent for metronidazole and levofloxacin sorption from pharmaceutical wastewater: Kinetics, isotherms and thermodynamic studies. *Sci. Am.,* **2020**, *7*, e00242.
[http://dx.doi.org/10.1016/j.sciaf.2019.e00242]

[57] Kuo, C.Y.; Wu, C.H.; Wu, J.Y. Adsorption of direct dyes from aqueous solutions by carbon nanotubes: Determination of equilibrium, kinetics and thermodynamics parameters. *J. Colloid Interface Sci.,* **2008**, *327*(2), 308-315.
[http://dx.doi.org/10.1016/j.jcis.2008.08.038] [PMID: 18786679]

[58] Ali, I.; Basheer, A. A.; Mbianda, X. Y.; Burakov, A.; Galunin, E.; Burakova, I.; Mkrtchyan, E.; Tkachev, A.; Grachev, V. Graphene based adsorbents for remediation of noxious pollutants from wastewater. In: *Environment International*; Elsevier, **2019**; pp. 160-180.
[http://dx.doi.org/10.1016/j.envint.2019.03.029]

[59] Saverini, M.; Catanzaro, I.; Sciandrello, G.; Avellone, G.; Indelicato, S.; Marcì, G.; Palmisano, L. Genotoxicity of citrus wastewater in prokaryotic and eukaryotic cells and efficiency of heterogeneous photocatalysis by TiO2. *J Photochem. Photobiol. B.,* **2012**, *108*, 8–15.
[http://dx.doi.org/10.1016/j.jphotobiol.2011.12.003]

[60] Jayakaran, P.; Nirmala, G. S.; Govindarajan, L. Qualitative and quantitative analysis of graphene-based adsorbents in wastewater treatment. In: *International Journal of Chemical Engineering*; Hindawi Limited, **2019**.
[http://dx.doi.org/10.1155/2019/9872502]

[61] Lim Teik Zheng, A.; Phromsatit, T.; Boonyuen, S.; Andou, Y. Synthesis of silver nanoparticles/porphyrin/reduced graphene oxide hydrogel as dye adsorbent for wastewater treatment. *FlatChem,* **2020**, *23*, 100174.
[http://dx.doi.org/10.1016/j.flatc.2020.100174]

[62] Sirajudheen, P.; Karthikeyan, P.; Ramkumar, K.; Meenakshi, S. Effective removal of organic pollutants by adsorption onto chitosan supported graphene oxide-hydroxyapatite composite: A novel reusable adsorbent. *J. Mol. Liq.,* **2020**, *318*, 114200.
[http://dx.doi.org/10.1016/j.molliq.2020.114200]

[63] Chaudhary, D.S.; Vigneswaran, S.; Jegatheesan, V.; Ngo, H.H.; Moon, H.; Shim, W.G.; Kim, S.H. *Granular activated carbon (GAC) adsorption in tertiary wastewater treatment: Experiments and Models; 2002.* https://iwaponline.com/wst/article-pdf/47/1/113/426552/113.pdf

[64] Asenjo, N.G.; Álvarez, P.; Granda, M.; Blanco, C.; Santamaría, R.; Menéndez, R. High performance activated carbon for benzene/toluene adsorption from industrial wastewater. *J. Hazard. Mater.,* **2011**, *192*(3), 1525-1532.
[http://dx.doi.org/10.1016/j.jhazmat.2011.06.072] [PMID: 21782335]

[65] Naseem, T.; Waseem, M. *International journal of environmental science and technology*; Springer Science and Business Media Deutschland GmbH, **2022**, pp. 2221-2246.
[http://dx.doi.org/10.1007/s13762-021-03256-8]

[66] Chou, T.W.; Gao, L.; Thostenson, E.T.; Zhang, Z.; Byun, J.H. An assessment of the science and technology of carbon nanotube-based fibers and composites. *Compos. Sci. Technol.,* **2010**, *70*(1), 1-19.
[http://dx.doi.org/10.1016/j.compscitech.2009.10.004]

[67]   Gu, Y.; Xia, K.; Wu, D.; Mou, J.; Zheng, S. Technical characteristics and wear-resistant mechanism of nano coatings: A review. *Coatings,* **2020**, *10*(3), 233.
[http://dx.doi.org/10.3390/coatings10030233]

[68]   Janas, D.; Liszka, B. *Copper Matrix nanocomposites based on carbon nanotubes or graphene. materials chemistry frontiers*; Royal Society of Chemistry, **2018**, pp. 22-35.
[http://dx.doi.org/10.1039/C7QM00316A]

[69]   Vilt, M.E.; Ho, W.S.W. Supported liquid membranes with strip dispersion for the recovery of Cephalexin. *J. Membr. Sci.,* **2009**, *342*(1-2), 80-87.
[http://dx.doi.org/10.1016/j.memsci.2009.06.026]

[70]   Zhong, L.S.; Hu, J.S.; Liang, H.P.; Cao, A.M.; Song, W.G.; Wan, L.J. Self-Assembled 3D flowerlike iron oxide nanostructures and their application in water treatment. *Adv. Mater.,* **2006**, *18*(18), 2426-2431.
[http://dx.doi.org/10.1002/adma.200600504]

[71]   Yu, L.; Ruan, S.; Xu, X.; Zou, R.; Hu, J. One-Dimensional nanomaterial-assembled macroscopic membranes for water treatment. In: *Nano Today*; Elsevier B.V., **2017**; pp. 79-95.
[http://dx.doi.org/10.1016/j.nantod.2017.10.012]

[72]   Nasir, S.; Hussein, M. Z.; Zainal, Z.; Yusof, N. A. Carbon-Based Nanomaterials/Allotropes: A Glimpse of Their Synthesis, Properties and Some Applications. *Materials.,* **2018**.
[http://dx.doi.org/10.3390/ma11020295]

[73]   Thirunavukkarasu, A.; Nithya, R.; Sivashankar, R. A review on the role of nanomaterials in the removal of organic pollutants from wastewater. *Reviews in Environmental Science and Biotechnology.,* **2020**, , 751-778.
[http://dx.doi.org/10.1007/s11157-020-09548-8]

[74]   Zhang, Z.L.; Hong, H.S.; Zhou, J.L.; Huang, J.; Yu, G. Fate and assessment of persistent organic pollutants in water and sediment from Minjiang River Estuary, Southeast China. *Chemosphere,* **2003**, *52*(9), 1423-1430.
[http://dx.doi.org/10.1016/S0045-6535(03)00478-8] [PMID: 12867172]

[75]   Liu, X.D.; Murayama, Y.; Matsunaga, M.; Nomizu, M.; Nishi, N. Preparation and characterization of DNA hydrogel bead as selective adsorbent of dioxins. *Int. J. Biol. Macromol.,* **2005**, *35*(3-4), 193-199.
[http://dx.doi.org/10.1016/j.ijbiomac.2005.01.008] [PMID: 15811474]

[76]   Miretzky, P.; Cirelli, A.F. Hg(II) removal from water by chitosan and chitosan derivatives: A review. *J. Hazard. Mater.,* **2009**, *167*(1-3), 10-23.
[http://dx.doi.org/10.1016/j.jhazmat.2009.01.060] [PMID: 19232467]

[77]   Guo, X.; Zhang, S.; Shan, X. Adsorption of metal ions on lignin. *J. Hazard. Mater.,* **2008**, *151*(1), 134-142.
[http://dx.doi.org/10.1016/j.jhazmat.2007.05.065] [PMID: 17587495]

[78]   Aydın, H.; Bulut, Y.; Yerlikaya, Ç. Removal of copper (II) from aqueous solution by adsorption onto low-cost adsorbents. *J. Environ. Manage.,* **2008**, *87*(1), 37-45.
[http://dx.doi.org/10.1016/j.jenvman.2007.01.005] [PMID: 17349732]

[79]   Koyuncu, I.; Sengur, R.; Turken, T.; Guclu, S.; Pasaoglu, M.E. advances in water treatment by microfiltration, ultrafiltration, and nanofiltration. In: *advances in membrane technologies for water treatment: materials, processes and applications*; Elsevier Inc., **2015**; pp. 83-128.
[http://dx.doi.org/10.1016/B978-1-78242-121-4.00003-4]

[80]   Abdel-Fatah, M. A. Nanofiltration systems and applications in wastewater treatment: Review article. In: *Ain Shams Engineering Journal*; Ain Shams University, **2018**; pp. 3077-3092.
[http://dx.doi.org/10.1016/j.asej.2018.08.001]

[81]   Han, Y.; Xu, Z.; Gao, C. Ultrathin graphene nanofiltration membrane for water purification. *Adv. Funct. Mater.,* **2013**, *23*(29), 3693-3700.

[http://dx.doi.org/10.1002/adfm.201202601]

[82]   Mostafavi, S.T.; Mehrnia, M.R.; Rashidi, A.M. Preparation of nanofilter from carbon nanotubes for application in virus removal from water. *Desalination,* **2009**, *238*(1-3), 271-280.
[http://dx.doi.org/10.1016/j.desal.2008.02.018]

[83]   Nair, R. R.; Wu, H. A.; Jayaram, P. N.; Grigorieva, I. V.; Geim, A. K. Unimpeded Permeation of Water through Helium-Leak-Tight Graphene-Based membranes. *Science (1979),* **2012**, *335*(6067), 442-444.
[http://dx.doi.org/10.1126/science.1211694]

[84]   Saravanan, R.; Gracia, F.; Stephen, A. *Basic Principles*; Mechanism, and Challenges of Photocatalysis, **2017**, pp. 19-40.
[http://dx.doi.org/10.1007/978-3-319-62446-4_2]

[85]   Bethi, B.; Sonawane, S. H.; Bhanvase, B. A.; Gumfekar, S. P. Nanomaterials-based advanced oxidation processes for wastewater treatment: a Review. In: *Chemical Engineering and Processing: Process Intensification*; Elsevier B.V., **2016**; pp. 178-189.
[http://dx.doi.org/10.1016/j.cep.2016.08.016]

[86]   Rueda-Marquez, J. J.; Levchuk, I.; Fernández Ibañez, P.; Sillanpää, M. A critical review on application of photocatalysis for toxicity reduction of real wastewaters. In: *Journal of Cleaner Production*; Elsevier, **2020**.
[http://dx.doi.org/10.1016/j.jclepro.2020.120694]

[87]   Mehrjouei, M.; Müller, S.; Möller, D. A review on photocatalytic ozonation used for the treatment of water and wastewater. *Chemical Engineering Journal.,* **2015**, 209-219.
[http://dx.doi.org/10.1016/j.cej.2014.10.112]

[88]   Feng, T.; Feng, G. S.; Yan, L.; Pan, J. H. One-dimensional nanostructured tio2 for photocatalytic degradation of organic pollutants in wastewater. In: *International Journal of Photoenergy*; Hindawi Limited, **2014**.
[http://dx.doi.org/10.1155/2014/563879]

[89]   Bai, L.; Wei, M.; Hong, E.; Shan, D.; Liu, L.; Yang, W.; Tang, X.; Wang, B. Study on the controlled synthesis of Zr/TiO2/SBA-15 nanophotocatalyst and its photocatalytic performance for industrial dye reactive red X–3B. *Mater. Chem. Phys.,* **2020**, *246*, 122825.
[http://dx.doi.org/10.1016/j.matchemphys.2020.122825]

[90]   Tahir, M.B.; Sohaib, M.; Sagir, M.; Rafique, M. Role of Nanotechnology in Photocatalysis. In: *Encyclopedia of Smart Materials*; Elsevier, **2021**; pp. 578-589.
[http://dx.doi.org/10.1016/B978-0-12-815732-9.00006-1]

[91]   Choi, H.; Khan, S.; Choi, J.; Dinh, D.T.T.; Lee, S.Y.; Paik, U.; Cho, S.H.; Kim, S. Synergetic control of band gap and structural transformation for optimizing TiO$_2$ photocatalysts. *Appl. Catal. B,* **2017**, *210*, 513-521.
[http://dx.doi.org/10.1016/j.apcatb.2017.04.020]

[92]   Esmaeili, A.; Saremnia, B. Synthesis and characterization of NaA zeolite nanoparticles from Hordeum vulgare L. husk for the separation of total petroleum hydrocarbon by an adsorption process. *J. Taiwan Inst. Chem. Eng.,* **2016**, *61*, 276-286.
[http://dx.doi.org/10.1016/j.jtice.2015.12.031]

[93]   Nassar, M.Y.; Abdelrahman, E.A.; Aly, A.A.; Mohamed, T.Y. A facile synthesis of mordenite zeolite nanostructures for efficient bleaching of crude soybean oil and removal of methylene blue dye from aqueous media. *J. Mol. Liq.,* **2017**, *248*, 302-313.
[http://dx.doi.org/10.1016/j.molliq.2017.10.061]

[94]   Talwar, S.; Sangal, V. K.; Verma, A. Feasibility of Using Combined TiO2 Photocatalysis and RBC Process for the Treatment of Real Pharmaceutical Wastewater. *J. Photochem. Photobiol. A Chem.,* **2018**, *353*, 263–270.
[http://dx.doi.org/10.1016/j.jphotochem.2017.11.013]

[95] Pandey, P.K.; Sharma, S.K.; Sambi, S.S. Removal of lead(II) from waste water on zeolite-NaX. *J. Environ. Chem. Eng.,* **2015**, *3*(4), 2604-2610.
[http://dx.doi.org/10.1016/j.jece.2015.09.008]

[96] Zhao, Y.; Zhang, B.; Zhang, X.; Wang, J.; Liu, J.; Chen, R. Preparation of highly ordered cubic NaA zeolite from halloysite mineral for adsorption of ammonium ions. *J. Hazard. Mater.,* **2010**, *178*(1-3), 658-664.
[http://dx.doi.org/10.1016/j.jhazmat.2010.01.136] [PMID: 20172651]

[97] Samarghandi, M.R.; Al-Musawi, T.J.; Mohseni-Bandpi, A.; Zarrabi, M. Adsorption of cephalexin from aqueous solution using natural zeolite and zeolite coated with manganese oxide nanoparticles. *J. Mol. Liq.,* **2015**, *211*, 431-441.
[http://dx.doi.org/10.1016/j.molliq.2015.06.067]

[98] Nyankson, E.; Adjasoo, J.; Efavi, J.K.; Amedalor, R.; Yaya, A.; Manu, G.P.; Asare, K.; Amartey, N.A. Characterization and evaluation of zeolite A/Fe$_3$O$_4$ Nanocomposite as a potential adsorbent for removal of eOrganic Molecules from wastewater. *J. Chem.,* **2019**, *2019*, 1-13.
[http://dx.doi.org/10.1155/2019/8090756]

[99] Gugushe, A.S.; Mpupa, A.; Nomngongo, P.N. Ultrasound-assisted magnetic solid phase extraction of lead and thallium in complex environmental samples using magnetic multi-walled carbon nanotubes/zeolite nanocomposite. *Microchem. J.,* **2019**, *149*, 103960.
[http://dx.doi.org/10.1016/j.microc.2019.05.060]

CHAPTER 2

Chemistry of Zero-Valent Metal Nanoparticles in Wastewater Treatment

Minakshi[1], Prabhjot Kaur[2], Neera Raghav[2] and Rashmi Pundeer[1,*]

[1] *Department of Chemistry, Indira Gandhi University, Meerpur, Rewari -122502, Haryana, India*

[2] *Department of Chemistry, Kurukshetra University, Kurukshetra-136119, Haryana, India*

Abstract: The chapter "Chemistry of Zero-Valent Metal Nanoparticle in Wastewater Treatment" offers a thorough glance at the synthesis, basic chemistry, and utilizations of zero-valent metal nanoparticles, explicitly nano zero-valent iron (nZVI), in the field of wastewater treatment. These nanoparticles have particular qualities, including a huge surface area, magnetic properties, redox potential, and reactivity, that make them an important asset for natural remediation attempts like soil remediation, groundwater treatment, and wastewater treatment. Dye wastewater is a major environmental and health concern because of its negative impacts on aquatic life. Dyes can pose serious health risks to humans and other animals due to their possible toxicity, carcinogenicity, and teratogenicity. Thus, it is essential to make industrial wastewater free from dyes before releasing it into the environment. Moreover, heavy metals are extremely harmful and can cause serious medical conditions when present in water. Inorganic anions in wastewater also pose significant environmental challenges, including the contamination of water resources, negative impacts on aquatic life, and potential risks to human health. This study features the particular utilization of nZVI in wastewater treatment, including the debasement of dyes, heavy metals, and inorganic anions subsequent to breaking down their major natural and well-being concern.

Keywords: Dyes, Environmental remediation, Sustainable development, Wastewater treatment, Zero-valent metal nanoparticles.

INTRODUCTION

Zero-valent metal nanoparticles (NPs) are nanoparticles made up of a metal in its zero-oxidation state, indicating that the metal atoms have not undergone oxidation and remain in their form with no charge. These nanoparticles are typically very small, ranging from 1 to 100 nanometers in diameter. Various methods can be used to develop zero-valent metal nanoparticles, including reduction, electrochemical techniques, and friendly approaches using plant extracts or

[*] **Corresponding author Rashmi Pundeer:** Department of Chemistry, Indira Gandhi University, Meerpur, Rewari - 122502, Haryana, India; E-mail: dr.rashmip@gmail.com

Anjaneyulu Bendi (Ed.)

microorganisms. The chosen synthesis method can impact the size, shape, and stability of the nanoparticles. Zero-valent metal nanoparticles possess properties that are different from those of their counterparts. Their high surface area, tiny size, and quantum effects all contribute to improved optical properties. These distinctive features make them highly valuable for a range of applications, such as catalysis, biomedical applications, environmental restoration, and electronics. Examples of zero-valent metal nanoparticles include zero-valent iron (ZVI), zero-valent copper (ZVCu), zero-valent silver (ZVAg), and others. Nano zero-valent iron (nZVI) is a kind of zero-valent metal nanoparticle that has gained attention due to its exceptional characteristics and diverse applications. The characteristics that make nZVI stand out among zero-valent metal nanoparticles, especially in specific situations, are as follows:

Reactivity

nZVI (nano-iron) has a high reactivity. Because of its small size and large surface area, it has greater numbers of active sites than other catalysts. Therefore, the environmental remediation factor of nZVI in various contaminants is very effective at catalyzing the reduction. It is especially used as a way to degrade chlorinated solvents and other organic pollutants.

Magnetic Properties

nZVI particles are usually magnetic, so it is easy not to lose them to reaction sites. This property is convenient for the removal of active nZVI particles, making their use especially advantageous in environmental applications.

Redox Potential

The redox potential of nZVI is suitable for the reduction of different pollutants. This property matters where we want to remove a particular contaminant from water or in other applications such as bioremediation.

Surface Modification

The surface of nZVI nanomaterial can be modified to improve its stability, reactivity, and selectivity. Surface modification techniques include coatings and attachment of functional units that allow the properties of nZVI to be tailored for different applications.

Synthesis Techniques

There are various methods of synthesizing nZVI for different purposes. nZVI synthesis can be optimized by researchers to control the size, shape, and surface properties of nZVI, depending on the planned application [1].

REDUCTION MECHANISM

Nano zero-valence iron (nZVI) can be reduced by transferring the valence from the iron zero particles to other chemical species, thus reducing them. In various applications, especially environmental remediation, this is a crucial mechanism for nZVI to work. The reduction reaction catalyzed by nZVI is dependent upon the contaminants in any one system. The general process for nano zero-valent iron reduction mechanisms is as follows:

- **Electron Transfer**:

The dominant part of the reduction mechanism is the allocation of electrons from zero-valent iron (Fe^0) nanoparticles to other chemical species. This process consists of the oxidation of iron, where iron donates electrons.

$$Fe^0 \longrightarrow Fe^{2+} + 2e^-$$

- **Reducing Contaminants**: For example

Chlorinated Compounds: In the context of environmental remediation, nZVI is often used to reduce chlorinated compounds, such as chlorinated solvents. As nZVI electrons pass between the chlorinated compounds, they are all reduced and dechlorinated.

Heavy Metals: nZVI has the ability to reduce certain heavy metal ions. For example, in the reduction of hexavalent chromium Cr (VI), the reaction entails the transfer of electrons to reduce Cr (VI) to less toxic Cr (III).

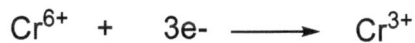

$$Cr^{6+} + 3e^- \longrightarrow Cr^{3+}$$

- **Formation of Iron Oxides/Hydroxides**:

Iron oxide/hydroxide species may form on the surface of nZVI particles upon reduction. These species have the potential to impact nZVI's stability and reactivity.

The first test of nZVI for the treatment of polluted water took place in the USA by Wang and Zhang from Lehigh University in 2006, who were attracted by its unique characteristics. After that, it was demonstrated that nZVI is highly efficient at destroying and removing numerous types of chemical pollutants, such as antibiotics containing nitroimidazole, nitroaromatic compounds, azo dyes, lactam, and chlorinated solvents, inorganic anions including perchlorate and nitrate, alkaline earth metals like beryllium and barium, transitions metals such as lead, chromium, vanadium, copper, cobalt, nickel, silver, *etc.*, metalloids like selenium and arsenic, actinides like uranium and plutonium, and post-transition metals like zinc and cadmium [2]. After meticulously analyzing the literature on nZVI and its application in environmental remediation, the authors considered it worthwhile to elaborate on the significance of nZVI in dye degradation and wastewater treatment in this chapter.

ZERO-VALENT IRON NANOPARTICLES IN DYE DEGRADATION

Dyeing, paper and pulp, textiles, plastics, leather, cosmetics, and food industries have been using dyes for a long time. The effluents from these industries contain colored substances that can cause environmental and health problems. These substances can reduce the sunlight reaching the water bodies and disturb the aquatic life. Dyes are usually composed of complex aromatic molecules that are resistant to biodegradation. Moreover, many dyes are harmful to some microorganisms and can damage or inhibit their enzymatic activities. These compounds are sophisticated organic molecules that can withstand light, washing, and microbial attacks. Therefore, they are not easily decomposed. If the dye-containing effluents are directly discharged into the municipal environment, they may produce toxic, carcinogenic by-products [3]. Many dyes and pigments are highly toxic, stable, resistant to biodegradation, and potentially carcinogenic. Industrial wastewater often contains a mixture of different organic dyes, which poses a serious challenge for pollution control and remediation. Therefore, it is necessary to develop effective approaches for the removal of various dyes from wastewater [4], and one of the major environmental challenges is the disposal of dye wastewater without proper treatment. Dye pollutants can block the sunlight from penetrating the deeper layers of water, which can adversely affect the growth of aquatic life. Hence, it is essential to eliminate dyes from wastewater before releasing them into the environment. Various methods, such as adsorption [5], flocculation [6], membrane filtration [7], electrolysis [8, 9], biological treatments [10, 11], and oxidation [12, 13], have been reported in the literature for the decolorization of dye wastewater [14]. However, none of these methods can fully remove the dye pollutants from the wastewater. Physical methods can only separate the dye molecules into different phases and are only feasible for small amounts of effluent. Chemical methods are too expensive to be widely applied and also generate chemical sludge that needs further treatment. Biological methods have low degradation rates and are ineffective for some types of dyes [15] (Fig. **1**).

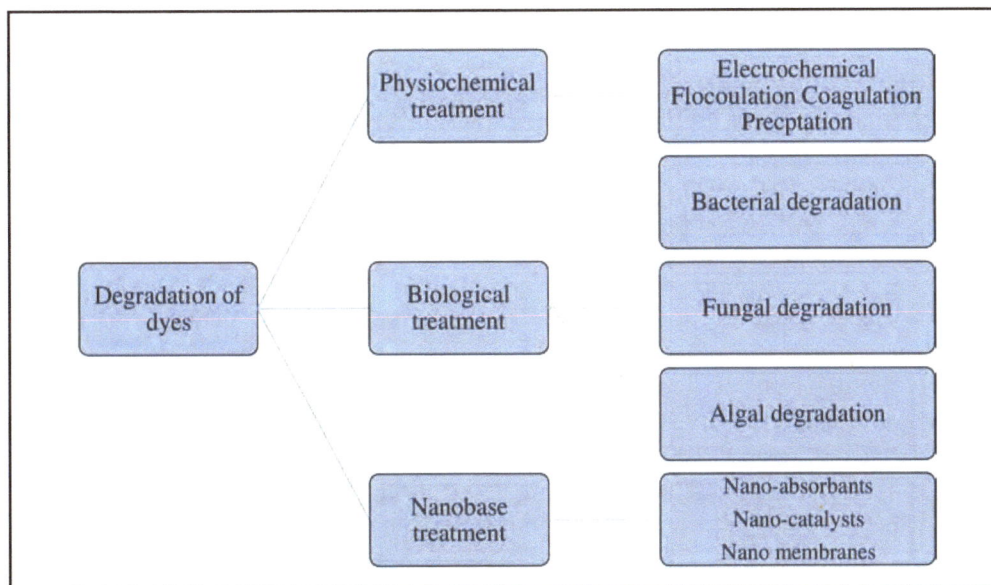

Fig. (1). A Figurative Dive into Dye Degradation Methods.

Types of Dyes

Azo Dyes

They are the most common type, constituting over 70% of synthetic dyes. Azo dyes are a type of synthetic organic dyes that have the azo chromophoric group (-N=N-) in their structure. This group connects two sp^2 hybridized carbon atoms to an aromatic or heterocyclic ring on one side and to another unsaturated molecule of carboxylic, heterocyclic, or aliphatic nature on the other side. No natural dyes have this group in their structure. Azo dyes are the most important and diverse class of organic dyes in the market. There are more than 10,000 Color Index generic names for commercial colorants, but only about 4,500 are used, and more than half of them are azo dyes [16]. Azo dyes have many applications in different industries, such as textile, paper, ink, leather, drug, and food processing. However, these industries produce effluents that contain dyes and organic pollutants. Many of these dyes are carcinogenic and toxic and reduce the light penetration in water bodies, which affects the natural growth of aquatic organisms [17]. Some of these dyes can also contaminate drinking water and pose a risk to people's health. This makes it necessary to eliminate or treat the water-containing dyes to minimize the environmental impact they pose [18, 19].

Anthraquinone Dyes

These dyes are another major group known for their bright colors and high stability, making them challenging to degrade. Anthraquinone dyes, which are classified as a class of synthetic dyes, contain anthraquinone chromophore. The anthraquinone structure is a fused aromatic ring system consisting of three benzene rings and two ketone groups. This structure imparts distinctive color properties to the dyes [20, 21].

Reactive Dyes

They covalently bond to fabrics during dyeing, making them difficult to remove from water waste. Reactive dyes are a class of synthetic dyes commonly used for coloring textiles, especially natural fibers like cotton, wool, and silk. Reactive dyes get their name from their chemical reaction with fibers, creating a strong covalent bond. This reaction takes place through a process known as dye fixation or dyeing, and with this reaction, a covalent bond is established between fiber and dye molecule [22, 23].

Research Methodology of Dye Degradation

We have seen that wastewater treatment is a challenge in the modern world, as water resources are becoming scarce and polluted by various contaminants, and these contaminants, synthetic dyes, organic materials, inorganic materials, and heavy metals cause consequential threats to human health and the ecosystem. To overcome this problem, researchers have explored novel methods based on nanotechnology, which is the manipulation of matter at the nanoscale. Nanomaterials have unique properties, such as high reactivity, overall larger surface area, the potential for functionalization, and size-dependent behaviour. These properties make them ideal options for treating wastewater and purifying water.

Suvanka Dutta *et al.* discussed that azo and anthraquinone dyes are common pollutants in textile wastewater and pose serious threats to the environment and human health. In this study, we evaluated the performance of nanoscale zero-valent iron (NZVI) particles in degrading reactive azo and anthraquinone dyes: reactive blue MR (RBMR) and remazol brilliant orange 3RID (RBO3RID). The parameters of nZVI dosage (0.15 - 0.3 g/L), initial dye concentration (100 - 500 mg/L), and pH (2 - 12) were varied to optimize dye removal efficiency. Researchers found that NZVI particles could degrade up to 97% of both dyes under optimal conditions, which were a pH of 8-12 and an NZVI dosage of 0.3 g/L. The degradation was rapid and occurred mostly throughout the first fifteen minutes of communication. The synthesized iron nanoparticles appeared spherical, with particle diameters under 35 nm and an average size of 16.46 ± 5.98 nm (n = 116). TEM images revealed particle aggregation, while the XRD plot displayed a prominent peak at 45° and smaller peaks at 64° and 82°, indicating Fe^0 in the nZVI particles [24] (Fig. **2**).

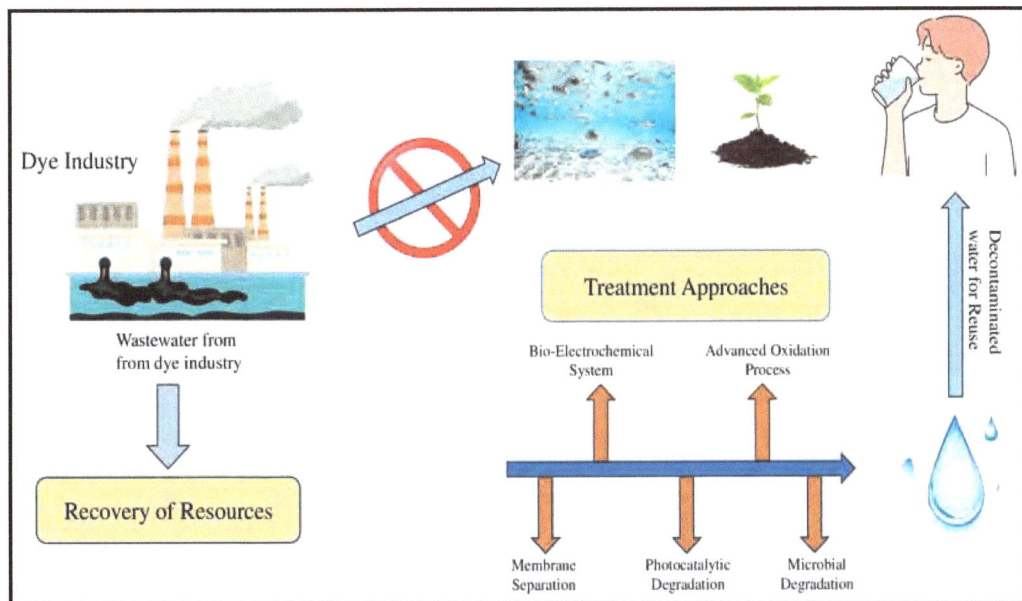

Fig. (2). From Contaminated Hues to Purified Waters: A Pictorial Exploration Odyssey through Dye Industry Effluents and Remediation Techniques:

Chandra Devi Raman *et al.* investigated the use of nano iron particles for decolorization of textile wastewater and mono-azo-methyl orange dye. The existing methods for treating textile dyeing wastewater are not effective enough to deal with persistent and toxic textile dyes in the environment. The nano iron particles presents a promising alternative because they are highly reactive, nontoxic, and cost-effective. The results show that the dye was decolorized by 98% in 30 min, and the total organic carbon was reduced by 54%. The analysis of the intermediate products like m/z 128, 149, 73, 84, 55, and 56 indicates that nano iron particles can degrade textile dyes to lower molecular compounds. Therefore, the reaction mechanism and dye decolorization pathways are suggested. The application of nano iron particles to real dyeing textile industry wastewater achieved 82% color removal and substantial reduction of inorganic pollutants such as sulfates and chlorides. Consequently, it is strongly advised to use nano iron particles to efficiently clean up wastewater from textile dyeing [25].

N. A. Zarime *et al.* developed the nZVI (Nano zero-valent iron), which has demonstrated significant potential in decolorizing anionic dyes, yet its tendency to aggregate can hinder its reactivity. The B-nZVI (bentonite-supported nZVI) composite was synthesized through a reduction process involving ferrous ions and sodium borohydride in the presence of bentonite. This study aimed to evaluate the effectiveness of the B-nZVI composite in removing acid orange II from aqueous solutions. The materials, nZVI, B-nZVI, and bentonite, underwent detailed characterization, including assessments of their physical, chemical, mineralogical, and morphological properties using techniques like BET surface area analysis, X-ray photoelectron spectroscopy combined with field emission scanning electron microscopy (FESEM), Auger electron spectroscopy (XPS-AES), and X-ray diffraction (XRD). Batch adsorption tests were performed on each material to assess dye removal efficiency, examining the effects of factors like initial dye concentration, adsorbent dose, pH, kinetics, and temperature. The results showed that B-nZVI (with an optimal adsorption capacity, qe opt, of 8.9286 mg/g) outperformed both nZVI (qe opt = 8.1224 mg/g) and bentonite alone (qe opt = 5.8469 mg/g). The increased efficiency of B-nZVI was attributed to improved dispersion of nZVI particles within the bentonite matrix, providing additional active sites for adsorption. This research indicates that B-nZVI can serve as an economical and effective adsorbent for the decolorization of anionic dyes in synthetic dye wastewater treatment. nZVI, B-nZVI, and bentonite display the morphology and textures (shape and size) of the nanoparticles by using field emission scanning electron microscopy (FESEM) analysis. It illustrates the nZVI particles, which have spherical shapes (ranging from 54.75 nm to 71.46 nm) and are aggregated into a chain-like structure [26].

Shoniya Thomas *et al.* investigated the degradation of an acid red 1 (AR1) and model azo dye by zero-valent iron nanoparticles (ZVINP), as well as their synergic effect with the ultrasound (US). The ZVINP treatment at pH 3 resulted in 53.0% color removal and 48.5% total organic carbon reduction of AR1 after 25 min. However, the combination of ZVINP and US enhanced the efficiency of the system, removing almost 95% of the color in under five minutes. The LC-Q-TOF analysis confirmed the formation of aromatic transformed products, which accounted for around 55% of total organic carbon reduction. The synergic effect of ZVINP and US was attributed to the different mechanisms operating at alkaline pH (pH 9) and acidic (pH 3). The products that were found revealed some information about the mechanism. In this study, the ZVINP utilized was easily recoverable; therefore, it may be regarded as an effective alternative to the standard Fenton's reagent. The ZVINP synthesized in this study was characterized using SEM, EDS, and TEM. SEM images reveal the morphology and distribution of ZVINP, showing chain-like aggregates ranging from 25–40 nm due to magnetic attraction between particles. EDS analysis confirms a core-shell structure, with 77.7% iron and 22.3% oxygen. TEM images show particles under 15 nm, forming larger spherical aggregates (30–50 nm) that link into chains with particle size distribution. The SAED pattern's discrete spots indicate a crystalline structure and bright-field imaging confirms nanoscale particle sizes [26].

G. Kumaravel Dinesh *et al.* presented the synthesis of amphoteric zero-valent copper (Cu^0) nanoparticles using hibiscus rosa-sinensis extract as a green reducing and stabilizing agent under ultrasound irradiation. The sonochemical method enhanced the crystallinity and morphology of the Cu^0 nanoparticles, which were characterized by XRD, SEM, EDX, PSA, BET and other techniques. The polyphenols in the hibiscus extract acted as both hard (–OH) and soft (–C=O) ligands, which reduced Cu^{2+} to Cu^0 and capped and stabilized the Cu^0 nanoparticles, respectively. The sonochemical degradation of 5-fluorouracil and lovastatin drugs using Cu^0 nanoparticles was investigated in the presence of different oxidants. The results indicated that the Cu^0 nanoparticles were effective in degrading both drugs, with degradation efficiencies of 91.3% and 93.2%, respectively. The degradation was attributed to the generation of ˙OH and O_2 radicals from the oxidation of Cu^+ released from Cu^0 by ultrasound. The Cu^0 nanoparticles exhibited amphoteric behavior and could degrade drugs that are resistant to conventional methods [27].

Chunwei Shi *et al.* identified that azo dyes, such as Congo red, can easily cause cancer when they come in contact with or are absorbed by the human body, so it is urgent to find a fast and simple method for degrading Congo red. In order to achieve this research goal better, an ultrasonic method was used to degrade Congo red solution in a rotating flow field. The concentration of hydroxyl radical in the solution was significantly increased under the action of ultrasonic cavitation, the chemical action of zero-valent iron, and mechanochemistry. Under the strong oxidation of hydroxyl radical and the reduction of nano zero-valent iron peeled off in the reaction process, the reaction speed is significantly accelerated and promotes the reaction. The effect of increasing stirring and adding iron powder particles on ultrasonic cavitation was studied by numerical simulation, and the yield of hydroxyl radical in the system was measured by fluorescence analysis. The experimental results show that, first, the rotating field formed by mixing increases the uniformity of ultrasonic sound field distribution and the amplitude of sound pressure, and it improves the cavitation intensity. In the effective dispersion area, the strong ultrasonic wave can form a temporary high-energy microenvironment in the suspension through cavitation, generate high-strength shockwaves and micro jets, and thus significantly deagglomerate the iron powder aggregates. The addition of iron powder particles then provides a complementary Fenton reagent for the degradation reaction. The concentration of hydroxyl radicals in the solution was significantly increased by the synergy of the two actions. The degradation rate of Congo red reached more than 99% after 30 min of reaction [28].

Iran Manesh *et al.* discussed the novel nanocomposite of nanoscale zero-valent iron-graphene (nZVI-G). This nanocomposite is used as an adsorbent for the removal and extraction of diazo direct red 81 dye from water. The nZVI-G was characterized by X-ray diffraction (XRD), scanning electron microscopy (SEM), Fourier-transform infrared (FT-IR), and energy-dispersive X-Ray spectroscopy (EDS) techniques. The particle size ranged from 20 nm to 35 nm. A study was conducted on the impact of several experimental factors on dye removal, including contact duration, adsorbent dosage, pH, temperature, and initial dye concentration. The maximum dye removal of 92% was achieved under the optimal conditions of pH = 3, contact time of 10 minutes, initial dye concentration of 10 mg/L, and adsorbent dose of 0.05 g. The adsorption data were well fitted by the Langmuir isotherm model ($R2$= 0.9773), having a maximum 29.07 mg/g adsorption capacity. The kinetic data followed the pseudo-first-order model ($R2 = 0.9838$). The adsorption process was shown to be exothermic and spontaneous by thermodynamic study. The nZVI-G nanocomposite showed a high adsorption performance and could be used as a cost-effective and efficient adsorbent to remove Direct Red 81 dye from water. The nZVI-G nanocomposite also exhibited reduction and degradation abilities for the dye due to the generation of radicals in a short reaction time [29].

Alexander J. Sutherland *et al.*described how electro-catalytic microfiltration membranes can be used to break down azo dyes in water. Chlorinated organic compounds (COCs) and azo dyes are harmful chemicals that can cause cancer and damage the environment. Traditional methods of treating these pollutants are not very effective, and they developed electro-catalytic nanocomposite membranes by embedding nZVI particles and double-walled and single-walled carbon nanotubes (SW/DWCNTs) into a polyethersulfone (PES) membrane. The nanocomposite membranes have a porous, conductive, and reactive surface that can degrade azo dyes electrochemically. In order to decompose methyl orange (MO) dye, charged membranes containing nZVI-SW/DWCNT nanocomposites on their surface were employed as cathodes in two distinct setups: a continuous dead-end filtering system and a batch system. The graphite anode system had a voltage of -2. The outcomes showed that in 2 to 3 hours, the charged membranes could eliminate all traces of the 0.25 mM MO (7.5 mMoles in 30 mL) in the batch system. The performance of nZVI-SW/DWCNT cathodes was better than the controls, indicating that the applied current could restore the reactivity of the iron nanoparticles in the nanocomposites. The continuous system with nZVI-SW/DWCNT membranes achieved approx. 87 mol % removal of MO and had a 2.7 times higher removal rate than the batch system and a 1.6 times higher removal rate than the continuous system with membranes without nZVI. Moreover, the applied voltage to the membrane surface nanocomposites enhanced the transmembrane flux, allowing for higher throughput at a given feed pressure.

These experiments demonstrated the potential of using reductive degradation by charged membranes with nZVI-SW/DWCNT nanocomposites to treat wastewater containing azo dyes and chlorinated organic compounds. A membrane-based system that can continuously regenerate the reactivity of nano zero-valent iron (nZVI) particles has been developed for the degradation of azo dyes and COC_s in water. This system overcomes the limitations of conventional methods that use nZVI as a single-use reagent and require frequent replenishment. The SEM image presents a cross-sectional view of an nZVI-SW/DWCNT membrane. The analysis of this image, along with similar images of membranes with the same specifications, revealed that the average thickness of the 10 mg nZVI-SW/DWCNT membrane composite layer is approximately 6 μm [30].

Aditya Kumar Jha *et al.* discussed that the conventional methods for dye removal face limitations in effectively addressing these pollutants. This study explored the photocatalytic breakdown of methyl orange (MO) dye using UV-assisted nanozerovalent iron particles (nZVI). The nZVI particles were synthesized using ferrous sulfate heptahydrate ($FeSO_4 \cdot 7H_2O$) and the leaf extract of *Shorea robusta* (Sal) as a reducing agent. The characterization of the synthesized nZVI was conducted using methods such as scanning electron microscopy (SEM), Fourier transform infrared spectroscopy (FTIR), dynamic light scattering (DLS), zeta potential (ZP), atomic force microscopy (AFM), X-ray diffraction (XRD), and transmission electron microscopy (TEM). The degradation of MO was then achieved under UV light, with optimization performed by adjusting dye concentration, nZVI dosage, pH, and exposure time, with degradation progress monitored *via* UV spectrophotometry at a peak wavelength of 465 nm. Characterization results confirmed the successful synthesis of nZVI, achieving an optimal degradation efficiency of 78% for MO. Additional analysis, including germination tests and FTIR of the dye pre- and post-degradation, further confirmed MO degradation [31].

Caiqi Gao *et al.* investigated the carbothermal reduction technology. Copper slag (CS) was transformed into porous silicate-supported micro nano zero-valent iron, which was used to activate persulfate (PS) for the removal of organic pollutants. The preparation conditions had a significant impact on the activator's characteristics. Calcinating for 20 minutes at 1100 °C with 20% anthracite was the most effective way to degrade orange G (OG). Higher doses of PS or micro-nano zero-valent iron (Psi-ZVI) and higher reaction temperatures accelerated the pace of OG breakdown. Psi-ZVI also performed exceptionally well in activating PS to degrade various organic dyes over a broad pH range. The safety and recyclability of Psi-ZVI were established. Experiments utilizing electron paramagnetic resonance and radical scavenging demonstrated that the most reactive species in the Psi-ZVI/PS system was the sulfate radical ($SO_4^{\cdot-}$). The results of X-ray diffraction showed that for the reduction of iron-bearing minerals to ZVI, the high calcination temperature (1100 °C) was favorable. The results from energy-dispersive spectroscopy and scanning electron microscopy showed that the silicate holes' surface was created by nano to micro-sized ZVI particles and that the Psi-ZVI had a porous structure. According to the X-ray photoelectron spectra, ZVI was first changed into Fe (II), which primarily stimulated PS and created Fe(III) in the Psi-ZVI/PS system. Additionally, gas chromatography-mass spectrometry was used to identify the intermediates of OG, and a potential OG breakdown mechanism was proposed. This work presents a unique strategy to reuse CS as a heterogeneous activator to successfully and efficiently activate the PS. The XRD and SEM-EDS results confirmed that fayalite and magnetite in CS were reduced to ZVI after modification, leading to good dispersion of micro-nano ZVI with the large specific surface area on the porous silicate holes in PSi@ZVI [32].

Tanima Nandia *et al.* reported that the synthesis and characterization of diethylenetriaminepentaacetic acid (DTPA) stabilized nano-scale zero-valent iron and its application for the degradation of various organic pollutants in water under ultrasound irradiation (US). The DTPA-stabilized nZVI was prepared by the coprecipitation method and characterized by different techniques. The results showed that the DTPA-stabilized nZVI had good dispersion and stability with an average particle size of about 50 nm. The degradation efficiency of the DTPA stabilized nZVI for methyl orange (MO), azo-based textile dyes, and phenols was investigated in the presence of US. The degradation rates were found to be 95%, 85%, and 70% for MO, azo dyes, and phenols, respectively, within a short time span. The ac conductivity of the polluted water was also measured during the degradation process, which revealed distinct patterns and provided an alternative way to monitor the degradation without using spectrophotometric analysis. The reusability of DTPA-stabilized nZVI has been tested for 3 cycles and has shown 85% degradation of MO in the last cycle. The FESEM images of bare and DTPA-stabilized nZVI particles reveal their spherical shape. The reduced particle size, averaging 27.3 nm, increases the surface area-to-volume ratio, enhancing reactivity and dye degradation efficiency. The analysis of over 250 nanoparticles showed that 90% were within 50 nm, though some larger aggregates around 100 nm were observed. This size reduction significantly boosts the catalytic properties of the DTPA-stabilized ZVI particles [33].

Faiza Lughmani *et al.*showed that MCC (microcrystalline cellulose) was chemically altered into OC (oxidized cellulose) and derivatives 6-deoxycellulose (N,N-diethyl)amine (MCC-DEM) and 6-deoxycellulose hydrazide (MCC-Hyd). These derivatives served as a scaffold for producing copper nanoparticles. Copper ions from an aqueous solution were first adsorbed onto the films of the modified cellulose and then reduced to ZVC (zero-valent copper nanoparticles) using sodium borohydride. Comprehensive characterization of the modified cellulose derivatives and Cu NP-embedded films was carried out using techniques such as elemental analysis, Fourier transform infrared spectroscopy, scanning electron microscopy, X-ray diffraction, nuclear magnetic resonance spectroscopy, ultraviolet-visible (UV–VIS) spectroscopy, and X-ray photoelectron spectroscopy, and was employed to study the degradation of contaminants, including 4-nitrophenol and azo dyes like methylene blue, Congo red, and methyl orange. The results indicated that effective degradation occurred only when ZVC NPs were present. Notably, oxidized cellulose, MCC-DEM, and MCC-Hyd showed high degradation efficiencies (over 85%) across all tests. This study suggests that MCC derivatives could be promising renewable materials for the removal of water pollutants [34].

Yanpeng Mao *et al.* discussed microwave radiation and "nanoscale zero-valent iron" (MW–nZVI) to study how fast and well soluble dyes can be degraded in different pH conditions ranging from 5.0 to 9.0. The nZVI particles, which have a size of 40 to 70 nm, were created using a liquid-phase chemical reduction process with starch used as a dispersant. The MW–nZVI method has shown better results than using only nZVI for removing solvents Reactive Yellow K-RN and Blue 36 and their total organic carbon. The dye's removal efficiency improved dramatically as pH decreased, but variation in the MW power has shown a negligible effect on the removal efficiency. The removal dye using 23 MW–nZVI followed both the pseudo-first-order model and an empirical equation, while the 24 kinetics of dye removal using nZVI could only be described by an empirical equation. The study concluded that heating of the dye solutions by microwave, as well as consumption of Fe(II) and acceleration of 26 corrosion of nZVI, was possibly the reason for the enhanced dye 27 degradation [35].

Thivaharan Varadavenkatesan *et al.* reported that the extract of Arachis hypogaea nuts, commonly known as peanuts, was used to synthesize silver nanoparticles in this study. The formation of the nanoparticles was confirmed by UV–visible spectroscopy, which showed a peak at 435 nm corresponding to the surface plasmon resonance of silver. The morphology, size, composition, and crystallinity of the nanoparticles were characterized by scanning electron microscopy, energy-dispersive X-ray spectroscopy, and X-ray diffraction, respectively. The role of phytochemicals in the reduction of silver ions to metallic silver was investigated by FT-IR spectroscopy. The nanoparticles exhibited a narrow size distribution and a negative zeta potential of -34.6 mV, indicating their stability and mono-dispersity. The catalytic potential of the nanoparticles was demonstrated by their ability to degrade two commercial fabric dyes, acid red 88 and acid blue 113, in the presence of $NaBH_4$. This study presents a simple, eco-friendly, rapid, and biocompatible method for the synthesis of silver nanoparticles, which is green synthesis, and their application in environmental remediation [36].

Burcu Ileri *et al.* discussed that the reactive dyes, widely used in textile manufacturing, pose carcinogenic and toxic risks to both ecosystems and human health. This study focused on assessing the removal of Reactive Blue 19 (RB19) from water and synthetic textile effluent by applying nanoscale zero-valent aluminum (nZVAl), an ultrasonic bath (US–40 kHz), and a combined US/nZVAl treatment while varying parameters such as pH, nZVAl dosage, contact time, and initial dye concentration. Acidic pH significantly enhanced RB19 degradation. Increasing the nZVAl dosage and contact time improved RB19 removal in both water and synthetic wastewater, with the combined US/nZVAl treatment showing high effectiveness. For instance, a dosage of 0.10g with a 30-minute exposure in the combined method achieved similar removal as 0.20 g of nZVAl alone in 60 minutes. The sono-degradation process stimulated nZVAl surface activation *via* ultrasonic cavitation and shock waves, enhancing RB19 removal with shorter contact times and lower nZVAl dosages. Higher initial dye concentrations, however, reduced removal efficiency. Reusability tests indicated that nZVAl particles could be reused for four cycles in the combined method and two cycles with nZVAl alone [37].

Mohammad Neaz Morshed *et al.* reported that the porous 3DG (three-dimensional graphene) structure, synthesized *via* chemical vapor deposition, served as a matrix to host nZVI particles. Methods were developed to control the morphology, dispersion, and size of nZVI particles on the 3DG surface, resulting in immobilized particles around 100 nm in size with dense distribution. The composite material (3DG-Fe) removed 94.5% of the orange IV azo dye within 60 minutes, outperforming free Fe nanoparticles in the solution, which achieved only 70.9% removal. Furthermore, the degradation rate of orange IV with 3DG-Fe was approximately five times faster than with free Fe nanoparticles. The study also explored the effects of 3DG-Fe dosage, dye concentration, pH, and temperature on the degradation efficiency. Results suggested that lower pH and higher temperature enhanced both the reaction rate and efficiency. Kinetic analysis indicated that the degradation of Orange IV follows pseudo-first-order kinetics, likely involving adsorption and redox reactions, with an activation energy of 39.2 kJ/mol. SEM analysis revealed that the untreated PET nonwoven has a network of randomly arranged, smooth fibers. Plasma treatment showed no significant morphological change. However, the PN@Dr nonwoven displayed irregular patches on the fiber surface, likely due to dendrimer branches forming layered clusters [38].

M.M. Naim *et al.* reported that the "chitosan nanoparticles (CS-NPs)" have demonstrated efficacy in the removal of dyes, phenols, and heavy metals from wastewater. Silver nanoparticles (Ag-NPs) are significant in the industry of electronics. Since "zero-valent iron (ZVI)" can adsorb a wide range of significant pollutants, such as metalloids and heavy metals, it has been thoroughly researched for environmental remediation. It also has a very strong reducing power. In the current study, chitosan, iron, and silver nanoparticles were prepared using chemical techniques. $AgNO_3$ served as the precursor for the production of Ag-NPs, which were then applied to the filter paper. Additionally, Ag-NPs and casting solution were combined to create a cellulose acetate (CA) membrane, and the efficacy of the membranes to destroy the germs was next tested by exposing both to Escherichia coli. The E. Coli was cultured using a normal protocol, and the colonies were enumerated using treated filter paper and membrane both before and after treatment. Notably, the current work includes a novel, straightforward technique for synthesizing Ag-NPs. The outcomes demonstrated that, in both filter paper and CA membrane, Ag-NPs totally eliminate E. coli germs at room temperature. Additionally, ferric chloride was reduced by sodium borohydride to create nano iron (nZVI), which was filtered and used to extract Cu (II) ions from an aqueous solution. Two procedures were used to prepare CS-NPs: first, low molecular weight chitosan (LWCS) was synthesized, and then, using varying concentrations of H_2O_2 solution, LWCS was broken down to produce CS-NPs. $AgNO_3$ was used to prepare CS-NPs in the second technique. Because the CS-NPs membrane contains a variety of functional groups in addition to amino and hydroxyl groups, it was able to biosorb 70.68 and 42.1% of NaCl from 9.38 and 15.2 g/L salt solutions, respectively. In contrast, biosorption of $CuSO_4$ was 59.8% from 12.5 g/L solution [39].

Shahid Ali Khan *et al.* outlined the purpose of supporting zero-valent metal nanoparticles, and a simple and extremely porous heterostructure nanocomposite was created. The ZnO/carbon black (ZnO/CB) nanocomposite, which is also referred to as the ZCA sheet, was embedded in cellulose acetate polymer (CA) and supported the bimetallic Cu Ag nanoparticles (NPs) and metallic Cu, Ag. ZCA-5, ZCA-4, and ZCA-3 are the names of the ZnO/CB incorporations made to the CA polymer host molecule in 5, 4, and 3 weight percent. The catalytic tunability for CuAg/ZCA-5 was evaluated by varying the molar ratio of Cu and Ag ion concentrations. Consequently, eight model pollutants were further chosen for removal utilizing CuAg/ZCA-5: m-nitrophenol (MNP), o-nitrophenol (ONP), p-nitrophenol (PNP), methyl orange (MO), 2,6-dinitrophenol (DNP), rhodamine B (RB), congo red (CR) and, methylene blue (MB). The Kapp value for MO was 9.0×10^{-2} min^{-1}, whereas it was 1.9×10^{-1} min^{-1} for PNP. Methyl orange (MO) breaks down faster than the other several nitrophenols, although the CuAg/ZCA-5 reaction rate followed the pattern: "PNP > ONP > MNP > DNP". The morphology, shape, elemental analysis, functional groups, and peak crystallinity were investigated using FESEM, EDS, ATR-FTIR, and XRD, respectively. The FESEM analysis revealed the distinct sheet-like morphology of ZnO/CB particles, while the ZCA-5 nanocomposite displayed a rough surface with numerous cavities, enhancing its ability to attract metal ions. The high polarity of oxygen and OH groups in ZnO/CB facilitated the effective capture of Ag^+ and Cu^{2+} ions, leading to monodispersed Ag^0 particles and dense Cu^0 and Cu^0Ag^0 structures [40].

Nilsun H. Ince *et al.* noted that the advanced oxidation processes (AOPs) and ultrasound can break down and degrade different organic contaminants that are harmful to the environment and human health. These pollutants include endocrine disruptors, anti-inflammatories, painkillers, and azo dyes. The paper focuses on the elimination of organic carbon and the treatability of these pollutants by ultrasound (US) and AOPs. The main findings of the paper are: i) under ideal circumstances, ultrasonic can oxidize these contaminants completely or partially, but it cannot mineralize them; ii) the hydroxyl radicals' (% OH) position in the solution or at the bubble-liquid interface, the pollutant's characteristics, the ultrasound's frequency and pH affect the degradation mechanism; iii) combining ultrasound with other AOPs such as UV/peroxide, ozone, UV/Fenton and Fenton enhances the degradation and mineralization performance significantly; iv) using catalysts such as alumina, TiO_2, and zero-valent iron using ultrasound can also improve the degradation and mineralization efficiency, especially at the catalyst surface where most reactions occur even if pollutants are highly soluble; v) catalysts modification by reducing their size and proportions (to nanoscale) or coating them with metal nanoparticles can increase their activity and stability under ultrasound, light or both [41].

Mohd Shaiful Sajab *et al.* investigated the process of Fenton oxidation, which uses Fe (III), Fe (II), and *in situ* creation of zero-valent iron (Fe(0)), for the discoloration of textile effluent from industrial wastewater from batik. The Fenton oxidation reaction was found to be optimal on effluent with pH 5, a color concentration of 4005 mg/L Pt-Co units, and a catalyst dosage of 0.5 mg/mL, which met the water standard regulation of Malaysian quality. The discoloration efficiency of precursors of Fe(0) was higher for Fe (II) than for Fe (III). The ex situ-synthesized Fe(0) particles formed chain-like aggregates and exhibited nano-spherical shapes with a diameter range of 20–70 nm. Ex situ Fe(0) had a discoloration performance of 97%, whereas Fe (II) had an 89% discoloration performance. Because *in situ* Fe (0) production in batik effluent is more reactive than partly oxidized ex situ Fe (0), it completely discolored the material. Moreover, *in situ* Fe (0) had a faster reaction time, which was shorter than 5 minutes. Fe(0), Fe(II), and Fe(III) regeneration showed the potential to be a recyclable catalyst for up to three Fenton oxidation cycles; however, the amount of effluent discoloration decreased somewhat to 62.1% [42].

ZERO VALENT IRON NANOPARTICLES IN THE TREATMENT OF INORGANIC POLLUTANTS FROM WASTEWATER

Treatment of Heavy Metals

Metallic elements that have a high atomic weight and a density that is at least five times greater than water are known as heavy metals. Lead, mercury, cadmium, arsenic, chromium, and other elements are among them. The main sources of these heavy metals in water are mining, agriculture, and industrial waste. Heavy metals in wastewater pose serious environmental hazards that affect ecosystems, human health, and the quality of water resources. Metals such as lead, mercury, cadmium, and arsenic are highly toxic to aquatic organisms, even at low levels, resulting in adverse effects on their growth, reproduction, and behavior. This toxicity can lead to bioaccumulation, where these metals build up in the tissues of organisms and reach higher concentrations in predators at the top of the food chain, disrupting aquatic food webs. Furthermore, heavy metals can settle into sediments, leading to long-term contamination problems that can harm benthic organisms and decrease biodiversity. They can also seep into groundwater, jeopardizing drinking water supplies and posing significant health risks to communities relying on these sources. When heavy metals are released onto land, they can contaminate soil, impair fertility, and hinder plant growth while also entering the food chain through plant absorption. The health implications for humans include a range of serious conditions, such as neurological disorders and developmental problems, particularly impacting vulnerable groups like children and pregnant women. One of the contaminants that can occur in water is chromium, which is most frequently found in groundwater and wastewater in the trivalent (Cr (III)) and hexavalent (Cr (VI)) forms. Because of its high solubility and mobility in aqueous conditions, Cr (VI) is more mobile and maybe more hazardous than Cr (III) in the pH range of 5-8. Cr (VI) has been linked to a number of health concerns, including respiratory disorders, skin irritation, gastrointestinal disorders, and kidney and liver damage, and has been categorized as potentially carcinogenic [43].

Industrial activities such as metallurgy, refractories, and chemical wastes are examples of manmade sources of Cr (VI) pollution in water [44]. It is crucial to remember, though, that recent research has also suggested that geogenic fonts may contribute to Cr (VI) groundwater contamination. Heavy metals can be aloof from water using a number of processes, including: 1. Chemical precipitation: In this technique, water is treated with a chemical reagent to create an insoluble precipitate that may be filtered or sedimented out. 2. Ion exchange: In this technique, the heavy metal ions present in water can be exchanged by other ions fixed on a resin or alternative exchange medium. 3. Adsorption: This technique entails removing the ions of heavy metal from the consumable water by adsorbing them using adsorbents like zeolites, activated carbon, or other materials. 4. Membrane filtration: In this technique, the heavy metal ions in the water are separated according to their size and charge using a membrane. 5. Electrochemical treatment: This technique uses electroplating or electrocoagulation with an electric current to remove heavy metals from water. 6. Biological treatment: This technique uses biosorption or bioremediation processes to extract heavy metals from the water. 7. Treatment based on nanotechnology: This technique uses carbon nanotubes and nZVI to eliminate heavy metals from the water *via* adsorption or reduction [45, 46].

Nano zero-valent iron (nZVI) has gained remarkable attention in recent years as a promising restoration technique in the removal of ions of heavy metals from water due to its versatile features, such as large surface area and reactivity. Electrons from the nZVI surface are transferred to the heavy metal ions during the reduction process, which causes the metals to immobilize or precipitate [47].

Nevertheless, there are a few limitations to treating groundwater with conventional nZVI: 1. Agglomeration and sedimentation: The mobility and efficacy of conventional nZVI particles in groundwater restoration are diminished due to their propensity for agglomeration and sedimentation. 2. Limited stability: The long-term efficacy of traditional nZVI particles in treating contaminated groundwater may be hampered by their limited stability. 3. Reduced removal efficiency under certain conditions: The removal efficiency of traditional nZVI particles may decrease under high pH conditions or when there is an imbalance in the iron to Cr (VI) ratio. These limitations highlight the need for improved nZVI composites, in combination with other materials such as the CNT-nZVI, SA-nZVI composite, and many more, which address these challenges and offer enhanced stability, mobility, and removal efficiency for groundwater treatment applications [48].

In this study, the removal capabilities of heavy metals with various composites of nZVI constituents are reported. The purpose of this compilation is to enrich the removal efficiency of heavy metal Cr (VI), Pb, Cd, and Zn from water at different pH values of the water system, temperature, and the initial concentration of the contaminant. This chapter also presents the synthetic procedure of different composites (biochar-CMC-nZVI, ACF-nZVI, nZVI/OMC, PEG-nZVI@BC, *etc.*) that are more stable and efficient than nZVI particles (Fig. **3**).

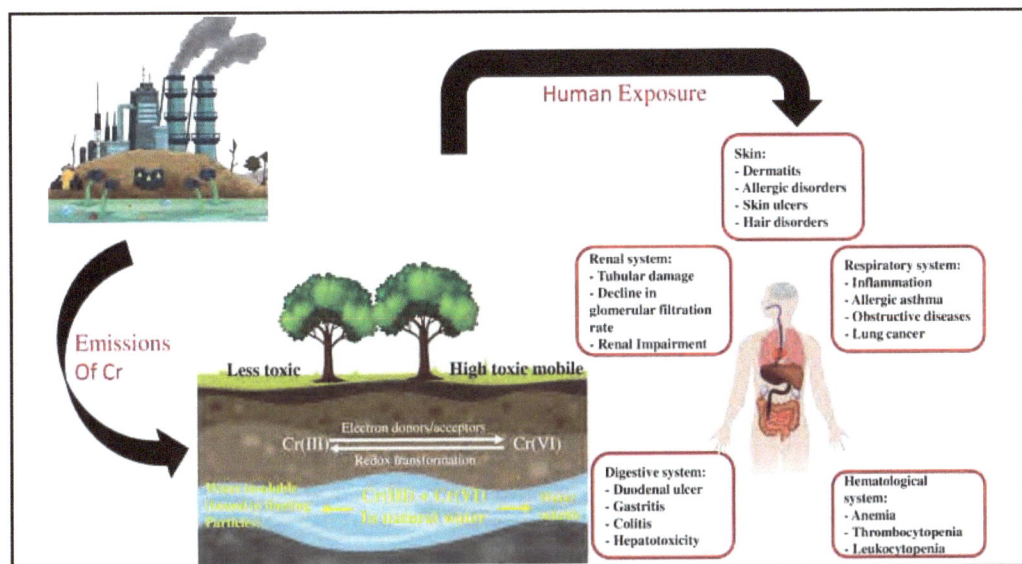

Fig. (3). Exploring the link between chromium emission from industry to human health.

Huihui Wu *et al.* presented a novel approach to remediate Cr (VI) with high efficiency using polyethylene glycol-stabilized nano zero-valent iron supported by biochar. The PEG-nZVI@BC composites were synthesized by impregnating biochar with polyethylene glycol (PEG) and $FeSO_4$ solution, followed by liquid-phase reduction using $NaBH_4$. Specifically, PEG was solubilized in a mixture solution of ethyl alcohol and deionized water in a three-necked flask. $FeSO_4 \cdot 7H_2O$ was then mixed in the PEG solution under nitrogen protection, and biochar was added with continuous stirring. Subsequently, a fresh $NaBH_4$ solution was added dropwise under vigorous stirring to fix and stabilize nano zero-valent iron (nZVI) on the surface of the biochar. The resulting solution was then centrifuged to obtain the PEG-nZVI@BC mixtures. Four different types of PEG-nZVI@BC mixtures were prepared with varying PEG quantities for comparative analysis. Resultant composites characterized by SEM, XRD, FTIR, and XPSwere found to have a great number of –OH functional groups, which enhance their performance. The XRD analysis of the 1.5PEG-nZVI@BC composite provides key insights into its crystalline structure. Distinct peaks in the diffraction pattern indicate the presence of biochar (BC), with significant reflections observed at approximately 24° and 43°, corresponding to the (002) and (001) planes of carbon fibers. The composite also displays prominent diffraction peaks near 44.7° and a less intense peak at approximately 82.5°, associated with the (110) and (211) planes of α-Fe^0. These results indicate the successful loading of nano zero-valent iron (nZVI) onto the biochar matrix. Notably, the lack of strong iron oxide peaks implies that PEG stabilization effectively minimized nZVI oxidation, maintaining its reactivity for

Cr (VI) remediation. SEM analysis revealed that the prepared composite has an irregular, porous biochar (BC) surface, enhancing its adsorption capacity. The nZVI particles, with uniform spherical shapes ranging from 50 to 350 nm, were more evenly distributed across the biochar surface in the 1.5PEG-nZVI@BC composite than in those with lower PEG content. The higher PEG concentration reduced particle agglomeration, improving nZVI dispersion. This enhanced distribution is expected to increase the composite's reactivity for Cr (VI) remediation, highlighting PEG's role in optimizing the composite's structure. XPS analysis of the composite revealed valuable information about the surface composition and chemical states involved in chromium (Cr) remediation. High-resolution scans in the Cr 2p region displayed distinct peaks for Cr (III) and Cr (VI), indicating that around 77.9% of the immobilized Cr was reduced to Cr (III), a less toxic form, emphasizing the reduction of hazardous Cr (VI). Additionally, the Fe 2p spectra showed peaks for Fe (II) and Fe (III), indicating partial oxidation of nZVI particles during remediation. These oxidation states, along with the even surface distribution of chromium, confirm the composite's efficiency in Cr (VI) removal through combined adsorption and reduction. The XPS results support the composite's effectiveness and its suitability for environmental remediation applications. The batch experiments in this study investigated the outcome of PEG concentration, pH, Cr (VI) concentration, and meld time on the efficiency of Cr (VI) removal using PEG-nZVI@BC composites. The findings of the research suggest that PEG-nZVI@BC had significantly greater Cr (VI) removal capabilities in comparison to biochar alone and nZVI@BC. The optimal PEG concentration was found to be 1.5 g, and the optimal pH was found to be 3.0. The kinetic experiment data showed that the Cr (VI) removal rate increased quickly within 60 minutes and, after that, it gradually slowed down. The Sips isotherm model was found to best describe the adsorption behavior of Cr (VI) onto PEG-nZVI@BC, with a maximum adsorption capacity of 125.22 mg/g at 25°C [49].

Hualin Chen *et al.*, in their study, explored the removal efficiency of hexavalent chromium from wastewater using starch-stabilized nanoscale zero-valent iron (S-nZVI). The synthesis of starch-stabilized nanoscale zero-valent iron (S-nZVI) involved a liquid-phase method with some modifications. In a flask containing 100 mL of 0.4 mol L^{-1} $FeSO_4$, 2 g of corn starch and 100 mL of ethanol were combined. The flask was then stirred in an ultrasonic bath for 30 minutes. Subsequently, sodium borohydride ($NaBH_4$) was added until the reaction was finished. A magnet was applied to facilitate nZVI precipitation. To prevent oxygen from dissolving in the aqueous phase, a nitrogen gas flow was used during the entire process. The S-nZVI slurries that had precipitated were shifted into centrifuge tubes, and the chamber was rapidly wrapped with parafilm. After centrifuging the nZVI particles for ten minutes, the supernatant was disposed of, and the remaining material in the centrifuge tubes was cleaned with pure ethanol. There were three rounds of this process. The composites underwent freeze-drying following the last centrifugation, and the finished products were packed bagged and kept at -4°C. SEM analysis of synthesized S-nZVI and commercial nZVI (C-nZVI) revealed that both materials consist mainly of spherical particles. S-nZVI showed uniform adherence to the starch chain, resulting in a consistent particle size distribution from 27.51 to 126.78 nm, with a mean size of 62.56 nm. In contrast, C-nZVI exhibited aggregation on larger particles, with sizes ranging from 30.89 to 187.2 nm and a mean size of 80.13 nm. This indicates that S-nZVI has a smaller, more uniform particle size, likely enhancing its reactivity and effectiveness in chromium removal. XPS analysis provided insights into the chemical composition and oxidation states of iron in S-nZVI and C-nZVI. The XPS spectra showed iron in multiple oxidation states, with prominent peaks for Fe^{2+} and Fe^{3+}. S-nZVI displayed a notable Fe^{2+} peak at 715.8 eV, while peaks at 710.2 eV and 712.0 eV indicated FeO and FeOOH, with Fe^{2+} and Fe^{3+} contents of 33.33% and 66.67%, respectively. C-nZVI showed similar Fe^{2+} and Fe^{3+} distributions, with a shoulder peak at 715.4 eV and a small Fe^{0} peak at 706.1 eV, confirming zero-valent iron in both samples. These oxidation states are critical for understanding nZVI's reactivity in environmental applications. SEM analysis revealed distinct morphological differences between S-nZVI and C-nZVI, both primarily spherical in shape. S-nZVI displayed uniform particle distribution and adhesion to the starch carrier, enhancing stability, with sizes ranging from 27.51 to 126.78 nm and a mean size of 62.56 nm. In contrast, C-nZVI showed significant aggregation on larger particles, with a size range of 30.89 to 187.2 nm and a mean of 80.13 nm. The smaller, more uniform particle size of S-nZVI may improve its reactivity and suitability for applications like chromium removal. Through batch sorption investigations, the factors impacting Cr (VI) reduction and adsorption by nZVIs were examined. Among the variables looked at were the solid-to-solution ratio, the starting pH, and the start chromium level. The 1.5-diphenyl carbohydrazide method (spectrophotometry) was utilized to quantify the Cr (VI) concentration in the experiment, while the atomic absorption spectrophotometer (PE AA800) was utilized to measure the total Cr concentration. The outcomes demonstrated that, even across a broad pH range (1–11), S-nZVI eliminated nearly 100% of Cr (VI). S-nZVI directly absorbed a minor amount of Cr (VI), but the bulk was eliminated by reduction to Cr^{3+} by S-nZVI and subsequent precipitation as $Cr(OH)_3$. Through co-precipitation, the precipitated Cr^{3+} was coupled to ferric hydroxide to create Cr-Fe hydroxide, such as $(Cr_xFe_{1-x})(OH)_3$ [50].

Xiaoshu Lv et al. developed nanoscale nZVI hybridized with the layered double (LDH) decorated reduced graphene oxide (Fe@LDH/rGO) for the elimination of Cr (VI) oxyanions from contaminated water. The synthesis of Fe@LDH/rGO involved several steps: Firstly, graphene oxide (GO) was developed from graphite powder by modified Hummer's process. Subsequently, LDH/rGO was prepared by mixing LDH and rGO under ultrasound, and then the mixture was washed to remove residual impurities. After that, Fe@LDH/rGO was found to be fictitious under an N_2 environment by mixing LDH/rGO and $FeSO_4 \cdot 7H_2O$ in water under ultrasound. Aq. $NaBH_4$ solution was added dropwise to reduce Fe^{2+}. The resulting composite was characterized by techniques such as SEM, TEM, and XRD to examine its structure. SEM analysis revealed key morphological features of the synthesized materials. Reduced graphene oxide (rGO) displayed a crumpled, paper-like structure, which is beneficial for increasing surface area and supporting interactions with nZVI nanoparticles. The layered double hydroxide (LDH) showed a clear, layered structure, confirming successful synthesis. Uniform distribution of iron nanoparticles (Fe° NPs) on the rGO and LDH surfaces was observed, maximizing the contact area for Cr (VI) adsorption. This confirms the effective integration of components in the Fe@LDH/rGO composite, enhancing its potential for contaminant removal applications. TEM images showed that the iron nanoparticles (Fe° NPs) were primarily spherical, with sizes ranging from 50 to 300 nm, reflecting successful nZVI synthesis for effective contaminant removal. The Fe° NPs were uniformly distributed across the rGO matrix, with some particles embedded within the LDH layers, enhancing interactions with Cr (VI) and overall removal efficiency. These TEM results align with SEM findings, confirming the successful integration of the composite's components. According to XRD analysis, reduced graphene oxide (rGO) showed a characteristic peak at $2\theta = 10.3°$, shifted from the graphite peak at $2\theta = 26.5°$, confirming successful reduction. The layered double hydroxide (LDH) exhibited sharp peaks at 2θ values of 11.6°, 23.5°, 34.9°, 39.5°, and 47.0°, indicating a well-crystallized hydrotalcite structure, consistent with JCPDS No. 70-2151. In the Fe@LDH/rGO composite, a peak at 44.5° confirmed the Fe° phase, although rGO and Fe° peaks were generally less distinct due to lower crystallinity. These XRD results confirm the effective synthesis and crystallinity of the composite, which is critical for environmental remediation uses. Trial experiments were led to evaluate the performance of Fe@LDH/rGO in hexavalent chromium elimination. The results showed that Fe@LDH/rGO had a higher removal tendency for Cr (VI) oxyanions compared to bare Fe° NPs, LDH, rGO, Fe@rGO, and other composites. The optimized composition proportion was found to be 80 mg/L Fe°, 0.15 g/L LDH, and 0.15 g/L rGO. The abolition efficiency was found to be influenced by the starting pH, initial Cr (VI) content, and temp. The highest removal efficiency was achieved at pH 7.0, and the removal efficiency decreased with increasing initial Cr (VI) concentration. Up to 35°C, there was an improvement in removal efficiency; beyond that, the efficiency declined. Overall, the batch trials showed how well Fe@LDH/rGO performed in removing Cr (VI) and offered information on the ideal conditions for its use [51].

Shuai Zhang *et al.* developed an innovative biochar-supported nano-scale zero-valent iron (biochar-CMC-nZVI) composite stabilized by carboxymethyl cellulose (CMC) to avoid the deposition of nZVI particles (about 80 nm) and used this composite in water treatment for the removal of Cr (VI). The authors added biochar to the new material because it has many attractive properties like large BET surface area, low bulk density, long-term stability, availability, cheapness, and great adsorption capacity. In addition, biochar is typically made from waste materials such as sawdust, rice husks, and straw and pyrolyzed at relatively low temperatures (<700 °C), with limited oxygen supply. CMC is used to stabilize nZVI because it is a non-toxic associated anionic polymer that prevents nZVI from being oxidized. Biochar-CMC-nZVI composite was made by dissolving $FeSO_4 \cdot 7H_2O$ into 1100 mL blue cover flask containing a certain volume of deionized water (purified nitrogen was purged for 30 min to make oxygen-free water) with continuous stirring and purified nitrogen conditions. After that, CMC solution in different dosages (1 g/100 mL) was added such that CMC concentration in the system was 0%, 0.05%, 0.1%, and 0.2%, and it was stirred for 20 minutes. Thereafter, biochar was added to the above mixture. Following that, a dropwise addition of 50 mL $NaBH_4$ solution took place. Afterward, under nitrogen purging for about thirty minutes, the reaction occurred. The final suspension (biochar: nZVI mass ratio = 1:1) was then aged for twelve hours. The composite biochar-CMC-nZVI was later separated using magnetic separation and freeze-dried prior to use. In this way, four kinds of biochar-CMC-nZVI composites (biochar-nZVI, biochar-0.05CMC-nZVI, biochar-0.1CMC-nZVI, and biochar-0.2CMC-nZVI) were synthesized just by varying the concentrations of CMC. SEM analysis highlighted the morphology of various materials. Biochar appeared irregular and rough, with a particle size under 1.5 μm, while nZVI showed gathered spherical particles around 100 nm. CMC-stabilized nZVI exhibited a nuclear shell structure with particles below 100 nm. The biochar-nZVI composite included nano-spherical particles and sheet-like α-Fe_2O_3 from nZVI oxidation. In biochar-CMC-nZVI composites, biochar-0.1CMC-nZVI displayed smaller, more uniform particles compared to biochar-0.05CMC-nZVI and biochar-0.2CMC-nZVI, where higher CMC (0.2%) caused irregular nZVI shapes, possibly reducing composite reactivity. XRD analysis was performed to examine the crystallinity and elemental composition of the samples before and after reacting with Cr (VI). The XRD patterns showed distinct peaks representing crystalline phases in the biochar-CMC-nZVI composite, with iron oxides like α-Fe_2O_3 indicating nZVI oxidation during the reaction. This analysis revealed changes in crystallinity and the formation of new phases resulting from Cr (VI) interaction. These results suggest that the biochar-CMC-nZVI composite effectively removes Cr (VI) while undergoing structural transformations that may enhance its reactivity and stability for environmental applications. XPS analysis provided insights into the surface composition and chemical states of the biochar-CMC-nZVI composite before and after Cr(VI) adsorption. The XPS spectra showed notable shifts in the binding energies of carbon (C 1s), oxygen (O 1s), iron (Fe $2p^{3/2}$), and chromium (Cr $2p^{3/2}$) post-reaction. The decreased intensity of C–O and O–C=O groups suggested their involvement in Cr (VI) reduction. Chromium was detected on the composite surface, confirming the conversion of Cr (VI) to Cr (III) and the formation of compounds like $Cr_xFe_{1-x}(OH)_3$ and $Fe_xCr_yO_4$. These findings highlight the composite's effectiveness in Cr (VI) reduction and emphasize the role of surface chemistry in its adsorption and reduction processes. Several batch experiments were conducted in sealed glass vials of forty milliliters to decontaminate Cr (VI) from fictitious Cr (VI)-contaminated groundwater. During experimentation, various parameters were studied, such as the effects of CMC concentration on Cr (VI) removal, adsorption kinetics, adsorption isotherms, effects of initial pH, and mechanisms of Cr (VI) removal by the composite. At a dose of 1.25 g/L and an initial pH of 5.6, the research showed that biochar-CMC-nZVI completely removed the entire 100 mg/L Cr (VI) in 18 hours. The acid condition facilitated better Cr (VI) elimination by biochar-CMC-nZVI. The adsorption of Cr (VI), on the other hand, was governed by three major mechanisms: surface complexation, reduction, and electrostatic attraction. Results indicate that the biochar-CMC-nZVI composite may be considered as a cheap "green" sorbent material that can be viable for removing Cr(VI) from environmental matrices [52].

The Following Reaction Represents the Transformation of Cr (VI) into Cr (III) and the Formation of $Cr_xFe_{1-x}(OH)_3$ and $Fe_xCr_yO_4$:

$$3Fe^0 + Cr_2O_7^{2-} + 14H^+ \longrightarrow 3Fe^{2+} + 2Cr^{3+} + 7H_2O \quad (1)$$

$$\text{Surface} - C - e^- \longleftrightarrow \text{Surface} - CO_xH \quad (2)$$

$$HCrO_4^- + 7H^+ + 3e^- \longleftrightarrow Cr^{3+} + 4H_2O \quad (3)$$

$$Fe^0 + H^+ \longleftrightarrow Fe^{2+} + H_2 \quad (4)$$

$$3Fe^{2+} + HCrO_4^- + 7H^+ \longrightarrow 3Fe^{3+} + Cr^{3+} + 4H_2O \quad (5)$$

$$Fe^{2+} + Cr_2O_7^{2-} \longleftrightarrow FeCr_2O_4 \text{ (S)} \quad (6)$$

$$2Cr^{3+} + 6OH^- \longleftrightarrow 2Cr(OH)_3 \ (S) \longleftrightarrow Cr_2O_3 \ (S) + 3H_2O \quad (7)$$

$$2Fe^{3+} + 6OH^- \longleftrightarrow 2Fe(OH)_3 \ (S) \longleftrightarrow Fe_2O_3 \ (S) + 3H_2O \quad (8)$$

$$xCr^{3+} + (1-x)Fe^{3+} + 3H_2O \longleftrightarrow (Cr_xFe_{1-x})(OH)_3 \ (S) + 3H^+ \quad (9)$$

$$xCr^{3+} + (1-x)Fe^{3+} + 3OH^- \longleftrightarrow (Cr_xFe_{1-x})(OH)_3 \ (S) \quad (10)$$

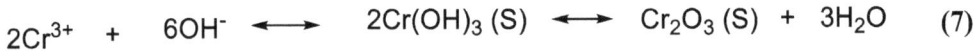

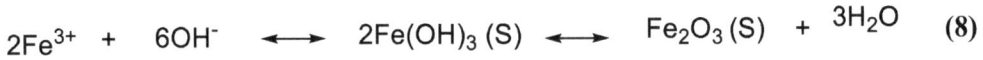

Chunli Zheng *et al.* developed zero-valent iron dispersed by sodium alginate (SA) and applied it to remove lead (Pb) from an aqueous solution. Sodium alginate-wrapped nanoscale zero-valent iron (nZVI) composites were synthesized through the following process: initially, 0.50 g of sodium alginate was dissolved in 25 mL of deionized, deoxygenated water. This solution was mixed continuously for 1.5 hours. Following this, nZVI particles were incorporated into the alginate solution with consistent stirring. Separately, a calcium chloride ($CaCl_2$) solution was prepared. The nZVI-alginate mixture was then added to the calcium chloride solution in a controlled dropwise manner under a nitrogen atmosphere. The resulting reaction was allowed to proceed for three hours. After the reaction, the gel particles were thoroughly rinsed with deoxygenated water and ethanol to remove impurities. The final step involved drying the composite at a temperature of -60°C for 24 hours using a vacuum freeze dryer to obtain the sodium alginate-wrapped nZVI composite (SN). Characterization techniques such as SEM-EDS, XRD, and FT-IR spectroscopy were employed to analyze the structure of the composite. The XRD patterns of sodium alginate (SA), nanoscale zero-valent iron (nZVI), and their composite (SN) revealed peaks at $2\theta = 44.6°$ and 82.3° for SN, indicating $\alpha\text{-}Fe^0(110)$ and $\alpha\text{-}Fe^0(211)$ reflections, confirming nZVI's successful incorporation into SA. Peaks at 20.7°, 26.2°, and 30.3° in nZVI indicated iron oxide formation, which was absent in SN, suggesting SA's role in preventing nZVI oxidation. FT-IR spectra of SA, nZVI, and SN showed hydroxyl group vibrations around 3500-3900 cm^{-1}. SA exhibited a broad weak band, while SN had a weak band between 3700-3900 cm^{-1}, linked to water vibrations. nZVI displayed a band at 3720 cm^{-1} and a broader band at 3124 cm^{-1}, attributed to hydroxyl stretching. Carboxylic group vibrations appeared at 1619 cm^{-1} and 1423 cm^{-1} for SA, while in SN, these shifted slightly to 1626 cm^{-1} and 1415 cm^{-1} due to nZVI embedding. SEM analysis of the sodium alginate-wrapped nanoscale zero-valent iron (nZVI) composite revealed a porous structure that enhances pollutant diffusion and increases the adsorption surface area. The composite's surface displayed irregularities like folds and bulges, contributing to its roughness and surface area. Unlike pure nZVI, the particles in the composite were well-dispersed due to the calcium ion (Ca^{2+}) cross-linking effect, preventing agglomeration. The integration of nZVI into the sodium alginate matrix suggests effective encapsulation and stabilization, making the composite suitable for adsorbing pollutants like lead ions Pb (II)). The adsorption behavior of Pb (II) on the composite was evaluated through adsorption kinetics and isotherm studies. Results indicated that the composite (SN) had a significantly higher specific surface area of 47.05 m^2/g compared to 7.56 m^2/g for pure nZVI, with reduced surface passivation. The maximum adsorption capacity of SN for Pb (II), determined through isotherm model fitting, was 70.92 mg/g. After five adsorption cycles, SN maintained a Pb (II) removal rate of 95.11%. The removal mechanism involved electrostatic adsorption, redox reactions, ion exchange, and coprecipitation. Even after 90 days, the Pb(II) removal rate by SN was still high at 95.39%, indicating its long-term stability and effectiveness as an adsorbent [53].

Jusu M. Ngobeh *et al.* developed a hydrochar-supported nanoscale zero-valent iron (nZVI) composite for the removal of cadmium from wastewater. The synthesis process started with hydrochar production from green tea leaves, which served as a sustainable base material. nZVI was also derived from these leaves, taking advantage of their natural properties to create nano-scale iron particles. The composite was formed by integrating nZVI onto the hydrochar, with an optimized mass ratio of 1:4 (nZVI to hydrochar), resulting in the highest adsorption capacity for cadmium removal. Characterization techniques such as XRD, FTIR, TEM, and SEM were used to confirm the successful formation and structural features of the composite. TEM analysis revealed that the nZVI particles had diameters between 10 and 200 nm, confirming their nanoscale nature. SEM images showed that the hydrochar had a dome-like shape, while the composite exhibited a more integrated structure with platy dendrite formations, indicating good electrical conductivity and favorable morphological features for pollutant adsorption. The XRD analysis confirmed the effective loading of nZVI onto the hydrochar, displaying distinct peaks of both materials and maintaining the structural integrity of the hydrochar. FTIR analysis identified key functional groups, particularly hydroxyl groups, that played a significant role in cadmium adsorption. The differences in FTIR spectra before and after adsorption highlighted the interaction between the composite and cadmium ions, demonstrating its efficiency in pollutant removal. Adsorption tests showed that the hydrochar-supported nZVI (H-nZVI) composite had a high cadmium removal capacity, with a maximum removal efficiency of 97.56%, indicating its strong potential for wastewater treatment applications [54].

Lu Yang *et al.* discussed carbon@nano-zero-valent iron (C@nZVI) for the reduction of Pb (II) and Zn (II) from mining wastewater. The carbon@nano-zero-valent iron (C@nZVI) composite is derived from biosynthesis through a process that utilizes biomass extracts, making it an environmentally friendly and economically viable method for producing iron nanoparticles. Initially, biomass, such as the leaves of Pinus massoniana Lamb, is collected and processed to extract organic compounds. This extracted biomass undergoes carbonization, a thermal process that transforms the biomolecules into carbon material, which not only produces carbon but also imparts strong reductive properties to the resulting material. The carbon then acts as a reducing agent for iron ions (Fe^{2+}/Fe^{3+}) present in the solution, facilitating their reduction to zero-valent iron (nZVI) under specific conditions. This nZVI is subsequently encapsulated within the carbon matrix, forming a composite that combines the adsorption capabilities of carbon with the reactivity of zero-valent iron. The resulting C@nZVI composite exhibits enhanced performance in removing heavy metals from wastewater due to the synergistic effects of the carbon layer, which provides a large surface area for adsorption and facilitates electron transfer during the reduction of heavy metal ions like Pb (II). This biosynthesis approach not only yields a stable and effective adsorbent for heavy metal removal but also promotes the reuse of biomass, contributing to sustainable environmental practices.

The study identified several mechanisms for the adsorption and reduction of heavy metal ions, specifically Pb (II) and Zn (II), using the carbon@nano-zer--valent iron (C@nZVI) composite. Key mechanisms included electrostatic interaction, where increased pH enhanced the negative charges on C@nZVI, improving attraction to positively charged metal ions (Pb^{2+} and Zn^{2+}). Complexation also played a role, as Fe-O bands and hydroxyl (-OH) groups on C@nZVI provided electrons for binding with the metal ions. Additionally, Pb (II) was reduced to Pb (0) by nZVI, with biomass-derived carbon facilitating electron transfer and enhancing reduction kinetics. Overall, these mechanisms—electrostatic interactions, complexation, and reduction—demonstrate the effective multifaceted approach of C@nZVI in treating heavy metal contamination in wastewater. The crystal structure of fresh and used C@nZVI were analyzed by XRD, Raman spectra, TEM, SEM-EDS, and XPS. XRD analysis of the freshly prepared material revealed peaks at 44.7° and 64.9°, corresponding to Fe^0 crystal planes (110) and (200), confirming the presence and stability of nZVI. Upon exposure to contaminates, additional peaks at 33.3°, 55.3°, and 62.3° appeared, indicating the reduction of Pb (II) to Pb^0. The removal of Zn^{2+} was attributed mainly to electrostatic attraction and complexation. Raman spectroscopy identified three peaks at 1365 cm^{-1} (D band), 1583 cm^{-1} (G band), and 2670 cm^{-1} (G′ band) after adsorption of lead and zinc, indicating the presence of defective multilayer graphitic carbon on the composite. An increase in the ID/IG ratio from 0.89 to 1.15 suggested enhanced defects in the carbon structure, promoting Pb and Zn adsorption through surface oxygen groups. The reduction in the G′ band intensity highlighted the role of graphite carbon in electron transfer during the reduction of Pb (II). SEM-EDS analysis of freshly prepared C@nZVI revealed well-dispersed spherical nanoparticles. After exposure, noticeable aggregation occurred due to interactions with contaminants *via* van der Waals forces, surface complexation, and electron transfer. Initial EDS results indicated a composition of C (14.82%), O (28.20%), and Fe (56.94%), mainly due to the pyrolysis of iron polyphenol complexes. After exposure, the presence of Pb (10.26%) and Zn (3.50%) confirmed their adsorption onto C@nZVI, while Fe content decreased to 29.33%, suggesting its role in the adsorption and reduction process. XPS analysis further verified that Pb(II) and Zn(II) were effectively removed by C@nZVI [55].

Treatments of Inorganic Anions

Inorganic anion contaminants in wastewater present significant environmental and health challenges due to their various sources and effects. Common anions include nitrates (NO_3^-), which primarily originate from agricultural runoff, wastewater treatment plants, and livestock operations. These can lead to issues like eutrophication and health risks such as methemoglobinemia, particularly in infants. Phosphates (PO_4^{3-}) are also frequently found in wastewater, coming from fertilizers, detergents, and effluents from treatment facilities. They contribute to algal blooms and oxygen depletion in aquatic environments. Sulfates (SO_4^{2-}) often result from industrial discharges and agricultural runoff, potentially leading to acid rain and corrosion of infrastructure. Chlorides (Cl^-), which enter water systems from road salt and various industrial processes, can raise salinity levels, negatively impacting freshwater ecosystems. Fluorides (F^-), derived from industrial effluents and water fluoridation, pose toxicity risks to aquatic life. To safeguard water resources, aquatic ecosystems, and public health, it is crucial to implement effective treatment and management practices for these inorganic anions in wastewater. As per the literature, nano-zero valent iron offers a promising approach for the treatment of various inorganic anion contaminants in wastewater through reduction reactions and adsorption processes [56 - 61]. Its application can lead to improved water quality and a reduction in environmental risks associated with these contaminants.

Yu-Kyung Jung *et al.* fabricated zero-valent iron on alginate-silicate polymer bead and used the resultant composition (nZVI-ASB) in the degradation of perchlorate (ClO^{4+}) ion. Perchlorate contamination is a significant health concern because it disrupts thyroid function by interfering with iodine uptake. Since iodine is crucial for producing thyroid hormones, exposure to perchlorate can impair the body's ability to maintain proper thyroid hormone levels, leading to potential health issues. The process for preparing nanoscale zero-valent iron-loaded alginate-silicate beads (nZVI-ASB) involved several critical stages. First, a hybrid polymer structure was created by combining sodium alginate with sodium silicate. The concentrations of these materials were varied to optimize the properties of the final beads. The alginate-silicate mixture was then cross-linked, typically by adding a cross-linking agent to stabilize the bead structure. After forming the beads, nanoscale zero-valent iron (nZVI) was introduced by treating the beads with an iron solution, allowing the iron to anchor onto the polymer matrix. A reduction step followed, converting the iron ions into their zero-valent form, which is crucial for effective perchlorate degradation. The nZVI-ASB was then characterized using methods such as SEM and FT-IR to confirm the incorporation of nZVI and assess the beads' physical properties. SEM was used to study the surface morphology of the beads, revealing differences in surface features depending on the alginate and silicate concentrations used during their preparation. Beads formed with lower silicate concentrations displayed a dense matrix with small pores, while higher silicate concentrations resulted in a rougher surface containing more small particles, which increased the surface area. FT-IR analysis verified the formation of the silica-alginate hybrid structure by identifying functional groups characteristic of both alginate and silicate. This confirmed successful cross-linking and the incorporation of nZVI into the polymer matrix. TGA was employed to evaluate the thermal stability of the beads, showing that the thermal stability varied with precursor concentrations. Beads produced from 2.0 wt% silicate and 1.0 wt% alginate exhibited the highest thermal stability. AAS analysis confirmed the presence of iron in the nZVI-ASB, validating the successful loading of nanoscale zero-valent iron onto the beads. The key findings on the catalytic activity of calcium-silicate-alginate beads loaded with nanoscale zero-valent iron (nZVI-ASB) indicate their promising potential for environmental remediation. The nZVI-ASB exhibited remarkable catalytic performance, successfully degrading 20 ppm of perchlorate within 4 hours at neutral pH, which suggests their effectiveness in typical natural water conditions. A reusability assessment showed that while the degradation efficiency diminished over time due to the oxidation of zero-valent iron (ZVI), it could be revitalized through iron supplementation. Mixing the used beads with an iron solution restored the degradation efficiency to over 95% within 4 hours, comparable to fresh nZVI-ASB. This demonstrates the beads' potential for effective reuse in remediation efforts [62].

Guanglong Liu *et al.* prepared nanoscale zero-valent iron confined in anion exchange resins and used this composition to selectively adsorp phosphate from wastewater. The preparation of nanoscale zero-valent iron (nZVI) encapsulated within anion exchange resins, specifically, mmberlite IRA-402 (Cl), was conducted through a series of well-defined steps. Initially, 10 grams of IRA-402 resin were treated with an acidic ethanol solution containing 0.005 mol/L of Fe^{3+} ions at 70 °C under continuous vigorous stirring for a duration of 12 hours.

Following this treatment, the mixture was filtered, and the resin was thoroughly rinsed with deionized water until the pH of the filtrate reached a stable value of 7.0. The rinsed resin was then introduced into a conical flask containing a 0.02 mol/L sodium borohydride ($NaBH_4$) solution and stirred at room temperature for 3 hours, which facilitated the reduction of Fe^{3+} ions to nZVI within the resin matrix. After reduction, the suspension was filtered, and the solid resin was immersed in a 5% sodium chloride (NaCl) solution to convert the resin from the −OH form to the −Cl form, thereby preventing fluctuations in pH due to hydroxide exchange. The final composite adsorbent (nZVI-402-Cl) was then dried at ambient temperature for 48 hours. This synthesized composite was designed to enhance the selective and efficient removal of phosphate from wastewater, demonstrating improved performance in pollutant adsorption applications. The XRD analysis of the nZVI-402-Cl composite revealed key structural features. A distinct peak for nZVI at $2\theta = 44.68°$ corresponded to the (110) crystal plane of zero-valent iron. Additionally, peaks at 14.14°, 26.38°, and 36.85° were identified, indicating the (020), (021), and (130) planes of lepidocrocite (γ-FeOOH), formed due to partial air oxidation of nZVI. While no prominent nZVI crystallization peaks were observed, likely due to the broad peak from the organic resin, the peak at $2\theta = 36.85°$ persisted in the composite. Lepidocrocite crystals suggested that the iron oxide layer protected nZVI from further oxidation and enhanced phosphorus adsorption, confirming the successful incorporation of nZVI and its role in phosphate removal. The SEM analysis of the nZVI-402-Cl composite revealed a rough, porous surface, enhancing its adsorption capacity. Lath-shaped crystals, attributed to lepidocrocite (γ-FeOOH), were observed, indicating iron oxide formation due to partial oxidation of nZVI. These structural features suggest that the increased surface area of the composite boosts its efficiency in phosphate removal. The SEM analysis highlighted the critical role of the composite's morphology in improving its phosphate adsorption-desorption performance, contributing to its effectiveness in wastewater treatment. The study on the nZVI-based composite, nZVI-402-Cl, highlighted its strong performance in phosphate removal from wastewater. The composite showed high adsorption capacity (56.27 mg P/g) at pH 7.2 and maintained effectiveness across a wide pH range (3.0–11.0). It demonstrated selectivity for phosphate even in the presence of competing ions (sulfate, nitrate) and humic acid. Additionally, it retained 95% of its efficiency after five regeneration cycles, showcasing its durability and cost-effectiveness. The phosphate removal mechanism involved electrostatic interactions, surface complexation, and chemical coprecipitation, enhanced by the resin's confinement of nZVI. In practical tests, the composite treated 1850 bed volumes of wastewater with phosphorus levels below 0.1 mg/L, outperforming the IRA-402 resin. These findings underscore nZVI-402-Cl as a promising material for efficient phosphate removal in wastewater treatment [63].

Yangyang Zhang *et al.* developed attapulgite loaded with nano-zero-valent iron (ATP-nZVI) for the removal of nitrate nitrogen from groundwater. To prepare attapulgite-loaded nano-zero-valent iron (ATP-nZVI), a chemical liquid deposition-coreduction approach was employed. First, 250 mL of deionized water was placed in a three-neck flask, and nitrogen gas was introduced for 30 minutes to eliminate any dissolved oxygen. Then, 4.8214 g of $FeCl_3 \cdot 6H_2O$ and 1 g of attapulgite (ATP) were added to the flask, followed by stirring at 500 rpm for 30 minutes. To trigger the reduction, 3 g of sodium borohydride ($NaBH_4$), dissolved in 100 mL of deionized water, was gradually introduced drop by drop into the flask using a peristaltic pump at a rate of 1 drop per second. After the addition, the mixture was stirred for an additional 40 minutes to allow the $NaBH_4$ to fully react with Fe^{3+} ions. After the reaction was complete, the resulting black suspension was washed two to three times with absolute ethanol. The solid particles of the ATP-nZVI composite were collected *via* vacuum-assisted suction filtration and characterized by SEM and XRD. The SEM analysis provided insights into the material morphologies. Attapulgite (ATP) exhibited rod- and block-like structures, while the nano-zero-valent iron (nZVI) particles were spherical but prone to aggregation due to their small size. After loading nZVI onto ATP, the dispersion improved, with nZVI particles distributed across the ATP surface and within its pores, leading to increased surface area and reaction sites. Following the reaction with nitrate, particle sizes increased, and aggregation indicated the adsorption of nitrate nitrogen. This resulted in the oxidation of nZVI, forming iron oxide and hydroxide, confirming the successful preparation of the ATP-nZVI composite for improved pollutant removal in groundwater. The XRD analysis provided essential insights into the crystallographic structure of the materials. The XRD patterns for attapulgite (ATP) showed distinct diffraction peaks at 2θ values of $19.82°$, $21.68°$, and $26.69°$, confirming its composition, mainly SiO_2. For nano-zero-valent iron (nZVI), a prominent peak at 2θ of $44.8°$ indicated the presence of $\alpha\text{-}Fe^0$. In the ATP-nZVI composite, characteristic peaks from both ATP and nZVI were visible, demonstrating the successful incorporation of nZVI into the ATP matrix. The nZVI loading increased the crystallinity of $\alpha\text{-}Fe^0$, as iron was adsorbed on ATP's aluminum silicate sites, enhancing the composite's stability. These XRD results also highlight the structural integrity of the ATP-nZVI composite, confirming its potential for nitrate removal in contaminated water. The study showed that pH had a significant impact on the removal of nitrate nitrogen (NO_3^-N) using ATP-nZVI composite. Tests were conducted at different initial pH values of 3, 5, 7, and 9, with removal rates of 79.77%, 79.58%, 78.61%, and 74.27%, respectively. The results indicated that pH had little effect on the final removal rates when equilibrium was reached, but equilibrium was attained more quickly at lower pH levels. At pH 3 and 5, the removal rates reached 79.77% and 79.58% within 50 minutes, while at pH 7, it was slightly lower at 78.61%, requiring 90 minutes to reach equilibrium. At pH 9, the removal rate dropped to 74.27%, with equilibrium reached after 120 minutes. This trend suggests that acidic conditions favor faster and more efficient NO_3^-N removal, with lower pH enhancing the reaction kinetics and overall nitrate removal efficiency in groundwater. Based on these findings, the attapulgite-loaded nano-zero-valent iron (ATP-nZVI) composite can show significant potential for wastewater treatment [64].

Nuntiya Paepatung *et al.*highlighted zero valent scrap iron for the anaerobic digestion of high-strength sulfate-rich wastewater and for producing methane from organic matter in the reactor. The inclusion of commercially available zero-valent iron (ZVI) significantly influences biogas and methane production in anaerobic digestion reactors, especially when treating sulfate-rich wastewater. ZVI acts as an electron donor, enhancing the metabolic activity of methanogenic archaea and facilitating the conversion of organic matter into methane. Its presence helps maintain optimal pH levels within the reactor, which is essential for effective methanogenesis. Reactors supplemented with ZVI have been observed to maintain higher pH levels, which can lead to improved methane production efficiency. Furthermore, ZVI mitigates the toxicity of sulfides produced during sulfate reduction, which can inhibit methane-producing microorganisms. By promoting sulfide precipitation and enhancing sulfate reduction, ZVI reduces competition between sulfate-reducing bacteria (SRB) and methanogens for available substrates. This reduction in competition allows methanogens to flourish, resulting in increased methane yields. Experimental findings indicate that reactors containing ZVI demonstrate higher methane production rates and greater methane concentrations in the biogas compared to control reactors lacking ZVI. For example, studies have shown that reactors with ZVI achieved maximum methane yields of 0.25 L CH_4/g COD added per day, with a methane composition in the biogas of 53%. In contrast, control reactors produced only 0.07 L CH_4/g COD added per day, with a methane composition of 27%. Additionally, the presence of ZVI has been linked to improved sulfate reduction efficiencies and increased levels of dissociated sulfide, further enhancing the anaerobic digestion process. This study establishes a framework for employing zero-valent iron (ZVI) to enhance the treatment of sulfate-containing wastewater. By improving methane production, increasing sulfate reduction efficiency, maintaining optimal pH levels, and minimizing inhibitory effects, ZVI can significantly enhance the overall effectiveness of wastewater treatment. These improvements promote more sustainable wastewater management practices and contribute to better environmental stewardship and resource recovery [65].

CONCLUSION

The chapter emphasizes the potential of zero-valent metal nanoparticles, particularly nZVI, in revolutionizing wastewater treatment and environmental restoration. The chapter outlines the various unique properties of nZVI, such as its high reactivity, magnetic properties, and versatility, which make it effective in catalyzing the reduction of various contaminants in environmental remediation. In this chapter, the authors highlight the requirement of degradation of dyes and elimination of heavy metals from water as they exert harmful effects on marine organisms and public health, as well as the synthesis of different nZVI composites to make better the stability of nZVI. They also discussed procedures for the removal of heavy metal and dye degradation. The use of nanoscale zero-valent iron (nZVI) and related materials has shown significant promise in treating inorganic anion contaminants in wastewater. These anions, including nitrates, phosphates, sulfates, chlorides, and fluorides, often originate from industrial and agricultural sources and pose considerable environmental and health risks. Studies demonstrate the efficacy of nZVI in removing these pollutants through reduction reactions, adsorption, and selective ion exchange processes. Overall, this compilation reinforces the significance of zero-valent metal nanoparticles, particularly nZVI, in advancing wastewater treatment and environmental remediation practices. Additionally, the need for more improved nZVI composites to address challenges such as agglomeration, limited stability, and reduced removal efficiency under certain conditions in groundwater treatment is also emphasized. The study also underscores the importance of surface modification and synthetic techniques to enhance the stability, reactivity, and selectivity of nZVI for specific applications.

LIST OF ABBREVIATIONS

AES	Auger Electron Spectroscopy
AOPs	Advanced Oxidation Processes
ATP	Attapulgite
AFM	Atomic Force Microscopy
B-nZVI	Bentonite-Supported nZVI
BC	Biochar
BET	Brunauer-Emmett-Teller
CA	Cellulose Acetate
CR	Congo Red
CB	Carbon Black
CMC	Carboxymethyl cellulose
COCs	Chlorinated Organic Compounds

3DG	3 Dimensional Graphene
DTPA	Diethylene Triamine Pentacetic Acid
DLS	Dynamic Light Scattering
DNP 2	6-dinitrophenol
EDS	Energy Dispersive Spectrum
FT-IR	Fourier Transform Infrared spectroscopy
FESEM	Field emission scanning electron microscopy
HRTEM	High-Resolution Transmission Electron Microscopy
LDH	Layered Double Hydroxide
LC-Q-TOF-MS	liquid chromatography quadrupole time-of-flight Mass spectrometry
MO	Methyl orange
MCC	Microcrystalline cellulose
nZVI	Nano zero-valent iron
ONP	o-nitrophenol
OG	Orange G
ONP	o-nitrophenol
PS	Persulfate
PSA	Polar Surface Area
PES	Polyethersulfone
PEG	Polyethylene glycol
RBMR	Reactive Blue MR
RBO3RID	Remazol Brilliant Orange 3RID
rGO	Reduced graphene oxide
SRB	Sulfate-Reducing Bacteria
SA	Sodium Alginate
SEM	Scanning Electron Microscopy
SAED	Selected Area Electron Diffraction
S-nZVI	Starch-Stabilized Nanoscale Zero-Valent Iron
SW/DWCNTs	Single-Walled and Double-Walled Carbon Nanotubes
TEM	Transmission Electron Microscopy
XPS	X-ray Photoelectron Spectroscopy
XRD	X-ray Diffraction
ZVCu	Zero-valent copper
ZVAg	Zero-valent silver
ZP	Zeta Potential

nZVI-ASB   Zero-Valent Iron-loaded Alginate-Silicate Beads

ACKNOWLEDGEMENTS

The authors would like to express their sincere thanks to the management of Indira Gandhi University, Meerpur, Rewari, Haryana, India, and Kurukshetra University, Kurukshetra -136119, Haryana, India, for providing the facilities to write and submit the book chapter for publication.

REFERENCES

[1]   Lu, H.J.; Wang, J.K.; Ferguson, S.; Wang, T.; Bao, Y.; Hao, H. Mechanism, synthesis and modification of nano zerovalent iron in water treatment. *Nanoscale,* **2016,** *8*(19), 9962-9975.
[http://dx.doi.org/10.1039/C6NR00740F] [PMID: 27128356]

[2]   Crane, R.A.; Scott, T.B. Nanoscale zero-valent iron: Future prospects for an emerging water treatment technology. *J. Hazard. Mater.,* **2012,** *211-212*, 112-125.
[http://dx.doi.org/10.1016/j.jhazmat.2011.11.073] [PMID: 22305041]

[3]   Arabi, S.; Sohrabi, M.R. Removal of methylene blue, a basic dye, from aqueous solutions using nano-zerovalent iron. *Water Sci. Technol.,* **2014,** *70*(1), 24-31.
[http://dx.doi.org/10.2166/wst.2014.189] [PMID: 25026575]

[4]   Xu, H.; Zhang, Y.; Cheng, Y.; Tian, W.; Zhao, Z.; Tang, J. Polyaniline/attapulgite-supported nanoscale zero-valent iron for the rival removal of azo dyes in aqueous solution. *Adsorpt. Sci. Technol.,* **2019,** *37*(3-4), 217-235.
[http://dx.doi.org/10.1177/0263617418822917]

[5]   Rashid, R.; Shafiq, I.; Akhter, P.; Iqbal, M.J.; Hussain, M. A state-of-the-art review on wastewater treatment techniques: the effectiveness of adsorption method. *Environ. Sci. Pollut. Res. Int.,* **2021,** *28*(8), 9050-9066.
[http://dx.doi.org/10.1007/s11356-021-12395-x] [PMID: 33483933]

[6]   Xiao, X.; Sun, Y.; Sun, W.; Shen, H.; Zheng, H.; Xu, Y.; Zhao, J.; Wu, H.; Liu, C. Advanced treatment of actual textile dye wastewater by Fenton-flocculation process. *Can. J. Chem. Eng.,* **2017,** *95*(7), 1245-1252.
[http://dx.doi.org/10.1002/cjce.22752]

[7]   Ejraei, A.; Aroon, M.A.; Ziarati Saravani, A. Wastewater treatment using a hybrid system combining adsorption, photocatalytic degradation and membrane filtration processes. *J. Water Process Eng.,* **2019,** *28*, 45-53.
[http://dx.doi.org/10.1016/j.jwpe.2019.01.003]

[8]   Yi, F.; Chen, S.; Yuan, C. Effect of activated carbon fiber anode structure and electrolysis conditions on electrochemical degradation of dye wastewater. *J. Hazard. Mater.,* **2008,** *157*(1), 79-87.
[http://dx.doi.org/10.1016/j.jhazmat.2007.12.093] [PMID: 18258359]

[9]   Kariyajjanavar, P.; Narayana, J.; Nayaka, Y.A. Degradation of textile wastewater by electrochemical method. *Hydrol. Curr. Res.,* **2011,** *2*(1), 2.
[http://dx.doi.org/10.4172/2157-7587.1000110]

[10]   Bhatia, D.; Sharma, N.R.; Singh, J.; Kanwar, R.S. Biological methods for textile dye removal from wastewater: A review. *Crit. Rev. Environ. Sci. Technol.,* **2017,** *47*(19), 1836-1876.
[http://dx.doi.org/10.1080/10643389.2017.1393263]

[11]   Piaskowski, K.; Świderska-Dąbrowska, R.; Zarzycki, P.K. Dye removal from water and wastewater using various physical, chemical, and biological processes. *J. AOAC Int.,* **2018,** *101*(5), 1371-1384.
[http://dx.doi.org/10.5740/jaoacint.18-0051] [PMID: 29669626]

[12] Raghu, S.; Lee, C.W.; Chellammal, S.; Palanichamy, S.; Basha, C.A. Evaluation of electrochemical oxidation techniques for degradation of dye effluents—A comparative approach. *J. Hazard. Mater.,* **2009**, *171*(1-3), 748-754.
[http://dx.doi.org/10.1016/j.jhazmat.2009.06.063] [PMID: 19592159]

[13] Ercan, Ö.; Deniz, S.; Yetimoğlu, E.K.; Aydın, A. Degradation of reactive dyes using advanced oxidation method. *Clean (Weinh.),* **2015**, *43*(7), 1031-1036.
[http://dx.doi.org/10.1002/clen.201400195]

[14] Wang, X.; Wang, P.; Ma, J.; Liu, H.; Ning, P. Synthesis, characterization, and reactivity of cellulose modified nano zero-valent iron for dye discoloration. *Appl. Surf. Sci.,* **2015**, *345*, 57-66.
[http://dx.doi.org/10.1016/j.apsusc.2015.03.131]

[15] Reza Sohrabi, M.; Mansouriieh, N.; Khosravi, M.; Zolghadr, M. Removal of diazo dye Direct Red 23 from aqueous solution using zero-valent iron nanoparticles immobilized on multi-walled carbon nanotubes. *Water Sci. Technol.,* **2015**, *71*(9), 1367-1374.
[http://dx.doi.org/10.2166/wst.2015.106] [PMID: 25945854]

[16] Zhang, C.; Chen, H.; Xue, G.; Liu, Y.; Chen, S.; Jia, C. A critical review of the aniline transformation fate in azo dye wastewater treatment. *J. Clean. Prod.,* **2021**, *321*, 128971.
[http://dx.doi.org/10.1016/j.jclepro.2021.128971]

[17] Sudha, M.; Saranya, A.; Selvakumar, G.; Sivakumar, N. Microbial degradation of azo dyes: A review. *Int. J. Curr. Microbiol. Appl. Sci.,* **2014**, *3*(2), 670-690.

[18] Kim, K.R.; Lee, B.T.; Kim, K.W. Arsenic stabilization in mine tailings using nano-sized magnetite and zero valent iron with the enhancement of mobility by surface coating. *J. Geochem. Explor.,* **2012**, *113*, 124-129.
[http://dx.doi.org/10.1016/j.gexplo.2011.07.002]

[19] Scott, T.B.; Popescu, I.C.; Crane, R.A.; Noubactep, C. Nano-scale metallic iron for the treatment of solutions containing multiple inorganic contaminants. *J. Hazard. Mater.,* **2011**, *186*(1), 280-287.
[http://dx.doi.org/10.1016/j.jhazmat.2010.10.113] [PMID: 21115222]

[20] Yeow, P.K.; Wong, S.W.; Hadibarata, T. Removal of azo and anthraquinone dye by plant biomass as adsorbent–a review. *Biointerface Res. Appl. Chem.,* **2021**, *11*, 8218-8232.

[21] Routoula, E.; Patwardhan, S.V. Degradation of anthraquinone dyes from effluents: a review focusing on enzymatic dye degradation with industrial potential. *Environ. Sci. Technol.,* **2020**, *54*(2), 647-664.
[http://dx.doi.org/10.1021/acs.est.9b03737] [PMID: 31913605]

[22] Asgher, M. Biosorption of reactive dyes: A review. *Water Air Soil Pollut.,* **2012**, *223*(5), 2417-2435.
[http://dx.doi.org/10.1007/s11270-011-1034-z]

[23] Arslan-Alaton, I. A review of the effects of dye-assisting chemicals on advanced oxidation of reactive dyes in wastewater. *Color. Technol.,* **2003**, *119*(6), 345-353.
[http://dx.doi.org/10.1111/j.1478-4408.2003.tb00196.x]

[24] Dutta, S.; Saha, R.; Kalita, H.; Bezbaruah, A.N. Rapid reductive degradation of azo and anthraquinone dyes by nanoscale zero-valent iron. *Environ. Technol. Innov.,* **2016**, *5*, 176-187.
[http://dx.doi.org/10.1016/j.eti.2016.03.001]

[25] Raman, C.D.; Kanmani, S. Decolorization of mono azo dye and textile wastewater using nano iron particles. *Environ. Prog. Sustain. Energy,* **2019**, *38*(s1), S366-S376.
[http://dx.doi.org/10.1002/ep.13063]

[26] Thomas, S.; Abraham, S.V.; Aravind, U.K.; Aravindakumar, C.T. Enhanced degradation of acid red 1 dye using a coupled system of zero valent iron nanoparticles and sonolysis. *Environ. Sci. Pollut. Res. Int.,* **2017**, *24*(31), 24533-24544.
[http://dx.doi.org/10.1007/s11356-017-0080-5] [PMID: 28905281]

[27] Dinesh, G.K.; Pramod, M.; Chakma, S. Sonochemical synthesis of amphoteric Cu^0-Nanoparticles

using Hibiscus rosa-sinensis extract and their applications for degradation of 5-fluorouracil and lovastatin drugs. *J. Hazard. Mater.,* **2020**, *399*, 123035.
[http://dx.doi.org/10.1016/j.jhazmat.2020.123035] [PMID: 32512280]

[28]   Ali Khan, S.; Bakhsh, E.M.; Asiri, A.M.; Bahadar Khan, S. Synthesis of zero-valent Au nanoparticles on chitosan coated NiAl layered double hydroxide microspheres for the discoloration of dyes in aqueous medium. *Spectrochim. Acta A Mol. Biomol. Spectrosc.,* **2021**, *250*, 119370.
[http://dx.doi.org/10.1016/j.saa.2020.119370] [PMID: 33412468]

[29]   Iran Manesh, M.; Sohrabi, M.R.; Mortazavi Nik, S. Nanoscale zero-valent iron supported on graphene novel adsorbent for the removal of diazo direct red 81 from aqueous solution: isotherm, kinetics, and thermodynamic studies. *Iran. J. Chem. Chem. Eng.,* **2022**, *41*(6), 1844-1855.

[30]   Sutherland, A.J.; Ruiz-Caldas, M.X.; de Lannoy, C.F. Electro-catalytic microfiltration membranes electrochemically degrade azo dyes in solution. *J. Membr. Sci.,* **2020**, *611*, 118335.
[http://dx.doi.org/10.1016/j.memsci.2020.118335]

[31]   Khaloo, S.S.; Fattahi, S. Enhancing decolorization of Eriochrome Blue Black R during nano-size zero-valent iron treatment using ultrasonic irradiation. *Desalination Water Treat.,* **2014**, *52*(16-18), 3403-3410.
[http://dx.doi.org/10.1080/19443994.2013.801322]

[32]   Gao, C.; Yu, W.; Zhu, Y.; Wang, M.; Tang, Z.; Du, L.; Hu, M.; Fang, L.; Xiao, X. Preparation of porous silicate supported micro-nano zero-valent iron from copper slag and used as persulfate activator for removing organic contaminants. *Sci. Total Environ.,* **2021**, *754*, 142131.
[http://dx.doi.org/10.1016/j.scitotenv.2020.142131] [PMID: 33254954]

[33]   Nandi, T.; De, A.; Sarkar, S.; Haldar, S. A study of nZVI DTPA induced degradation of selective organic pollutants by the help of ac conductivity measurement. *Environ. Nanotechnol. Monit. Manag.,* **2020**, *14*, 100296.
[http://dx.doi.org/10.1016/j.enmm.2020.100296]

[34]   Shu, H.Y.; Chang, M.C.; Chen, C.C.; Chen, P.E. Using resin supported nano zero-valent iron particles for decoloration of Acid Blue 113 azo dye solution. *J. Hazard. Mater.,* **2010**, *184*(1-3), 499-505.
[http://dx.doi.org/10.1016/j.jhazmat.2010.08.064] [PMID: 20833471]

[35]   Mao, Y.; Xi, Z.; Wang, W.; Ma, C.; Yue, Q. Kinetics of Solvent Blue and Reactive Yellow removal using microwave radiation in combination with nanoscale zero-valent iron. *J. Environ. Sci. (China),* **2015**, *30*, 164-172.
[http://dx.doi.org/10.1016/j.jes.2014.09.030] [PMID: 25872723]

[36]   Varadavenkatesan, T.; Pai, S.; Vinayagam, R.; Selvaraj, R. Characterization of silver nano-spheres synthesized using the extract of Arachis hypogaea nuts and their catalytic potential to degrade dyes. *Mater. Chem. Phys.,* **2021**, *272*, 125017.
[http://dx.doi.org/10.1016/j.matchemphys.2021.125017]

[37]   Chen, Z.; Jin, X.; Chen, Z.; Megharaj, M.; Naidu, R. Removal of methyl orange from aqueous solution using bentonite-supported nanoscale zero-valent iron. *J. Colloid Interface Sci.,* **2011**, *363*(2), 601-607.
[http://dx.doi.org/10.1016/j.jcis.2011.07.057] [PMID: 21864843]

[38]   Rahman, N.; Abedin, Z.; Hossain, M.A. Rapid degradation of azo dyes using nano-scale zero valent iron. *Am. J. Environ. Sci.,* **2014**, *10*(2), 157-163.
[http://dx.doi.org/10.3844/ajessp.2014.157.163]

[39]   Naim, M.M.; El-Shafei, A.A.; Elewa, M.M.; Moneer, A.A. Application of silver-, iron-, and chitosan-nanoparticles in wastewater treatment. *Int. Conf. Eur. Desalin. Soc. Desalin. Environ. Clean Water Energy,* **2016**, *Vol. 73*, pp. 268-280.

[40]   Khan, S.A.; Khan, S.B.; Farooq, A.; Asiri, A.M. A facile synthesis of CuAg nanoparticles on highly porous ZnO/carbon black-cellulose acetate sheets for nitroarene and azo dyes reduction/degradation. *Int. J. Biol. Macromol.,* **2019**, *130*, 288-299.
[http://dx.doi.org/10.1016/j.ijbiomac.2019.02.114] [PMID: 30797005]

[41]   Ince, N.H. Ultrasound-assisted advanced oxidation processes for water decontamination. *Ultrason. Sonochem.,* **2018**, *40*(Pt B), 97-103.
[http://dx.doi.org/10.1016/j.ultsonch.2017.04.009] [PMID: 28552350]

[42]   Sajab, M.S.; Ismail, N.N.N.; Santanaraj, J.; Mohammad, A.W.; Abu Hassan, H.; Chia, C.H.; Zakaria, S.; Mohamed Noor, A. Insight observation into rapid discoloration of batik textile effluent by *in situ* formations of zero valent iron. *Sains Malays.,* **2019**, *48*(2), 393-399.
[http://dx.doi.org/10.17576/jsm-2019-4802-17]

[43]   Ihsanullah, ; Al-Khaldi, F.A.; Abu-Sharkh, B.; Abulkibash, A.M.; Qureshi, M.I.; Laoui, T.; Atieh, M.A. Effect of acid modification on adsorption of hexavalent chromium (Cr(VI)) from aqueous solution by activated carbon and carbon nanotubes. *Desalination Water Treat.,* **2016**, *57*(16), 7232-7244.
[http://dx.doi.org/10.1080/19443994.2015.1021847]

[44]   Saha, R.; Nandi, R.; Saha, B. Sources and toxicity of hexavalent chromium. *J. Coord. Chem.,* **2011**, *64*(10), 1782-1806.
[http://dx.doi.org/10.1080/00958972.2011.583646]

[45]   Lee, C.G.; Kim, S.B. Removal of arsenic and selenium from aqueous solutions using magnetic iron oxide nanoparticle/multi-walled carbon nanotube adsorbents. *Desalination Water Treat.,* **2016**, *57*(58), 28323-28339.
[http://dx.doi.org/10.1080/19443994.2016.1185042]

[46]   Bakather, O. Y.; Kayvani Fard, A.; Khraisheh, M.; Nasser, M. S.; Atieh, M. A. Enhanced adsorption of selenium ions from aqueous solution using iron oxide impregnated carbon nanotubes. *Bioinorg. Chem. Appl.,* **2017**, *2017*
[http://dx.doi.org/10.1155/2017/4323619]

[47]   Qu, X.; Alvarez, P.J.J.; Li, Q. Applications of nanotechnology in water and wastewater treatment. *Water Res.,* **2013**, *47*(12), 3931-3946.
[http://dx.doi.org/10.1016/j.watres.2012.09.058] [PMID: 23571110]

[48]   Zhang, Y.; Wu, B.; Xu, H.; Liu, H.; Wang, M.; He, Y.; Pan, B. Nanomaterials-enabled water and wastewater treatment. *NanoImpact,* **2016**, *3-4*, 22-39.
[http://dx.doi.org/10.1016/j.impact.2016.09.004]

[49]   Wu, H.; Wei, W.; Xu, C.; Meng, Y.; Bai, W.; Yang, W.; Lin, A. Polyethylene glycol-stabilized nano zero-valent iron supported by biochar for highly efficient removal of Cr(VI). *Ecotoxicol. Environ. Saf.,* **2020**, *188*, 109902.
[http://dx.doi.org/10.1016/j.ecoenv.2019.109902] [PMID: 31704325]

[50]   Chen, H.; Xie, H.; Zhou, J.; Tao, Y.; Zhang, Y.; Zheng, Q.; Wang, Y. Removal efficiency of hexavalent chromium from wastewater using starch-stabilized nanoscale zero-valent iron. *Water Sci. Technol.,* **2019**, *80*(6), 1076-1084.
[http://dx.doi.org/10.2166/wst.2019.358] [PMID: 31799951]

[51]   Lv, X.; Qin, X.; Wang, K.; Peng, Y.; Wang, P.; Jiang, G. Nanoscale zero valent iron supported on MgAl-LDH-decorated reduced graphene oxide: Enhanced performance in Cr(VI) removal, mechanism and regeneration. *J. Hazard. Mater.,* **2019**, *373*, 176-186.
[http://dx.doi.org/10.1016/j.jhazmat.2019.03.091] [PMID: 30921568]

[52]   Zhang, S.; Lyu, H.; Tang, J.; Song, B.; Zhen, M.; Liu, X. A novel biochar supported CMC stabilized nano zero-valent iron composite for hexavalent chromium removal from water. *Chemosphere,* **2019**, *217*, 686-694.
[http://dx.doi.org/10.1016/j.chemosphere.2018.11.040] [PMID: 30448748]

[53]   Zheng, C.; Ren, J.; He, F.; Yong, Y.; Tu, Y.; Wang, Z. Nanoscale zero-valent iron dispersed by sodium alginate enables highly efficient removal of lead (Pb) from Aqueous solution. *Adsorpt. Sci. Technol.,* **2023**, *2023*, 1829725.
[http://dx.doi.org/10.1155/2023/1829725]

[54] Ngobeh, J. M.; Nan, X.; Sorathiya, V. Removal of a cadmium from wastewater using hydrochar-supported nanoscale zero-valent iron (NZVI) on composite materials. **2024**.

[55] Yang, L.; Jin, X.; Lin, Q.; Owens, G.; Chen, Z. Enhanced adsorption and reduction of Pb(II) and Zn(II) from mining wastewater by carbon@nano-zero-valent iron (C@nZVI) derived from biosynthesis. *Separ. Purif. Tech.,* **2023**, *311*, 123249.
[http://dx.doi.org/10.1016/j.seppur.2023.123249]

[56] Foltynowicz, Z.; Maranda, A.; Czajka, B.; Wachowski, L.; Sałaciński, T. The effective removal of organic and inorganic contaminants using compositions based on zero-valent iron nanoparticles (n-ZVI). *Materiały Wysokoenergetyczne / High Energy Materials,* **2020**, *12*(1), 37-74.
[http://dx.doi.org/10.22211/matwys/0172E]

[57] Pasinszki, T.; Krebsz, M. Synthesis and application of zero-valent iron nanoparticles in water treatment, environmental remediation, catalysis, and their biological effects. *Nanomaterials (Basel),* **2020**, *10*(5), 917.
[http://dx.doi.org/10.3390/nano10050917] [PMID: 32397461]

[58] Liu, J.; Ding, Y.; Qiu, W.; Cheng, Q.; Xu, C.; Fan, G.; Song, G.; Xiao, B. Enhancing anaerobic digestion of sulphate wastewater by adding nano-zero valent iron. *Environ. Technol.,* **2023**, *44*(26), 3988-3996.
[http://dx.doi.org/10.1080/09593330.2022.2077137] [PMID: 35546259]

[59] Lin, D.; Hu, L.; Lo, I.M.C.; Yu, Z. Size distribution and phosphate removal capacity of nano zero-valent iron (nZVI): Influence of pH and ionic strength. *Water,* **2020**, *12*(10), 2939.
[http://dx.doi.org/10.3390/w12102939]

[60] Jahin, H.S. Fluoride removal from water using nanoscale zero-valent iron (NZVI). *Int. Water Technol. J,* **2014**, *4*, 173-182.

[61] Peng, X.; Chen, N.; Wei, K.; Li, S.; Shang, H.; Sun, H.; Zhang, L. Zero-valent iron coupled calcium hydroxide: A highly efficient strategy for removal and magnetic separation of concentrated fluoride from acidic wastewater. *Sci. Total Environ.,* **2022**, *838*(Pt 3), 156336.
[http://dx.doi.org/10.1016/j.scitotenv.2022.156336] [PMID: 35654177]

[62] Jung, Y.K.; Narendra Kumar, A.V.; Jeon, B.H.; Kim, E.Y.; Yum, T.; Paeng, K.J. Exploration of zero-valent iron stabilized calcium–silicate–alginate beads' catalytic activity and stability for perchlorate degradation. *Materials (Basel),* **2022**, *15*(9), 3340.
[http://dx.doi.org/10.3390/ma15093340] [PMID: 35591672]

[63] Liu, G.; Han, C.; Kong, M.; Abdelraheem, W.H.M.; Nadagouda, M.N.; Dionysiou, D.D. Nanoscale zero-valent iron confined in anion exchange resins to enhance selective adsorption of phosphate from wastewater. *ACS ES T Eng.,* **2022**, *2*(8), 1454-1464.
[http://dx.doi.org/10.1021/acsestengg.1c00506]

[64] Zhang, Y.; Tan, Y.; Zu, B.; Zhang, X.; Zheng, C.; Lin, Z.; He, F.; Chen, K. Removal of nitrate nitrogen in groundwater by attapulgite loaded with nano-zero-valent iron. *Adsorpt. Sci. Technol.,* **2023**, *2023*, 5594717.
[http://dx.doi.org/10.1155/2023/5594717]

[65] Paepatung, N.; Songkasiri, W.; Yasui, H.; Phalakornkule, C. Enhancing methanogenesis in fed-batch anaerobic digestion of high-strength sulfate-rich wastewater using zero valent scrap iron. *J. Environ. Chem. Eng.,* **2020**, *8*(6), 104508.
[http://dx.doi.org/10.1016/j.jece.2020.104508]

CHAPTER 3

Metal Oxides Nanoparticles in Wastewater Treatment

Taruna[1], Rajni[1] and Lakhwinder Singh[2,*]

[1] *Department of Chemistry, Faculty of Basic and Applied Sciences, SGT University, Gurugram-122505, Haryana, India*

[2] *IIMT University, Meerut-250001, Uttar Pradesh, India*

Abstract: A rise in industrial and human activity degrades the quality of the water. This implies that wastewater from residential and commercial properties needs to have pollutants removed. The effectiveness of water filtration and decontamination can be greatly enhanced using nanotechnology. Nanomaterials are effective at killing microbes and eliminating heavy metals and organic and inorganic pollutants from wastewater. Because of their special qualities, namely their vast surface areas and low concentration, metal oxide nanoparticles offer a lot of potential for treating contaminated water. Metal oxide nanoparticles and their uses for eliminating heavy metals and dyes from wastewater, as well as their antibacterial properties, are covered in this chapter.

Keywords: Anti-microbial activity, Dye degradation, Heavy metal, Nanocomposite, Nanoparticles, Wastewater treatment.

INTRODUCTION

The demand for clean water has increased significantly worldwide in recent years due to technology and population development. Seven hundred and eighty three million people lack access to safe water worldwide [1 - 5]. This has led to an urgent search for environmentally acceptable and energy-efficient wastewater treatment techniques. Diverse methodologies have been devised to eliminate these contaminants. Nonetheless, the selection of suitable media is largely responsible for the efficacy of these water treatment methods. Because harmful compounds seep into the process water, the majority of commercially available and chemically synthesized materials used in water treatment also pose extra risks. The selection of materials and optimal techniques for combining them to mitigate the pollution caused by dyes and heavy metals are the main points of emphasis.

* **Corresponding author Lakhwinder Singh:** IIMT University, Meerut-250001, Uttar Pradesh, India; E-mail: lakhwinder.pharma@gmail.com

Anjaneyulu Bendi (Ed.)

The biological approach is a cost-effective option for producing nanoparticles, including zinc oxide, titanium, aluminum, silver, and nickel oxide nanoparticles. The production of non-toxic end products, low energy consumption, and environmental friendliness make biological synthesis of nanoparticles the preferred method. One of the safest, most straightforward, and least expensive methods for removing dyes, heavy metals, and organic compounds from water contaminated by homes and businesses is photocatalytic degradation, which is based on heterogeneous semiconductors [6 - 10]. Numerous metal oxide nanoparticles (NPs), including CuO, TiO_2, and ZnO, are of concern for their ability to remove heavy metals and photodegrade toxic colors [11 - 21] (Fig. **1**).

CHEMISTRY OF METAL OXIDES FOR THE REMOVAL OF HEAVY METALS

Muhammad Kaleem et al. synthesized iron oxide nanoparticles by using cyanobacteria extract. Initially, cyanobacteria were first mixed with water and heated at 373k for 24 h to prepare the extract, followed by the addition of Nostoc sp. MK-11 and $FeCl_3$. Finally, the mixture was heated for 2 h at 343k to obtain nanoparticles.

Synthesized particles were characterized using various techniques. XRD confirmed the crystalline nature with a crystallite size of 18.21 nm, and SEM images showed a spherical or cubic shape. EDX technique showed the presence of constituent elements, and FTIR peaks confirmed the presence of cyanide, C=O, S=O, O-H, C=C, C-H, C-Cl, and Fe-O bonds. UV spectra peak at 348 nm with strong intensity showed effective synthesis of nanoparticles.

All the above characteristics enable nanoparticles to have a large adsorption surface area, due to which they are used to remove heavy metals Cd and Pb with the efficiency of 85 and 95%, respectively, *via* pseudo-second-order kinetics (Scheme **1**) [22].

Nazia Hossain et al. introduced a novel thermos-chemical conversion technique to prepare solvochars of silver nanoparticles as represented in the scheme.

Several spectroscopic techniques were used to analyze crystalline nature, morphology, and functional groups. The BET technique analyzed the porosity and roughness of particles, and XRD confirmed that the FCC structure had an irregular shape and crystalline nature. FTIR data showed the presence of C=C, C-O-C, C-OH, S=O, P=O, C-F, and Si-O-Si bonds. XPS confirmed the presence of constituent elements Mg, Si, Al, *etc*. These nanoparticles were also tested for antibacterial activity and showed strong inhibition against *E. Coli* and *S. aureus*.

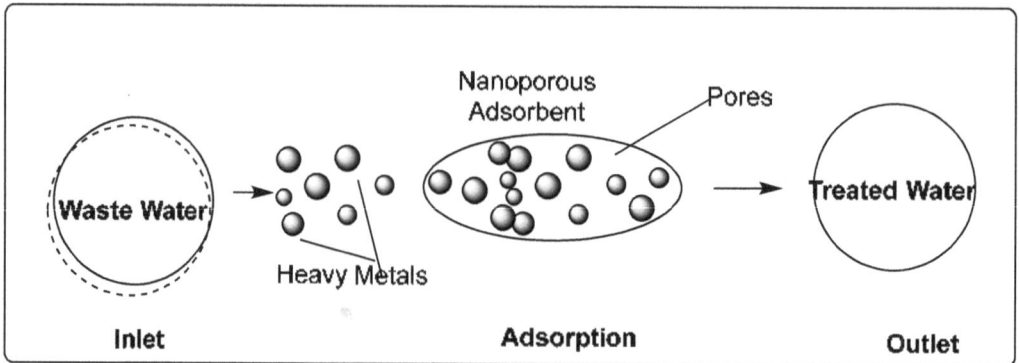

Fig. (1). General mechanism of heavy metal removal from wastewater.

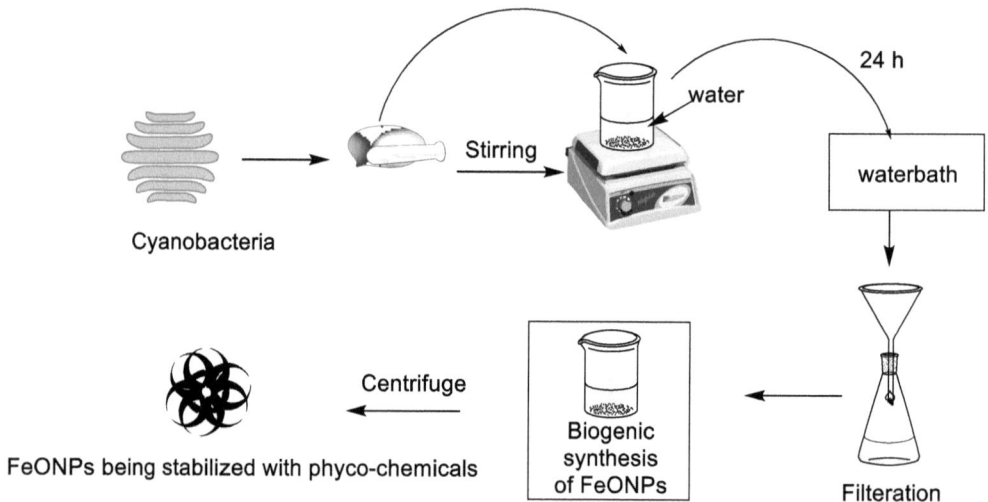

Scheme 1. Synthesis of iron oxide nanoparticles.

These nanoparticles, due to their antibacterial activity and large surface area, were used for water treatment and tested for the removal of heavy metals like copper, iron, lead, zinc, and manganese. (Scheme **2**) [23].

Omolbanin Hosseinkhani *et al.* synthesized graphene oxide-ZnO nanocomposites by mixing ZnO nanoparticles in an ethanol-water mixture followed by the addition of graphene oxide paste and NaOH and heated at 85^{0}C for 5 h. Synthesized nanocomposites were characterized using techniques like XRD and SEM. XRD showed the formation of pure particles without any impurities, and SEM images confirmed the uniform distribution of particles without their agglomeration. These nanocomposites, due to their large surface area, were used as degrading agents and tested for the removal of heavy metals from wastewater. These nanocomposites showed high efficiency in the removal of

Cd^{2+} and Pb^{2+} metals. (Scheme **3**) (Fig. **1**) [24].

Ziguo *et al.* prepared $Fe_3O_4@SiO_2$-(-NH_2/-COOH). Initially, $Fe_3O_4@SiO_2$-NH_2 was dissolved in deionized water along with nitrogen gas, followed by the addition of N,N-methylene bisacrylamide, and heated for 2 h at 25^0C. Secondly, ammonium persulphate, sodium sulfite, ethylenediaminetetraacetic acid, and di-isopropyl azodicarboxylate solutions were added and stirred for 4h to get the desired precipitates.

Characterization of synthesized particles was carried out using various techniques. XRD showed an inverse spinel structure of the nanocomposite with a particle size of 24 nm. TEM images showed a core-shell structure.

The synthesized nanocomposite was used to remove heavy metals from wastewater, and it degraded three heavy metals, Pb^{2+}, Cd^{2+}, and Zn^{2+}, with good efficiency. These nanoparticles showed maximum adsorption of lead metal, followed by cadmium and zinc. (Scheme **4**) [25].

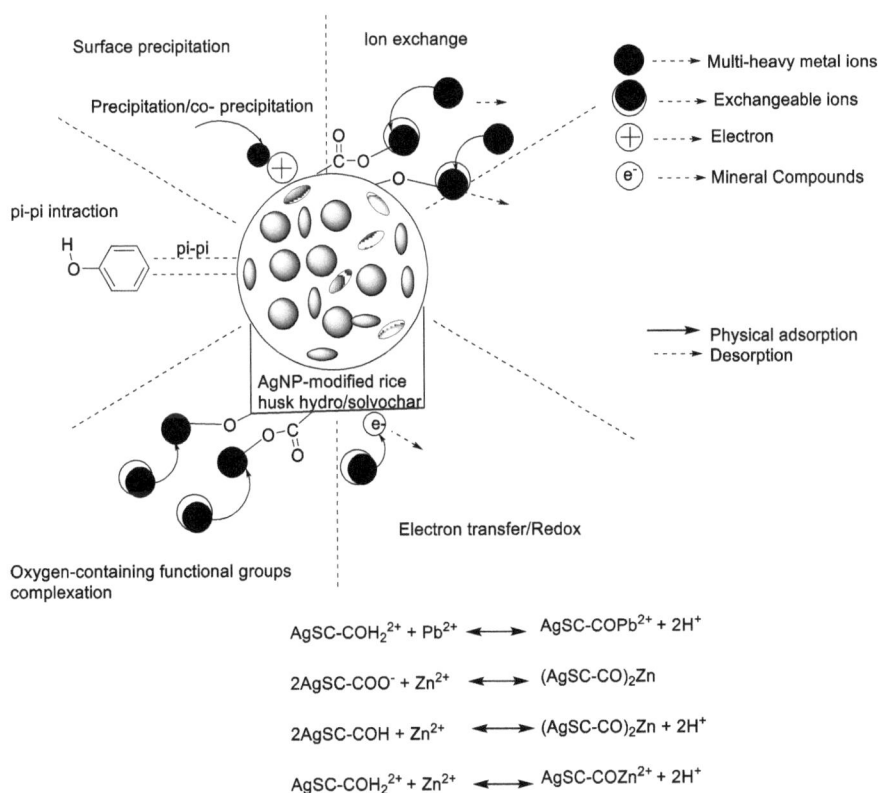

Scheme 2. Adsorption of Cu^{2+}, Fe^{3+}, Pb^{2+}, and Zn^{2+} by AgNPs.

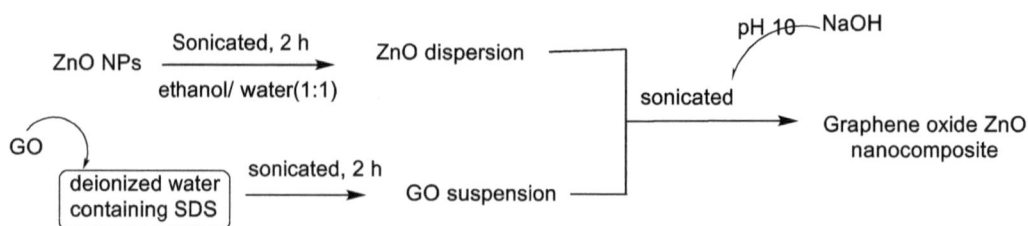

Scheme 3. Synthesis of GO-ZnO nanocomposite.

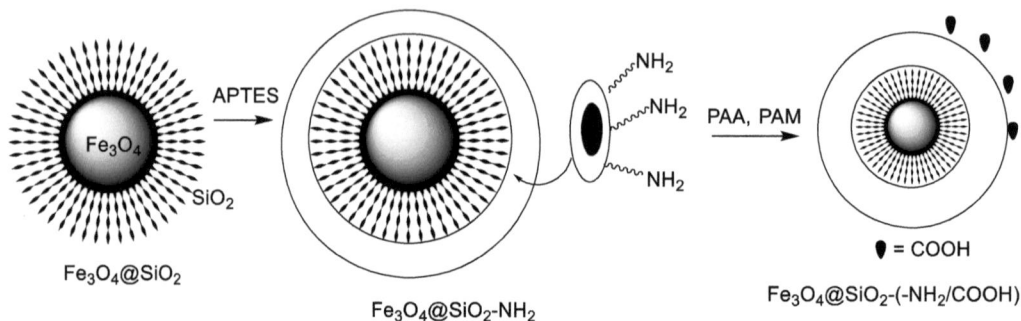

Scheme 4. Synthesis of Fe$_3$O$_4$@SiO$_2$-(-NH$_2$/-COOH) NPs.

Andrea Basso Peressut *et al.* synthesized nanocomposite by combining reduced graphene oxide, titanium oxide, and thallium oxide, as represented in the scheme.

Synthesized nanocomposites were characterized using XRD, EDX, and SEM techniques. XRD showed a crystalline nature with anatase and rutile structure, and SEM images showed a rough surface with a uniform distribution of particles along with their agglomeration. EDX showed the presence of constituent elements oxygen, Ti, and Tl. UV absorption spectra showed that these nanoparticles were found to have good absorption properties.

These nanocomposites were used to remove heavy metals like iron and copper from wastewater, and due to their strong adsorption ability, they were removed to a greater extent. (Scheme **5**) [26].

V. Masindi *et al.* synthesized magnesium oxide nanoparticles. Initially, pulverization of magnesite was carried out using a ball-milling process, which was passed through a sieve to make it crystalline. Secondly, magnesite was calcined for 1 h at 1100^{0}C to prepare magnesium oxide, followed by passing it through a sieve to convert it into desired nanoparticles.

Synthesized nanoparticles were characterized using various spectroscopical techniques. XRD data analysis showed crystalline and brucite structures of particles. HR-FESEM images showed particles in the form of spherical nanosheets, and EDX confirmed the presence of Mg, O, and C. FTIR showed the presence of C=O, O-H, Mg-O, and C-O. XRF technique showed a higher percentage of magnesium in nanoparticles.

Synthesized nanoparticles were tested as degrading agents for water treatment, and they removed heavy metals like Fe, U, Cr, Mn, *etc.*, in the form of hydroxides, sulfides, carbonates, and oxides *via* precipitation, co-adsorption, and co-precipitation. (Scheme **6**) [27].

Ahmed M. Eid *et al.* used Portulaca oleracea leaves to synthesize copper oxide nanoparticles by mixing $Cu(CH_3COO)_2.H_2O$ with leaves and heating at 200^0C.

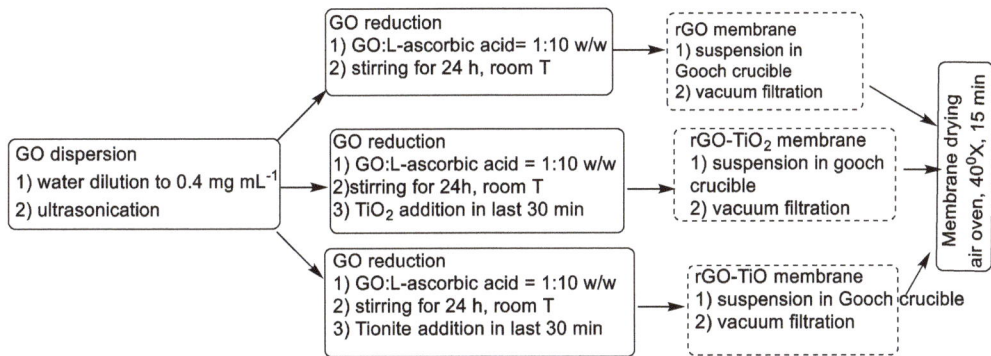

Scheme 5. Synthesis of rGO, rGO-TiO_2 and rGO-TiO membrane.

Scheme 6. Synthesis of MgO-NPs.

The crystalline nature and morphology of particles were measured using SEM, TEM, XRD, and EDX. SEM images showed smooth and spherical surfaces without aggregation, and EDX showed main constituents, such as Cu and O. XRD data showed an FCC structure with crystalline nature and purity. FTIR confirmed the presence of O-H, N-H, C-H, C=O, C=N, C-O, and cyanide groups.

Antimicrobial activity was also measured against S. aureus, B. subtilis, P. aeruginosa, E. Coli, and C. albicans and found to show strong activity with ZOI values in the range of 12.6-19.3 mm.

These antimicrobial activities and other data enabled particles to show good degrading efficiency of heavy metals from wastewater and were tested for the removal of nickel, cobalt, lead, cadmium, and chromium with an efficiency of 72, 73, 80, 64, and 90%, respectively. (Scheme 7) [28].

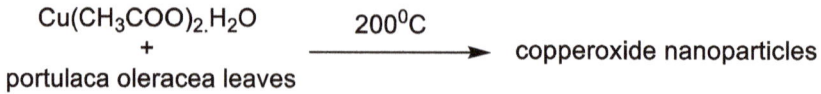

$$Cu(CH_3COO)_2.H_2O + portulaca\ oleracea\ leaves \xrightarrow{200^0C} copperoxide\ nanoparticles$$

Scheme 7. Synthesis of copper oxide NPs.

Emile Salomon Massima Mouele *et al.* synthesized nanoparticles by mixing gliadin protein from wheat with $AgFe_2O_3$, TiO_2, and $AgFe–TiO_2$ *via* chemical precipitations, as represented in the scheme.

Various spectroscopical techniques were used to assess these nanoparticles. SEM analysis of gliadin blended composites was conducted. The SEM images showed the non-porous nature of particles, and EDX confirmed the presence of constituent elements Fe, Ti, Ag, C, N, and O. XRD showed particles to have BCC tetragonal lattice, and FTIR peaks confirmed the presence of C–N, C–O, N–H, C-C, Ti – O –Ti, O–Ti–O, Ag–O, Fe–O, Ag–Fe–O, Ag–O–Fe, and Ag–Fe. The TGA curve showed particles to be stable at low temperatures, and the adsorption process was found to be spontaneous and endothermic. Antibacterial activity was measured against E. coli, and good activity was observed in dark and light conditions.

Synthesized nanoparticles were tested for the removal of Co(II) from wastewater showed good removal efficiency.(Scheme 8) [29].

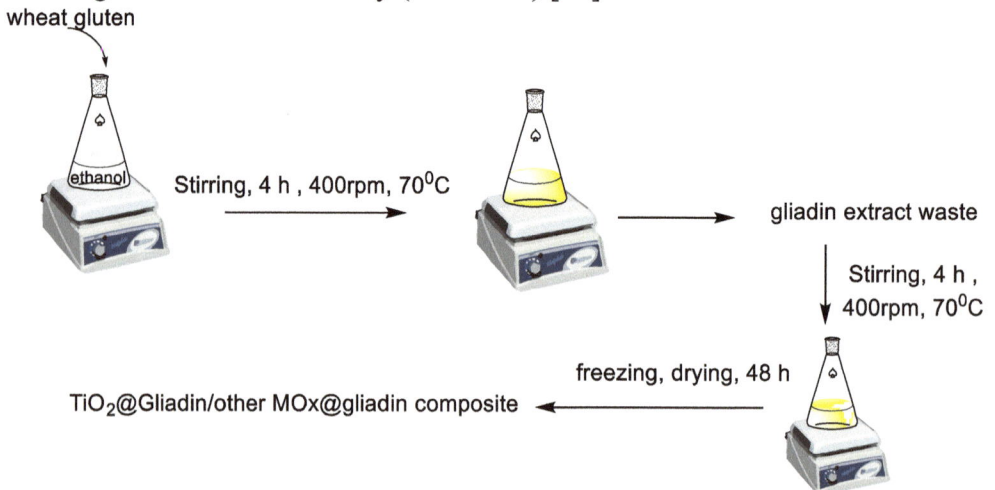

Scheme 8. Synthesis of nanoparticles by mixing gliadin protein from wheat with $AgFe_2O_3$, TiO_2, and AgFe–TiO_2.

Virendra Kumar Yadav *et al.* synthesized iron oxide nanoparticles from coal fly ash (CFA). Initially, ferrous particles were taken out from CFA and added with HCl, followed by sonication at 60-70°C. Secondly, the mixture obtained was heated at 60-70°C, followed by the addition of NaOH to form the desired precipitate.

Nanoparticles were subjected to various characterizations to measure their physical and morphological structure and the presence of functional groups. FTIR data confirmed the presence of O-H, Fe-O-Fe, and Fe-O-OH bonds. XRD data showed both magnetite and hematite phases in particles. Raman spectroscopy confirmed the Fe-O-Fe bond and crystalline nature of nanoparticles. TEM and SEM images showed particles to be spherical, along with some rod-shaped particles with aggregation. These nanoparticles also showed magnetic properties. A UV-visible peak at 210 nm showed the presence of ferrous ions and also the successful synthesis of nanoparticles.

Synthesized nanoparticles, due to the above characteristics, were used to remove heavy metals from wastewater with an efficiency of 40-70%. (Scheme **9**) [30].

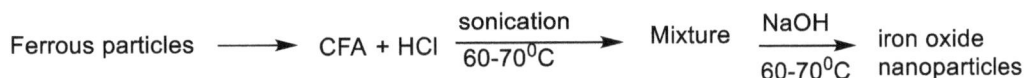

Ferrous particles $\longrightarrow$ CFA + HCl $\xrightarrow[\text{60-70}^0\text{C}]{\text{sonication}}$ Mixture $\xrightarrow[\text{60-70}^0\text{C}]{\text{NaOH}}$ iron oxide nanoparticles

Scheme 9. Synthesis of iron oxide nanoparticles.

CHEMISTRY OF METAL OXIDES FOR THE REMOVAL OF DYES

Han Fu *et al.* prepared metal oxide nanoparticles fabricated with crumpled graphene oxide using a nano-spray drying technique (Fig. **2**).

Synthesized nanoparticles were characterized using various spectroscopical techniques. SEM images of particles showed irregular or crumpled structures with many twists. XPS data confirmed the presence of constituent elements Si, C, and oxygen and also confirmed the presence of C=O, C-O-C, and COO. FTIR peaks confirmed the presence of OH, C=O, Si-O, and C-O-C.

These nanoparticles were used as dye-removing agents, and due to good adsorption capacity, they decomposed dye particles with an efficiency of 86%. (Scheme **10**) [31].

Tarmizi Taher *et al.* reported a technique to synthesize montmorillonite-mixed metal oxide (MMT-MMO) composite by the facile coprecipitation method. Initially, MMT suspension was prepared with MMT powder in distilled water, followed by vigorous stirring under ultrasonic irradiations for 1 h. Secondly, a solution containing $(Al(NO_3)_3.9H_2O)$ and $Zn(NO_3)_2.6H_2O$ was slowly added to the

MMT suspension, followed by the addition of NaOH dropwise to make the pH of the mixture 7.5 and stirring for 25 h at 60^{0}C to synthesize MMT-LDH composite. Finally, calcination of MMT-LDH was carried out for 1h at 400^{0}C to produce the final product.

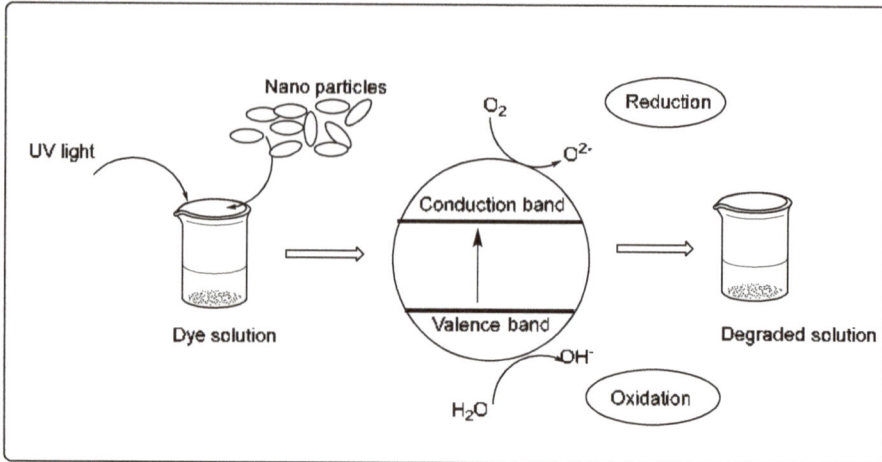

Fig. (2). General mechanism for the removal of dyes.

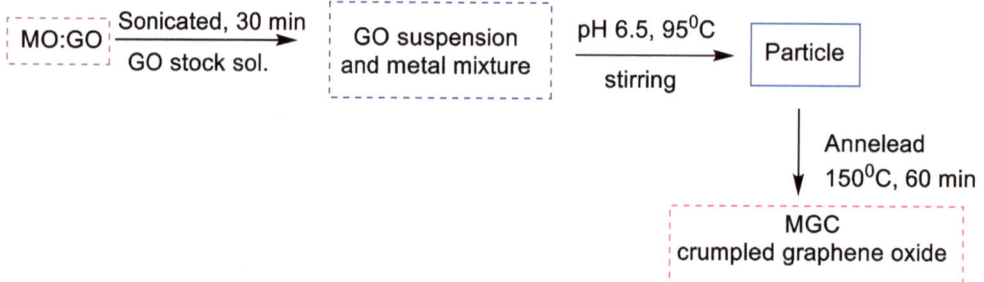

Scheme 10. Synthesis of MGC.

The synthesized composite was characterized by several characterization techniques like XRD, FTIR, TEM, and SEM. The absorption capacity was tested for the removal of both methylene blue (MB) as a cationic dye and congo red (CR) as the anionic dye from polluted water. XRD data showed well-structured particles with a crystalline nature. The SEM and TEM characterization of synthesized MMT-MMO was carried out to analyze their morphology and microstructure, and it showed that nanosized ZnO particles were uniformly distributed on the flat surface of MMT. The TGA graph analysis of the MMT_MMO composite showed that the MMT-LDH formed converted completely into MMT-MMO after calcination at 400^{0}C.

FTIR spectra confirmed the presence of NO^{3-} and CO^{3-} ions. Adsorption capacity studies showed that this composite was found to be capable of the removal of methylene blue and congo red dye *via* pseudo-second-order kinetics. These studies also confirmed the capacity of composite to remove both cationic and anionic dyes from water waste. UV absorption studies showed that these nanoparticles absorbed the anionic dye congo red more efficiently than the cationic dye methylene blue.

The synthesized composite was found to have a high surface area as compared to soil and coal, which was advantageous for the adsorption of dyes to a greater extent. (Scheme **11**) (Fig. **3**) [32].

Scheme 11. Synthesis of MMT-MMO nanocomposite.

V. Perumal et al. synthesized SnO_2-CuO nanocomposites using a simple solution processes method. Initially, copper acetate tetrahydrate and tin chloride pentahydrate were added to 100 mL distilled water and stirred for 1 h, followed by the addition of PVP. Secondly, sodium hydroxide solution was added to the above mixture. Finally, the precipitate was washed with ethanol and distilled water, followed by heating at 400^0C for 3 h to obtain SnO_2 – CuO nanocomposites.

The physical properties of prepared nanocomposites were analyzed by XRD and μ-Raman spectra, and it was observed that nanoparticles were found to have a tetragonal rutile structure. The presence of SnO_2 and CuO in the nanocomposite was confirmed by analyzing the vibration mode of the SnO_2 - CuO composite using μ-Raman spectroscopy. FTIR analysis confirmed the presence of Sn-O and Cu–O bonds. The rock-like structure of SnO_2 - CuO heterostructures was analyzed by FESEM techniques, and in TEM analysis, the clear lattice fringes were observed, which was due to the good crystalline nature of SnO_2 - CuO nanocomposite. The absorption peak at 500 nm showed good absorption capacity, as analyzed by UV absorption spectra.

The photodegradation activity of SnO_2 – CuO nanocomposites was studied using an MB model pollutant, and it was observed that SnO_2 - CuO heterostructure degraded 90.3% of highly reactive MB in 180 min. (Scheme **12**) (Fig. **3**) [33].

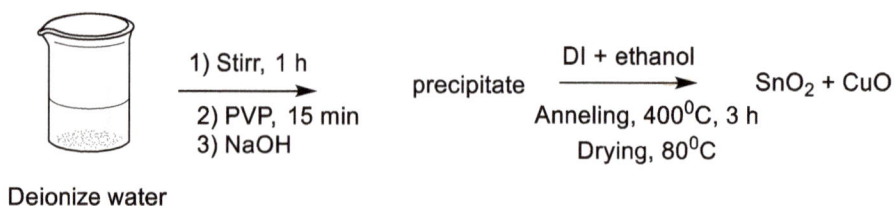

Scheme 12. Synthesis of SnO_2-CuO nanocomposites.

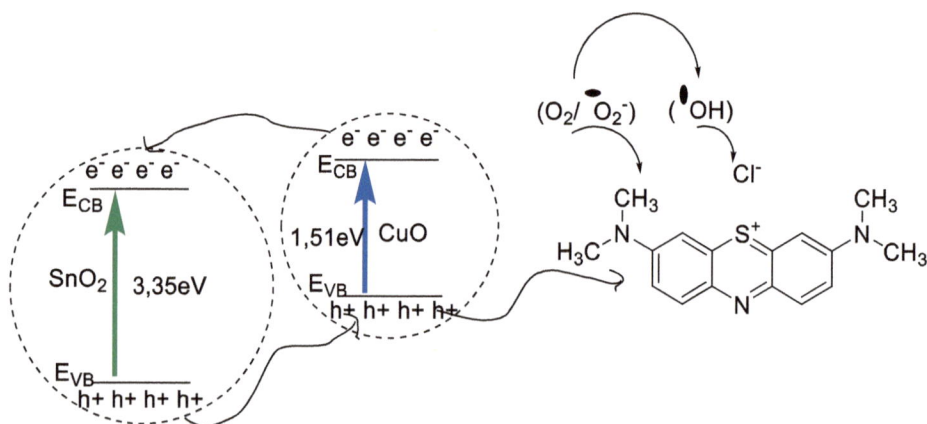

Fig. (3). Photocatalytic mechanism of SnO2-CuO nanocomposites.

Abbas Aziz *et al.* reported a green approach for the synthesis of zinc nanoparticles using Grewia asiatica extract. Initially, the zinc acetate solution was stirred at 40°C for 40 min, followed by the addition of NaOH to keep the pH at 12. Secondly, Grewia extract was added to the solution with continuous stirring. Finally, a color change from colorless to light orange was observed, which confirmed the formation of desired nanoparticles.

Synthesized nanoparticles were characterized by using techniques like scanning electron microscope (SEM), UV-visible spectroscopy (Uv-vis), Fourier transform infrared spectroscopy (FTIR), and energy dispersive X-ray spectroscopy (EDX).

SEM image analysis showed the spherical shape of ZnO NPs, and the particle size was found to be in the range of 28–42 nm. EDX technique showed that a high concentration of Zn and oxygen was present in the product, and only a negligible amount of K, Ca, P, and Mg was present, which could be from fruit extract.

FTIR studies of nanoparticles confirmed the presence of C-O, C=O, C-H, and O-H.

Photocatalytic degradation studies were conducted to evaluate photocatalytic activity through the absorbance of methylene blue azo dye, and it was observed that it was decomposed by 90% in 70 min *via* pseudo-first-order kinetics.

The importance of these nanoparticles lies in the fact that they can be recycled easily from the reaction mixture and can be used efficiently. (Scheme **13**) [34].

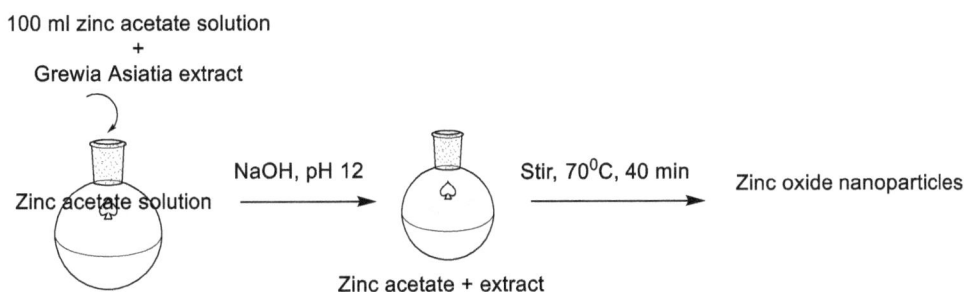

Scheme 13. Synthesis of ZnO nanoparticles.

Scanning electron microscopy (SEM) was used to evaluate the morphology of the synthesized nanoparticles, and it showed an asymmetrical tetragonal shape with some imperfections and a particle size of 20-48 nm in agglomerated form. TEM and HRTEM techniques were used to study the effect of Ag doping on Fe_2O_3 nanoparticles and $Ag-Fe_2O_3$ NPs, and the crystalline structure of $Ag-Fe_2O_3$ was observed with a lattice spacing of 2.66 nm.

X-ray photoelectron spectroscopy (XPS) analysis showed the presence of Fe, O, and Ag atoms in the pattern of $Ag-Fe_2O_3$ and showed good binding of Fe^{3+} in $Ag-Fe_2O_3$ nanoparticles, which was responsible for the photodegradation process.

Synthesized nanoparticles were tested for degradation of RR120 and O-II dyes, and it was observed that $Ag-Fe_2O_3$ was found to be most effective against RR120 and O-II dyes with 99.2% and 98.8% efficiency. (Scheme **14**) [35].

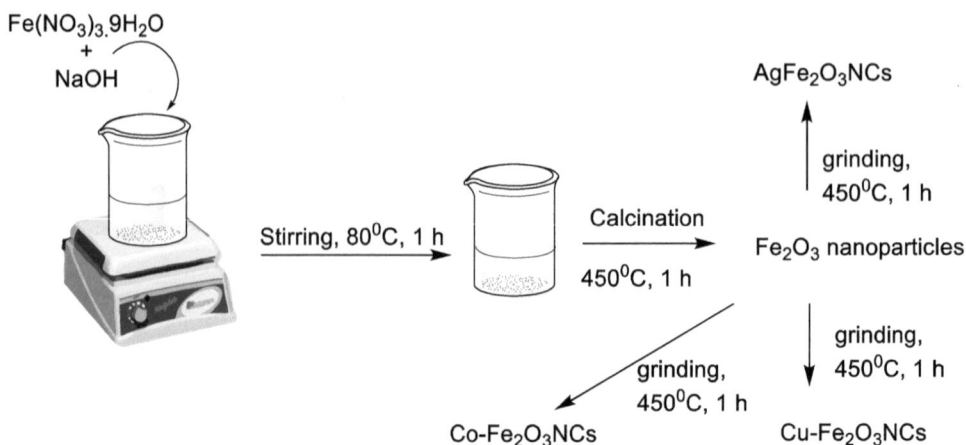

Scheme 14. Synthesis of Ag, Co, Cu-doped Fe_2O_3 nanoparticles.

Mohammed N. Almarri *et al.* prepared new $CuO/La_2O_3/GO$-based nanocomposites. CuO, La_2O_3, and GO were added separately as well as together in a ratio in deionized water to prepare binary nanocomposites like CuO/GO, La_2O_3/GO, and CuO/La_2O_3 as well as ternary nanocomposite $CuO/La_2O_3/GO$.

The physical properties of prepared nanocomposites were evaluated *via* techniques such as XRD, SEM, TEM, FTIR, and EDX. X-ray diffraction of prepared nanocomposites confirmed hexagonal structure with crystalline nature. SEM and TEM images showed roughness of structure with a nanorod size of 250 nm. FTIR peaks showed the presence of La-O and La-OH bonds.

SEM and TEM were used to determine the morphology of the prepared ternary nanocomposite, and it showed a rough surface with a particle size of 200 nm. EDX confirmed the presence of constituent elements without any impurities.

Thermal gravimetric analysis (TGA) showed high stability and a greater temperature range.

Synthesized nanocomposites were tested for the degradation of methylene blue dye, and it was observed that La_2O_3, CuO, La_2O_3/GO, CuO/GO, La_2O_3/CuO, and $La_2O_3/CuO/GO$ degraded the dye with an efficiency of 47, 49, 56, 58, 67, and 85%, respectively, after 240 min *via* pseudo first-order kinetics. (Scheme **15**) [36].

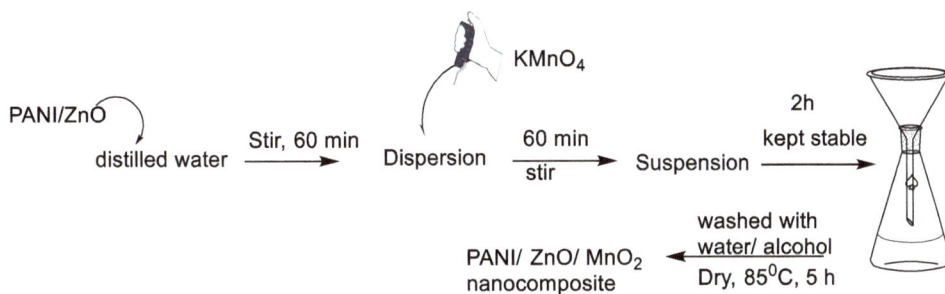

Scheme 15. Synthesis of CuO/La$_2$O$_3$/GO-based nanocomposite.

Twinkle *et al.* developed reduced iron oxide(rIOD) nanoparticles from waste iron dust which was used for the adsorption of azo dyes[RG(reactive green 19), RO16(reactive orange 16)] and one cationic dye[MG(malachite green)]. These r-IOD were synthesize from iron oxide dust, NaBH4 methanol and water. Then the solution was chemically reduced from Fe(III) to Fe(0) and turned black completely. The precipitate were collected with magnet washed with ethanol, water and CH$_3$COCH$_3$ to remove unreacted NaBH$_4$, and dried at room temperature.

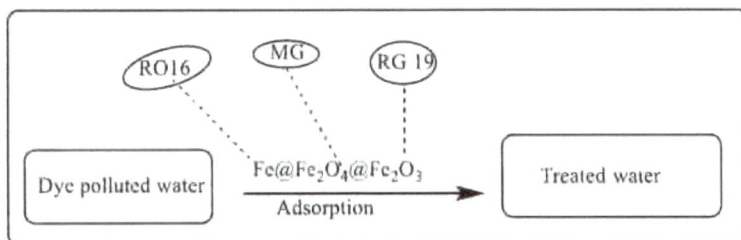

Fig. (4). Adsorption of pollutant from water.

SEM and EDX of r-IOD confirmed the presence of Fe and O and HRTEM showed flake-like morphology having interplanar spacing 0.24nm with (311) plane of Fe3O4 and 0.20nm with (110) plane of Fe wheras PXRD showed diffraction peaks of alpha-Fe2O3 planes(104)(012)(113)(300) and (024). (Scheme **16**) (Fig. **4**) [37].

Scheme 16. Synthesis of rIOD from IOD.

Parmeshwar Lal Meena *et al.* developed a protocol to synthesize novel PANI/ZnO/MnO$_2$ nanocomposite by dissolving prepared PANI/ZnO nanocomposite in distilled water and stirring for 60 min, followed by the addition of KMnO$_4$ dropwise.

Synthesized nanocomposite was characterized by various analytical tools like XRD, FTIR, FESEM, EDS, XPS, and TGA.

XRD spectra showed hexagonal crystal structure and crystal size, which was found to be 1.47–6.79 nm. The FTIR confirmed the presence of ZnO, O-Mn-O, and Mn-O. FESEM analysis was done to assess microstructural and textural properties, and it was observed that nanocomposite particles were found to have porous, gritty, and highly fibrous microstructures having a diameter of 73-86 nm with a rough surface, which enhanced their adsorption capability. Energy-dispersive X-ray spectroscopy (EDS) was used to determine the elemental composition of the PANI/ZnO/MnO$_2$ TNC, and the spectrum showed peaks of Zn, Mn, C, N O, and Cl elements. TGA showed thermal stability. XPS analysis showed the presence of Mn and Zn in the form of Mn^{4+} and Zn^{2+}. UV-vis absorption spectra showed peaks corresponding to good absorption capacity.

The synthesized nanocomposite was tested for CV dye degradation, and it was observed that due to large surface area, rough surface, and good porosity, it degraded CV by 96.13% in 60 min as compared to pure ZnO and PANI with values of 40.68 and 45.85, respectively. Due to its high adsorption capacity, this nanocomposite found great application in removing dye effluents from water waste. (Scheme **17**) [38].

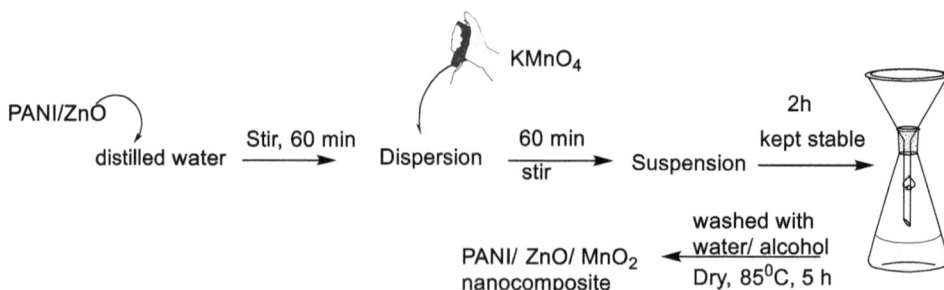

Scheme 17. Synthesis of PANI/ZnO/MnO$_2$ nanocomposite.

Akbar K. Inamdar *et al.* reported the synthesis of copper oxide transition metal oxide nanoparticles by using the flame pyrolysis method with some modifications. These nanoparticles were synthesized by dissolving copper chloride dihydrate in ethanol and burnt at high temperatures.

The synthesized CuO NPs were characterized by using various analytical tools. X-ray diffraction [XRD] and Fourier transform infrared [FT–IR] spectroscopy were used to determine crystallinity, size of particles, phase, and their structure, which confirmed the monoclinic structure of CuO nanoparticles. FTIR confirmed the presence of Cu-O and their monoclinic structure. SEM and TEM images showed a spherical shape with agglomeration. Adsorption capacity was measured using UV spectroscopy, and high values were observed.

The synthesized nanoparticles were tested for dye removal from water effluents, and it was observed that these particles were able to adsorb congo red dye with an efficiency of 99.90%. (Scheme **18**) [39].

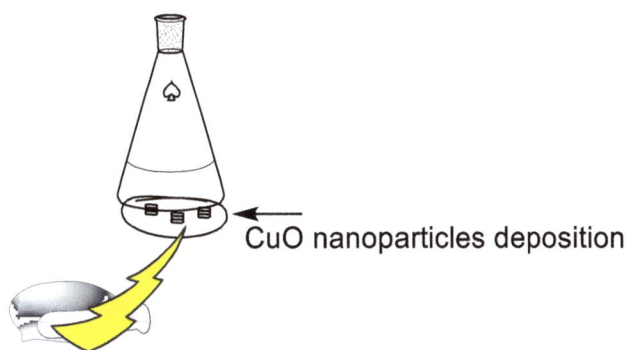

CuO nanoparticles deposition

ethanolic copper chloride solution in crusible with flame

Scheme 18. Synthesis of CuO nanoparticles.

Wei Wang *et al.* synthesized NiO/g-C$_3$N$_4$ for wastewater treatment. Initially, NiO nanospheres and graphite-phase carbon nitride (g-C$_3$N$_4$) nanosheets were synthesized by solvothermal and thermal polymerization methods. Secondly, these nanospheres and nanosheets were mixed in the requisite ratio and heated at 400^{0}C for 1 h to prepare the final nanocomposite.

Characterization of prepared nanocomposite was carried out by using various techniques like scanning electron microscope (SEM), transmission electron microscope (TEM), X-ray diffractometer (XRD), Fourier transform infrared spectrometer (FT-IR), and X-ray photoelectron spectroscopy (XPS).

SEM analysis showed that nanocomposite was found to have many small-size nanosheets without aggregation. XRD analysis showed FCC geometry with good crystallinity. BET analysis showed a mesoporous nature, and FTIR analysis showed the presence of N-H, O-H, and Ni-O bonds. XPS spectra analysis showed the presence of C, N, Ni, and O in nanocomposite Ni^{2+} in NiO.

The prepared nanocomposite was tested for the removal of MO dye from water waste, and it was observed that the compound showed an efficiency of more than 90% degradation in 100 min *via* pseudo first-order kinetics. This nanocomposite was found to be good for dye degradation as combining NiO and g-C_3N_4 increased the surface area and photocatalytic activity of the nanocomposite. (Scheme **19**) [40].

Scheme 19. Synthesis of NiO/g-C_3N_4X nanocomposite.

Sachin *et al.* reported a green synthesis of zinc oxide (ZnO) nanoparticles by using Lychee peel extract by reacting zinc acetate dihydrate and lychee peel extract and heating them at 70⁰C for 3 h. Synthesized particles were characterized by XRD, FTIR, SEM, and EDS spectroscopic techniques.

The X-ray diffraction analysis confirmed anisotropic and crystalline morphology along with hexagonal wurtzite structure. The average size of nanoparticles was calculated to be 2.18 pm. SEM and EDS images confirmed that nanoparticles were found to have a spherical shape and were formed without any impurity. FTIR analysis of prepared nanoparticles showed peaks due to O-H, C-H, C-OH, and Zn-O bonds. Synthesized nanoparticles were also tested for antimicrobial activity against *Pseudomonas aeruginosa, Bacillus subtilis S. aureus*, and *E. coli*, and it was found that nanoparticles showed activity with the zone of inhibition values of 22, 17, 15, and 23 mm, respectively.

These nanoparticles, due to their large surface area and high adsorption capacity, were tested for the removal of Congo red dye, and it was observed that the dye was degraded by 98.4% for 120 min. (Scheme **20**) [41].

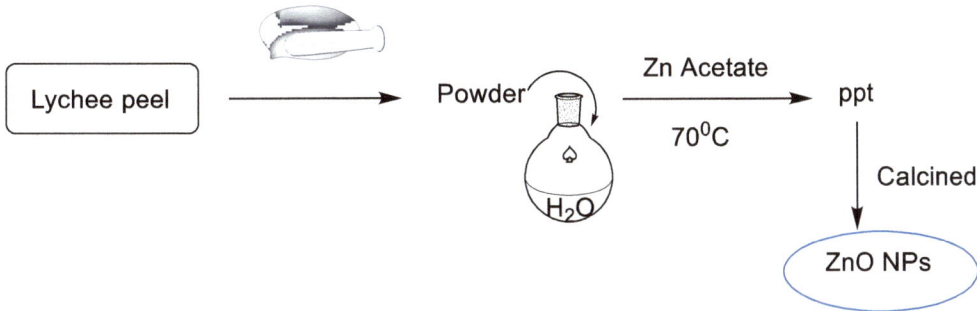

Scheme 20. Synthesis of ZnO NPs.

Some real-world applications of metal oxide nanoparticlesused in industrial wastewater treatment are mentioned below in Table **1**:

Table 1. List of companies working on wastewater treatment.

Company	Description
LANXESS	Offers a product utilizing iron nanoparticles designed to remove arsenic from water effectively.
American Elements	Develops iron nanoparticles aimed at addressing pollutants in groundwater, focusing on remediation.
Lehigh Nanotech	It specializes in the use of iron nanoparticles to treat groundwater contaminants.
SiREM	Provides solutions featuring iron nanoparticles for the remediation of various groundwater pollutants.
NanoH₂O	Focuses on membranes enhanced by nanotechnology to improve water desalination processes.
NanoOasis	Creates reverse osmosis membranes incorporating carbon nanotubes to enhance desalination efficiency.
Campbell Applied Physics	Engages in research on capacitive deionization technologies using carbon aerogel for purification.
Porifera	Develops nanotube-enhanced membranes aimed at advancing desalination techniques.
Itn Nanovation	Manufactures filters composed of nanoparticles to improve water filtration effectiveness.

ENVIRONMENTAL AND HUMAN HEALTH IMPLICATIONS

Metal oxides play a crucial role in the degradation of dyes and heavy metal particles, which helps in cleaning toxic materials from waste treatment. Although they are beneficial, they also have long-term adverse effects on the environment and human health if used without any protocols. These should be used in limited quantities by keeping in mind their adverse effects. Metal oxides should be assessed and regulated for proper usage with known potential risks.

CONCLUSION

Metal oxide nanoparticles have gained significant importance in water treatment due to their large surface area, high adsorption capacity, and many other properties. In order to help researchers create new nanomaterials and explore their chemistry for water and wastewater cleanup, this chapter offered invaluable information about synthetic processes along with their application for the removal of harmful heavy metals and dyes.

LIST OF ABBREVIATIONS

CR-	Congo Red
DI-	Deionized
E. Coli	Escherichia Coli
EDX-	Energy-Dispersive X-Ray
FTIR-	Fourier Transform Infrared Spectrometer
GO-	Graphene Oxide
MMT-MMO-	Montmorillonite-Mixed Metal Oxides
MB-	Methylene Blue
NPs-	Nanoparticles
PSO-	Pseudo-Second Order
PANI-	Polyaniline
PANI-EB-	PANI-Emeraldine Base
S. aureus	Staphylococcus aureus
SEM-	Scanning Electron Microscope
TEM-	Transmission Electron Microscope
XRD-	X-Ray Diffractometer
XPS-	X-Ray Photoelectron Spectroscopy
ZnO-	Zinc Oxide

ACKNOWLEDGEMENTS

The authors would like to express their sincere thanks to the management of SGT University, Gurugram, Haryana-122505, India, for providing the facilities to write and submit this chapter for publication.

REFERENCES

[1] Naseem, T.; Durrani, T. The role of some important metal oxide nanoparticles for wastewater and antibacterial applications: A review. *Environmental Chemistry and Ecotoxicology,* **2021**, *3*, 59-75.
[http://dx.doi.org/10.1016/j.enceco.2020.12.001]

[2] Call, T.P.; Carey, T.; Bombelli, P.; Lea-Smith, D.J.; Hooper, P.; Howe, C.J.; Torrisi, F. Platinum-free, graphene based anodes and air cathodes for single chamber microbial fuel cells. *J. Mater. Chem. A Mater. Energy Sustain.,* **2017**, *5*(45), 23872-23886.
[http://dx.doi.org/10.1039/C7TA06895F] [PMID: 29456857]

[3] Zhang, Y.; Liu, L.; Van der Bruggen, B.; Yang, F. Nanocarbon based composite electrodes and their application in microbial fuel cells. *J. Mater. Chem. A Mater. Energy Sustain.,* **2017**, *5*(25), 12673-12698.
[http://dx.doi.org/10.1039/C7TA01511A]

[4] Oyewo, O. A.; Elemike, E. E.; Onwudiwe, D. C.; Onyago, M. S. Metal oxide-cellulose nanocomposite for the removal of toxic metals and dyes from wastewater. *Int. J. Biolog. Macromol.,* **2020**, 1-69.

[5] Yuan, H.; Hou, Y.; Abu-Reesh, I.M.; Chen, J.; He, Z. Oxygen reduction reaction catalysts used in microbial fuel cells for energy-efficient wastewater treatment: a review. *Mater. Horiz.,* **2016**, *3*(5), 382-401.
[http://dx.doi.org/10.1039/C6MH00093B]

[6] Mashkour, M.; Rahimnejad, M. Effect of various carbon-based cathode electrodes on the performance of microbial fuel cell. *Biofuel Research Journal,* **2015**, *2*(4), 296-300.
[http://dx.doi.org/10.18331/BRJ2015.2.4.3]

[7] Elakkiya, E.; Matheswaran, M. Comparison of anodic metabolisms in bioelectricity production during treatment of dairy wastewater in Microbial Fuel Cell. *Bioresour. Technol.,* **2013**, *136*, 407-412.
[http://dx.doi.org/10.1016/j.biortech.2013.02.113] [PMID: 23567709]

[8] Ng, I.S.; Hsueh, C.C.; Chen, B.Y. Electron transport phenomena of electroactive bacteria in microbial fuel cells: a review of Proteus hauseri. *Bioresour. Bioprocess.,* **2017**, *4*(1), 53.
[http://dx.doi.org/10.1186/s40643-017-0183-3]

[9] He, W.; Zhang, X.; Liu, J.; Zhu, X.; Feng, Y.; Logan, B.E. Microbial fuel cells with an integrated spacer and separate anode and cathode modules. *Environ. Sci. Water Res. Technol.,* **2016**, *2*(1), 186-195.
[http://dx.doi.org/10.1039/C5EW00223K]

[10] Hidalgo, D.; Tommasi, T.; Bocchini, S.; Chiolerio, A.; Chiodoni, A.; Mazzarino, I.; Ruggeri, B. Surface modification of commercial carbon felt used as anode for microbial fuel cells. *Energy J. (Camb. Mass.),* **2016**, *99*, 193-201.

[11] Sorbiun, M.; Shayegan Mehr, E.; Ramazani, A.; Taghavi Fardood, S. Green synthesis of zinc oxide and copper oxide nanoparticles using aqueous extract of oak fruit hull (jaft) and comparing their photocatalytic degradation of basic violet 3. *Int. J. Environ. Res.,* **2018**, *12*(1), 29-37.
[http://dx.doi.org/10.1007/s41742-018-0064-4]

[12] Hindatu, Y.; Annuar, M.S.M.; Gumel, A.M. Mini-review: Anode modification for improved performance of microbial fuel cell. *Renew. Sustain. Energy Rev.,* **2017**, *73*, 236-248.
[http://dx.doi.org/10.1016/j.rser.2017.01.138]

[13] Taghavi Fardood, S.; Ramazani, A.; Moradi, S.; Azimzadeh Asiabi, P. Green synthesis of zinc oxide nanoparticles using arabic gum and photocatalytic degradation of direct blue 129 dye under visible light. *J. Mater. Sci. Mater. Electron.,* **2017**, *28*(18), 13596-13601.
[http://dx.doi.org/10.1007/s10854-017-7199-5]

[14] Singh, J.; Dutta, T.; Kim, K.H.; Rawat, M.; Samddar, P.; Kumar, P. 'Green' synthesis of metals and their oxide nanoparticles: applications for environmental remediation. *J. Nanobiotechnology,* **2018**, *16*(1), 84.

[http://dx.doi.org/10.1186/s12951-018-0408-4] [PMID: 30373622]

[15] Gajda, I.; Greenman, J.; Santoro, C.; Serov, A.; Melhuish, C.; Atanassov, P.; Ieropoulos, I.A. Improved power and long term performance of microbial fuel cell with Fe-N-C catalyst in air-breathing cathode. *Energy,* **2018**, *144*, 1073-1079.
[http://dx.doi.org/10.1016/j.energy.2017.11.135] [PMID: 29456285]

[16] Singh, J.; Kumar, V.; Kim, K.H.; Rawat, M. Biogenic synthesis of copper oxide nanoparticles using plant extract and its prodigious potential for photocatalytic degradation of dyes. *Environ. Res.,* **2019**, *177*, 108569.
[http://dx.doi.org/10.1016/j.envres.2019.108569] [PMID: 31352301]

[17] Yin, Y.; Huang, G.; Zhou, N.; Liu, Y.; Zhang, L. Increasing power generation of microbial fuel cells with a nano-CeO_2 modified anode. *Energy Sources A Recovery Util. Environ. Effects,* **2016**, *38*(9), 1212-1218.
[http://dx.doi.org/10.1080/15567036.2014.898112]

[18] Atrak, K.; Ramazani, A.; Taghavi Fardood, S. Green synthesis of amorphous and gamma aluminum oxide nanoparticles by tragacanth gel and comparison of their photocatalytic activity for the degradation of organic dyes. *J. Mater. Sci. Mater. Electron.,* **2018**, *29*(10), 8347-8353.
[http://dx.doi.org/10.1007/s10854-018-8845-2]

[19] Alatraktchi, F.A.; Zhang, Y.; Angelidaki, I. Nanomodification of the electrodes in microbial fuel cell: Impact of nanoparticle density on electricity production and microbial community. *Appl. Energy,* **2014**, *116*, 216-222.
[http://dx.doi.org/10.1016/j.apenergy.2013.11.058]

[20] Haidri, I.; Shahid, M.; Hussain, S.; Shahzad, T.; Mahmood, F.; Hassan, M.U.; Al-Khayri, J.M.; Aldaej, M.I.; Sattar, M.N.; Rezk, A.A.S.; Almaghasla, M.I.; Shehata, W.F. Efficacy of biogenic zinc oxide nanoparticles in treating wastewater for sustainable wheat cultivation. *Plants,* **2023**, *12*(17), 3058.
[http://dx.doi.org/10.3390/plants12173058] [PMID: 37687305]

[21] Hidalgo, D.; Tommasi, T.; Bocchini, S.; Chiolerio, A.; Chiodoni, A.; Mazzarino, I.; Ruggeri, B. Surface modification of commercial carbon felt used as anode for Microbial Fuel Cells. *Energy,* **2016**, *99*, 193-201.
[http://dx.doi.org/10.1016/j.energy.2016.01.039]

[22] Kaleem, M.; Anjum Minhas, L.; Zaffar Hashmi, M.; Umer Farooqi, H.M.; Waqar, R.; Kamal, K.; Saad Aljaluod, R.; Alarjani, K.M.; Samad Mumtaz, A. Biogenic synthesis of iron oxide nanoparticles and experimental modeling studies on the removal of heavy metals from wastewater. *J. Saudi Chem. Soc.,* **2024**, *28*(1), 101777.
[http://dx.doi.org/10.1016/j.jscs.2023.101777]

[23] Hossain, N.; Nizamuddin, S.; Ball, A.S.; Shah, K. Synthesis, performance and reaction mechanisms of Ag-modified multi-functional rice husk solvochar for removal of multi-heavy metals and water-borne bacteria from wastewater. *Process Saf. Environ. Prot.,* **2024**, *182*, 56-70.
[http://dx.doi.org/10.1016/j.psep.2023.11.058]

[24] Hosseinkhani, O.; Hamzehlouy, A.; Dan, S.; Sanchouli, N.; Tavakkoli, M.; Hashemipour, H. Graphene oxide/ZnO nanocomposites for efficient removal of heavy metal and organic contaminants from water. *Arab. J. Chem.,* **2023**, *16*(10), 105176.
[http://dx.doi.org/10.1016/j.arabjc.2023.105176]

[25] Liu, Z.; Lei, M.; Zeng, W.; Li, Y.; Li, B.; Liu, D.; Liu, C. Synthesis of magnetic $Fe_3O_4@SiO_2$-(-NH_2/-COOH) nanoparticles and their application for the removal of heavy metals from wastewater. *Ceram. Int.,* **2023**, *49*(12), 20470-20479.
[http://dx.doi.org/10.1016/j.ceramint.2023.03.177]

[26] Basso Peressut, A.; Cristiani, C.; Dotelli, G.; Dotti, A.; Latorrata, S.; Bahamonde, A.; Gascó, A.; Hermosilla, D.; Balzarotti, R. Reduced graphene oxide/waste-derived TiO_2 Composite membrane: Preliminary study of a new material for hybrid wastewater treatment. *Nanomaterials (Basel),* **2023**,

13(6), 1043.
[http://dx.doi.org/10.3390/nano13061043]

[27]   Masindi, V.; Tekere, M.; Foteinis, S. Treatment of real tannery wastewater using facile synthesized magnesium oxide nanoparticles: Experimental results and geochemical modeling. *Water Resour. Ind.,* **2023**, *29*, 100205.
[http://dx.doi.org/10.1016/j.wri.2023.100205]

[28]   Eid, A.M.; Fouda, A.; Hassan, S.E.D.; Hamza, M.F.; Alharbi, N.K.; Elkelish, A.; Alharthi, A.; Salem, W.M. Plant-based copper oxide nanoparticles; biosynthesis, characterization, antibacterial activity, tanning wastewater treatment and heavy metals sorption. *Catalysts,* **2023**, *13*(2), 348.
[http://dx.doi.org/10.3390/catal13020348]

[29]   Massima Mouele, E.S.; Bediako, J.K.; El Ouardi, Y.; Anugwom, I.; Butylina, S.; Mukaba, J.L.; Petrik, L.F.; Zar Myint, M.T.; Kyaw, H.H.; Al-Abri, M.; Al Belushi, M.A.; Dobretsov, S.; Laatikainen, K.; Repo, E. Sustainable gliadin - Metal oxide composites for efficient inactivation of Escherichia coli and remediation of cobalt (II) from water. *Environ. Pollut.,* **2024**, *340*(Pt 2), 122788.
[http://dx.doi.org/10.1016/j.envpol.2023.122788] [PMID: 37879550]

[30]   Yadav, V.K.; Amari, A.; Gacem, A.; Elboughdiri, N.; Eltayeb, L.B.; Fulekar, M.H. Treatment of Fly-ash-contaminated wastewater loaded with heavy metals by using fly-ash-synthesized iron oxide nanoparticles. *Water,* **2023**, *15*(5), 908.
[http://dx.doi.org/10.3390/w15050908]

[31]   Fu, H.; Cai, H.; Gray, K.A. Metal oxide encapsulated by 3D graphene oxide creates a nanocomposite with enhanced organic adsorption in aqueous solution. *J. Hazard. Mater.,* **2023**, *444*(Pt A), 130340.
[http://dx.doi.org/10.1016/j.jhazmat.2022.130340] [PMID: 36402105]

[32]   Taher, T.; Munandar, A.; Mawaddah, N.; Syamsuddin Wisnubroto, M.; Siregar, P.M.S.B.N.; Palapa, N.R.; Lesbani, A.; Wibowo, Y.G. Synthesis and characterization of montmorillonite – Mixed metal oxide composite and its adsorption performance for anionic and cationic dyes removal. *Inorg. Chem. Commun.,* **2023**, *147*, 110231.
[http://dx.doi.org/10.1016/j.inoche.2022.110231]

[33]   Perumal, V.; Uthrakumar, R.; Chinnathambi, M.; Inmozhi, C.; Robert, R.; Rajasaravanan, M.E.; Raja, A.; Kaviyarasu, K. Electron-hole recombination effect of SnO_2 – CuO nanocomposite for improving methylene blue photocatalytic activity in wastewater treatment under visible light. *J. King Saud Univ. Sci.,* **2023**, *35*(1), 102388.
[http://dx.doi.org/10.1016/j.jksus.2022.102388]

[34]   Aziz, A.; Memon, Z.; Bhutto, A. Efficient photocatalytic degradation of industrial wastewater dye by Grewia asiatica mediated zinc oxide nanoparticles. *Optik (Stuttg.),* **2023**, *272*, 170352.
[http://dx.doi.org/10.1016/j.ijleo.2022.170352]

[35]   Huang-Mu, L.; Devanesan, S.; Farhat, K.; Kim, W.; Sivarasan, G. Improving the efficiency of metal ions doped Fe_2O_3 nanoparticles: Photocatalyst for removal of organic dye from aqueous media. *Chemosphere,* **2023**, *337*, 139229.
[http://dx.doi.org/10.1016/j.chemosphere.2023.139229] [PMID: 37354953]

[36]   Almarri, M.N.; Khalaf, M.M.; Gouda, M.; El-Taib Heakal, F.; Elmushyakhi, A.; Abou Taleb, M.F.; Abd El-Lateef, H.M. Chemical, surface, and thermal studies of mixed oxides cupric oxide (CuO), lanthanum oxide (La2O3), and graphene oxide for dye degradation from aqueous solution. *J. Mater. Res. Technol.,* **2023**, *23*, 2263-2274.
[http://dx.doi.org/10.1016/j.jmrt.2023.01.152]

[37]   Twinkle, Tewatia, H.; Kaushik, J.; Sahu, A.; Chaudhary, S. K.; Sonkar, S. K. Waste iron dust deried iron oxide nanoparticles for efficient adsorption of multiple azo dyes. *ACS Sustainable Resour. Manage.,* **2024**, *1*, 278-288.
[http://dx.doi.org/10.1016/j.arabjc.2023.105176]

[38]   Lal Meena, P.; Kumar Saini, J. Synthesis of polymer-metal oxide (PANI/ZnO/MnO2) ternary

nanocomposite for effective removal of water pollutants. *Results in Chemistry,* **2023**, *5*, 100764.
[http://dx.doi.org/10.1016/j.rechem.2023.100764]

[39]   Inamdar, A.K.; Rajenimbalkar, R.S.; Thabet, A.E.; Shelke, S.B.; Inamdar, S.N. Environmental applications of flame synthesized CuO nanoparticles through removal of Congo Red dye. *Mater. Today Proc.,* **2023**, *92*, 515-521.
[http://dx.doi.org/10.1016/j.matpr.2023.03.698]

[40]   Wang, W.; Lv, B.; Tao, F. NiO/g-C$_3$N$_4$ composite for enhanced photocatalytic properties in the wastewater treatment. *Environ. Sci. Pollut. Res. Int.,* **2022**, *30*(10), 25620-25634.
[http://dx.doi.org/10.1007/s11356-022-24121-2] [PMID: 36413264]

[41]   Sachin; Jaishree; Singh, N.; Singh, R.; Shah, K.; Pramanik, B. K. Green synthesis of zinc nanoparticles using lychee peel and its application in anti-bacterial properties and CR dye removal. *Chemosphere,* **2023**, *327*, 138497.

CHAPTER 4

Carbon Nanotubes in Wastewater Treatment

Shubhika Goyal[1], **Vishaka Chauhan**[1], **Chinmay Mittal**[1] and **Allu Udayasri**[2,*]

[1] *Department of Chemistry, SGT University, Gurugram-122001, Haryana, India*

[2] *Department of Chemistry, Aditya Institute of Technology and Management, Tekkali-532201, Andhra Pradesh, India*

Abstract: The growing worldwide water problem in recent years has highlighted the critical need for creative and effective wastewater treatment solutions. Since carbon nanotubes (CNTs) exhibit special structural, mechanical, electrical, and chemical capabilities, they have become more attractive options. This chapter offers a thorough summary of the latest developments, difficulties, and potential uses of CNTs and their composites in wastewater treatment. We go over the synthesis processes, characterization methodologies, and the effects of many parameters on treatment efficiency, including CNT type, shape, functionalization, and composite formulations. Furthermore, we investigate the regulatory landscape, toxicological issues, and environmental consequences related to the broad use of CNT-based technology in wastewater treatment. To fully realize the promise of CNTs and their composites for sustainable and effective wastewater treatment, this study attempts to compile the state-of-the-art, highlight research gaps, and offer guidance for future studies.

Keywords: Adsorption, Carbon nanotubes, Catalytic degradation, Environmental implications, Membrane filtration, Photocatalysis, Wastewater treatment.

INTRODUCTION

Wastewater treatment is one of the most important problems facing environmental science and engineering today. Wastewater creation has reached previously unheard-of heights due to growing urban populations and industrial activity, endangering both public health and water supplies. Though they have clearly made progress in reducing pollution, conventional wastewater treatment methods frequently fail to meet strict water quality criteria and manage newly emergent pollutants. In this regard, the application of nanotechnology has shown promise as a means of transforming wastewater treatment procedures.

* **Corresponding author Allu Udayasri:** Department of Chemistry, Aditya Institute of Technology and Management, Tekkali-532201, Andhra Pradesh, India; E-mail: alluudayasri.udayasri@gmail.com

Anjaneyulu Bendi (Ed.)

Recent years have seen a significant increase in interest in carbon nanotubes (CNTs), which have the potential to revolutionize wastewater treatment due to their exceptional physicochemical features and multifunctionalities [1 - 4]. The remarkable mechanical strength, chemical stability, large surface area, and nanoscale dimensions of these materials make them very suitable for a range of environmental remediation applications, including wastewater treatment. To better understand carbon nanotubes' mechanics, applications, difficulties, and potential uses, this chapter will present a thorough overview of the state-of-the-art in this field.

The basic characteristics and methods of carbon nanotube production will be covered in detail in this chapter. To maximize their effectiveness and customize them for particular wastewater treatment applications, it is essential to comprehend the synthesis pathways and structural complexities. Engineering new materials with improved functions is made possible by the variety of morphologies and characteristics of carbon nanotubes, ranging from single-walled to multi-walled forms [5 - 11].

Explaining the fundamental relationship between carbon nanotubes and pollutants in wastewater will be the next area of emphasis. A wide range of contaminants found in wastewater may be efficiently sequestered, broken down, or transformed by CNTs because of their special physicochemical characteristics, which also include their high adsorption capacity, catalytic activity, and capability to promote electron transfer [12 - 16]. For CNT-based treatment systems to be designed effectively and function optimally in a variety of environmental circumstances, it is essential to comprehend these mechanisms.

The broad range of treatment processes that carbon nanotubes may be used in, from membrane filtration and photocatalysis to adsorption and catalysis, demonstrate how versatile these materials are in wastewater treatment [17 - 23]. Their remarkable ability to adsorb permits the elimination of developing contaminants, heavy metals, and organic pollutants from wastewater. Additionally, their catalytic characteristics facilitate the breakdown of resistant pollutants by means of sophisticated oxidation procedures. In addition, the incorporation of carbon nanotubes into membranes presents a significant opportunity to improve the effectiveness and specificity of membrane-based separation procedures, thereby tackling the problem of water shortage and resource retrieval [24 - 29].

The extensive use of carbon nanotubes in wastewater treatment is hampered by a number of issues, including cost-effectiveness, scalability, environmental impact, and regulatory problems, despite their enormous potential. Interdisciplinary research projects, including materials science, engineering, environmental science, and policy-making, are required to address these issues [30 - 35]. In addition, it is crucial to guarantee the sustainable manufacture and disposal of carbon nanotubes in order to avoid any unexpected environmental effects and to guarantee their long-term viability as a wastewater treatment option.

With so much potential, carbon nanotubes have a bright future ahead of them in wastewater treatment. The development of next-generation CNT-based treatment technologies is expected to reach new heights thanks to ongoing developments in functionalization methods, process engineering, and nanomaterial synthesis. Furthermore, combining biotechnology, nanotechnology, and artificial intelligence might lead to the creation of self-governing, adaptive treatment systems that can be optimized and monitored in real time, ushering in a new era of intelligent and sustainable wastewater management [36 - 42] (Fig. **1**).

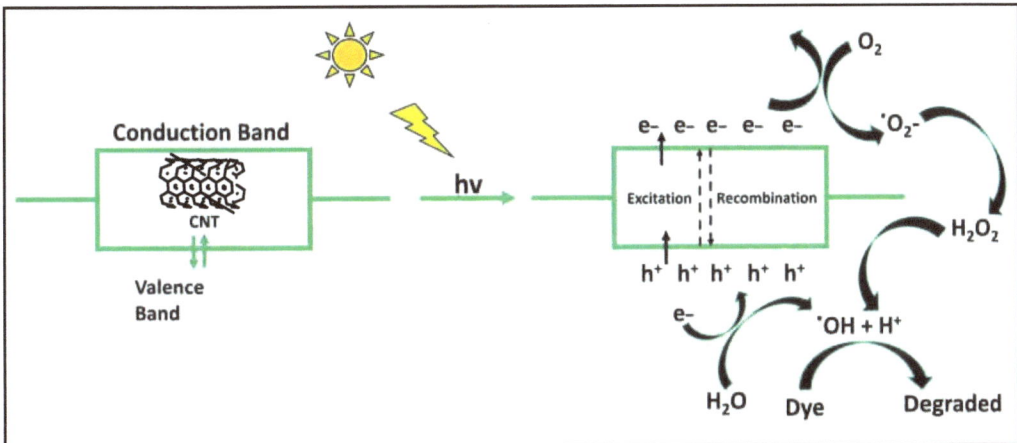

Fig. (1). Degradation of dyes present in wastewater. [37].

To sum up, carbon nanotubes are an innovative technology that might transform wastewater treatment and tackle the growing issues associated with water. This study aims to provide thorough knowledge of the role of carbon nanotubes in advancing state-of-the-art wastewater treatment by clarifying their fundamental features, processes, applications, problems, and future possibilities. Carbon nanotubes have the potential to become essential instruments in the pursuit of sustainable development and clean water *via* coordinated scientific endeavors and tactical partnerships.

CHEMISTRY OF CARBON NANOTUBE COMPOUNDS FOR THE ADSORPTION OF POLLUTANTS PRESENT IN WASTEWATER BODIES

Wang, Z. et al. synthesized the recyclable adsorbent modified with magnetic nanoparticles by a solvothermal method. This modification improves the overall performance of MWCNTs for heavy metal adsorption and recovery [43]. (Schemes **1 - 3**).

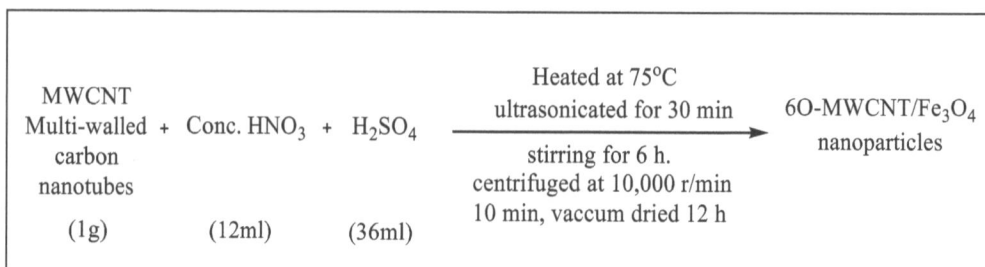

MWCNT
Multi-walled + Conc. HNO_3 + H_2SO_4 → 6O-MWCNT/Fe_3O_4 nanoparticles
carbon nanotubes

Heated at 75°C ultrasonicated for 30 min
stirring for 6 h.
centrifuged at 10,000 r/min 10 min, vaccum dried 12 h

(1g)   (12ml)   (36ml)

Scheme 1. Synthesis of 6O-MWCNTs.

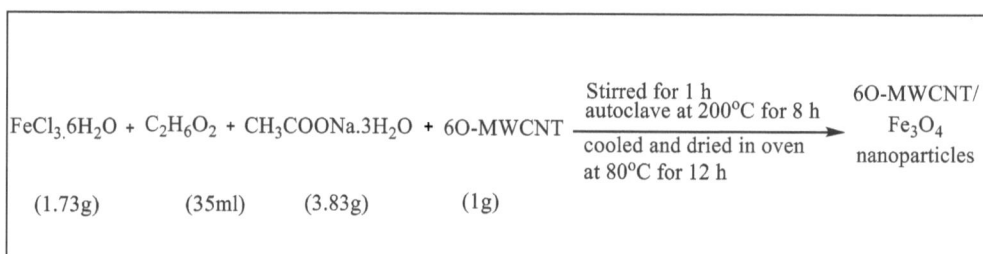

$FeCl_3.6H_2O$ + $C_2H_6O_2$ + $CH_3COONa.3H_2O$ + 6O-MWCNT → 6O-MWCNT/Fe_3O_4 nanoparticles

Stirred for 1 h
autoclave at 200°C for 8 h
cooled and dried in oven at 80°C for 12 h

(1.73g)   (35ml)   (3.83g)   (1g)

Scheme 2. Functionalization of magnetic nanoparticles on the surface of 6O-MWCNTs.

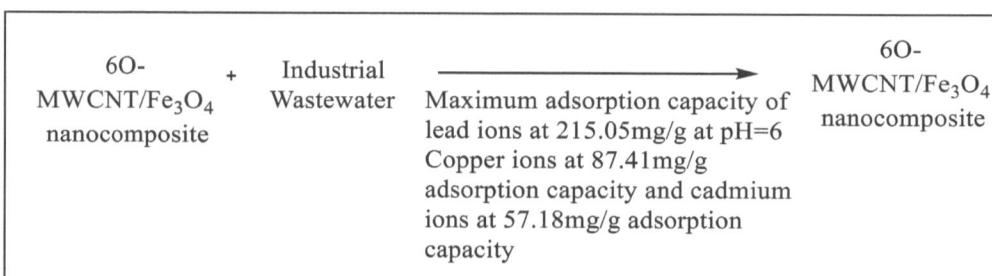

6O-MWCNT/Fe_3O_4 nanocomposite + Industrial Wastewater → 6O-MWCNT/Fe_3O_4 nanocomposite

Maximum adsorption capacity of lead ions at 215.05mg/g at pH=6 Copper ions at 87.41mg/g adsorption capacity and cadmium ions at 57.18mg/g adsorption capacity

Scheme 3. Adsorption of copper and cadmium.

A maximum adsorption capacity of 215.05 mg/g for lead ions is shown by the adsorption assays to indicate the composite's strong adsorption capacity. When compared to current adsorbents of the same kind, this is substantially greater. Comparing lead to copper and cadmium, it is discovered that lead shows a higher attraction to the lone pair of electrons in oxygen atoms. When carbon nanotubes are magnetically modified, a new reusable adsorbent is introduced for the effective removal of heavy metals from wastewater. In summary, the study contributed to the field's advancement and provided a useful method for effectively removing heavy metal ions from wastewater.

Polydopamine was used by ***Ghasemi, S. S. et al.*** to functionalize the surface and magnetize single-walled carbon nanotubes. The adsorbent carbon nanotube is appropriate for eliminating heavy metal ions from aqueous solutions [44] (Schemes **4** - **6**).

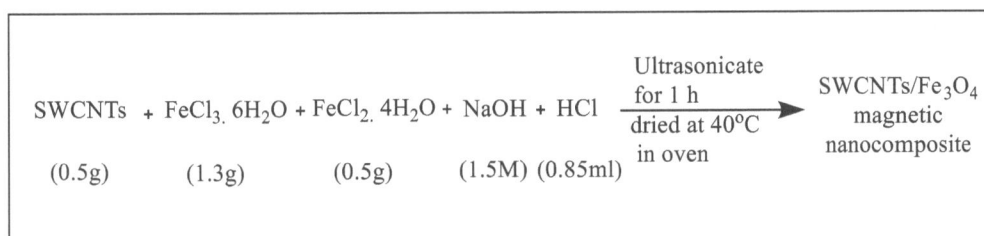

SWCNTs + $FeCl_3.6H_2O$ + $FeCl_2.4H_2O$ + NaOH + HCl $\xrightarrow[\substack{\text{dried at } 40^\circ C \\ \text{in oven}}]{\text{Ultrasonicate for 1 h}}$ SWCNTs/Fe_3O_4 magnetic nanocomposite

(0.5g)   (1.3g)   (0.5g)   (1.5M) (0.85ml)

Scheme 4. Synthesis of single-walled carbon nanotubes magnetic composite.

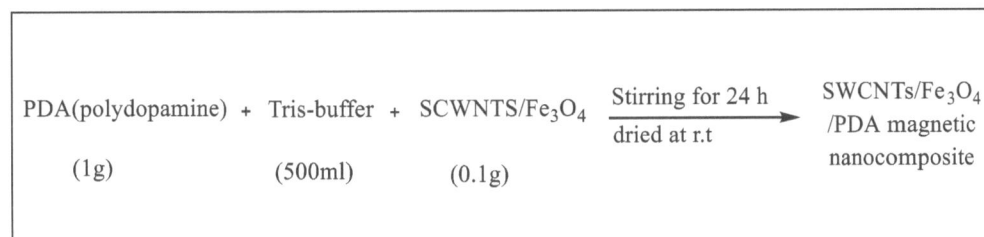

PDA(polydopamine) + Tris-buffer + SCWNTS/Fe_3O_4 $\xrightarrow[\text{dried at r.t}]{\text{Stirring for 24 h}}$ SWCNTs/Fe_3O_4/PDA magnetic nanocomposite

(1g)   (500ml)   (0.1g)

Scheme 5. Synthesis of polydopamine magnetic nanocomposite (SWCNTs/Fe_3O_4/PDA).

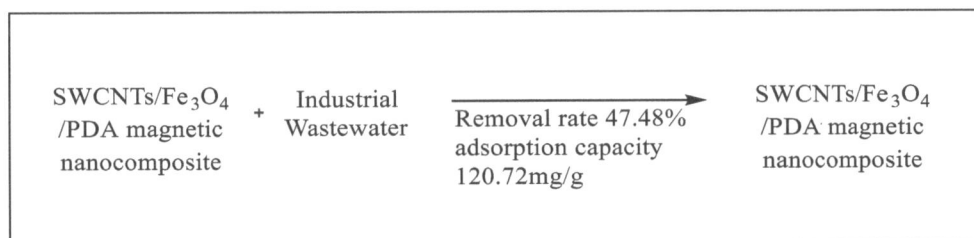

SWCNTs/Fe_3O_4/PDA magnetic nanocomposite + Industrial Wastewater $\xrightarrow[\substack{\text{adsorption capacity} \\ 120.72mg/g}]{\text{Removal rate 47.48\%}}$ SWCNTs/Fe_3O_4/PDA magnetic nanocomposite

Scheme 6. Adsorption of cadmium.

The value of the pseudo-second-order kinetic model's adsorption capacity was quite similar to that of the experiment. Reducing the starting concentration of metal ions also enhanced the adsorption capacity, and it was found that increasing the amount of adsorbent utilized improved the effectiveness of adsorption. As stated differently, more adsorbent was used, and less metal ion concentration in the solution resulted in improved metal ion adsorption onto the adsorbent material. The Langmuir model demonstrated the nanocomposite adsorbent's good adsorption capacity, with a maximum adsorption capacity of 186.48 mg/g. Therefore, when used in the best possible circumstances, the adsorbent is a good choice for eliminating cadmium ions from aqueous solutions.

Maryam, F. et al. fabricated an S-coated magnetic multi-walled carbon nanotube composite. The process used during the synthesis of composite is a simple heating process. Mercury was subsequently extracted from aqueous solutions using the composite as an adsorbent [45] (Schemes **7** - **9**).

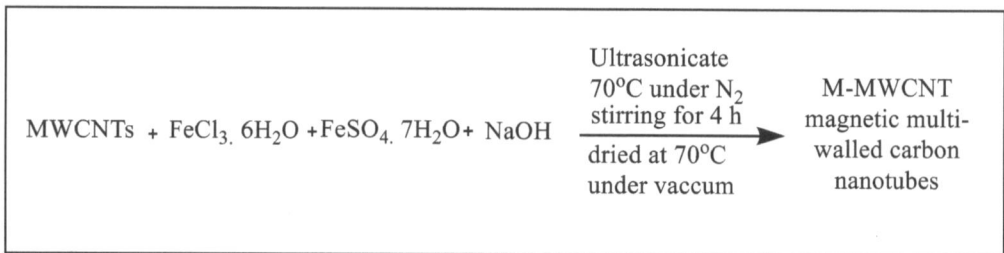

MWCNTs + FeCl$_3$. 6H$_2$O +FeSO$_4$. 7H$_2$O+ NaOH $\xrightarrow[\text{dried at 70°C under vaccum}]{\substack{\text{Ultrasonicate} \\ \text{70°C under N}_2 \\ \text{stirring for 4 h}}}$ M-MWCNT magnetic multi-walled carbon nanotubes

Scheme 7. Synthesis of M-MWCNT carbon nanotubes.

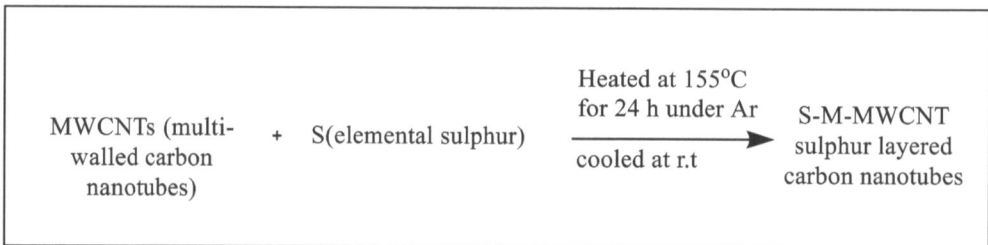

MWCNTs (multi-walled carbon nanotubes) + S(elemental sulphur) $\xrightarrow[\text{cooled at r.t}]{\substack{\text{Heated at 155°C} \\ \text{for 24 h under Ar}}}$ S-M-MWCNT sulphur layered carbon nanotubes

Scheme 8. Synthesis of sulfur-layered magnetic multi-walled nanotubes.

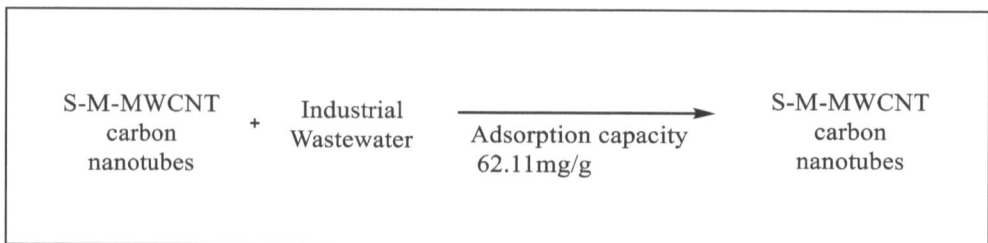

S-M-MWCNT carbon nanotubes + Industrial Wastewater $\xrightarrow{\text{Adsorption capacity 62.11mg/g}}$ S-M-MWCNT carbon nanotubes

Scheme 9. Adsorption of mercury metal ions.

Its ability to be readily separated from the solution by magnetic methods, without the need for filtering or centrifugation processes, is a benefit of this adsorbent. Furthermore, because of the magnetic iron oxide nanoparticles, the composite has magnetic characteristics that may be used for targeted medication administration and magnetic separation. Mercury was shown to adsorb onto the composite in a pH-dependent manner, with higher pH values showing greater adsorption. The adsorbent was found to have a maximum adsorption capacity of 62.11 mg/g. Because it could be utilized for further mercury removal without suffering a substantial loss of adsorption capability, the composite demonstrated high reusability.

Konczyk, J. et al. utilized functionalized multi-walled carbon nanotube composite groups to remove lead ions from aqueous and synthetic solutions. The adsorbent exhibited high selectivity for lead ion removal compared to other metal ions [46] (Scheme **10**).

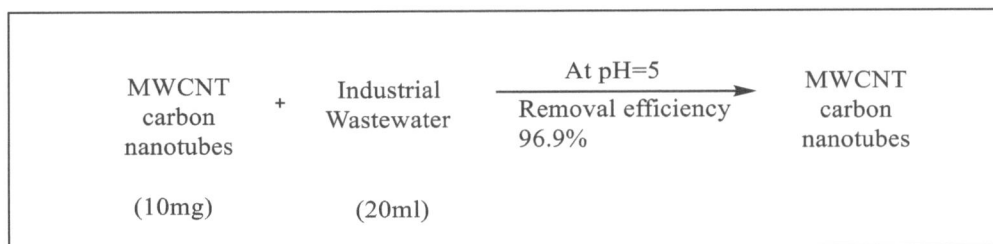

MWCNT carbon nanotubes + Industrial Wastewater → At pH=5 Removal efficiency 96.9% → MWCNT carbon nanotubes

(10mg) (20ml)

Scheme 10. Adsorption of lead metal ions.

The batch experiments provided valuable insights into the adsorption efficiency, reusability, kinetics, mechanism, and selectivity of the functionalized carbon nanotubes for removing lead ions from aqueous solutions. The greatest removal effectiveness at pH 5.0 and 313K was observed. The researchers found that the adsorbent can be effectively reused four times without noticeably diminishing its adsorption capabilities after conducting reusability trials. This suggests that the functionalized carbon nanotubes are long-lasting and may be used again before they need to be changed. The development of efficient and ecologically friendly techniques for removing heavy metals from polluted water sources may be impacted by these findings.

An environmentally beneficial nanocomposite based on carbon nanotubes grafted with lignin was created by ***Li, Z. et al.*** The generated nanocomposite can be applied as a remedy for contaminated water and the demand for a more environmentally friendly atmosphere [47] (Schemes **11 - 14**).

The nanocomposite has a lead ion distribution coefficient that is high compared to benchmark materials like metal-organic frameworks and mesoporous carbon. Our nanocomposite's wide surface area and tridimensional structure allow for excellent oil droplet removal effectiveness from water. In order to handle oil spills and other problems with oil-related water pollution, this is crucial. The researchers discovered that oil droplets can be effectively removed from water using the nanocomposite. Using a relatively low concentration of 0.5g/L of nanocomposite, a high removal efficiency of an oil-in-water emulsion is attained, surpassing 98.3%. By using the inexpensive lignin, the cost of the nanocomposite is reduced. The natural polymer layer of the nanocomposite keeps it inexpensive and environmentally friendly while improving its adsorption capacity.

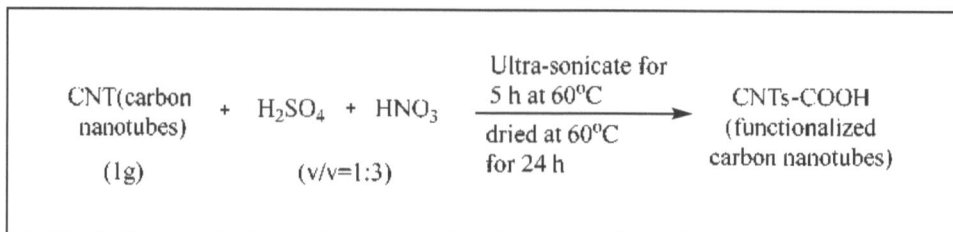

Scheme 11. Synthesis of functionalized carbon nanotubes (CNTs-COOH).

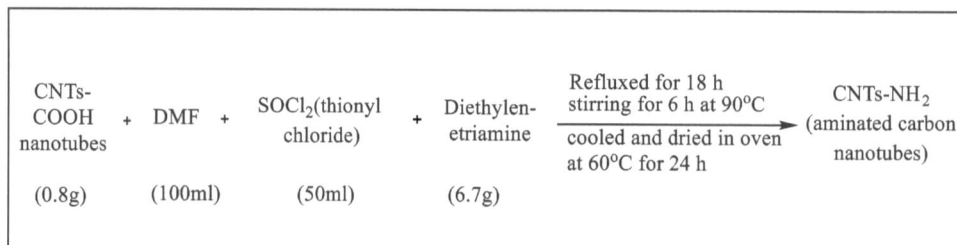

Scheme 12. Synthesis of amination of carbon nanotubes (CNTs-NH$_2$).

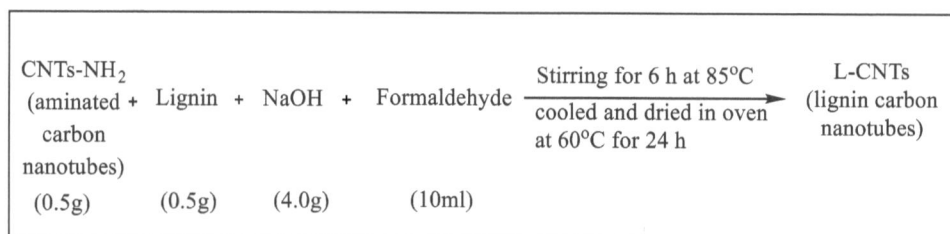

Scheme 13. Synthesis of lignin-CNTs.

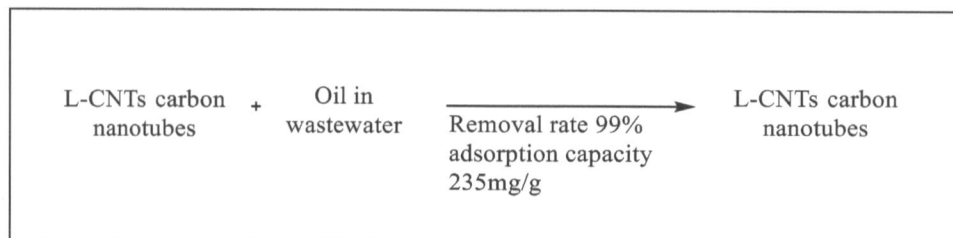

Scheme 14. Adsorption of lead ions from oil in wastewater.

Saleh, T. A. et al. synthesized a nanocomposite material consisting of multi-wall carbon nanotubes [48]. (Schemes **15 - 17**).

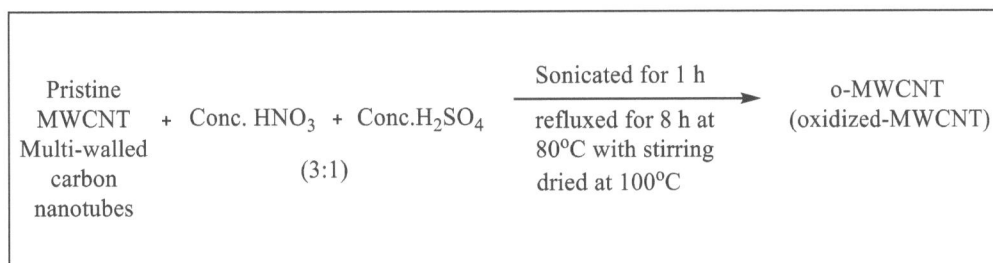

Pristine
MWCNT   + Conc. HNO$_3$ + Conc.H$_2$SO$_4$ Sonicated for 1 h → o-MWCNT
Multi-walled   refluxed for 8 h at   (oxidized-MWCNT)
carbon   (3:1)   80°C with stirring
nanotubes   dried at 100°C

Scheme 15. Functionalization of MWCNT with oxygen-containing groups.

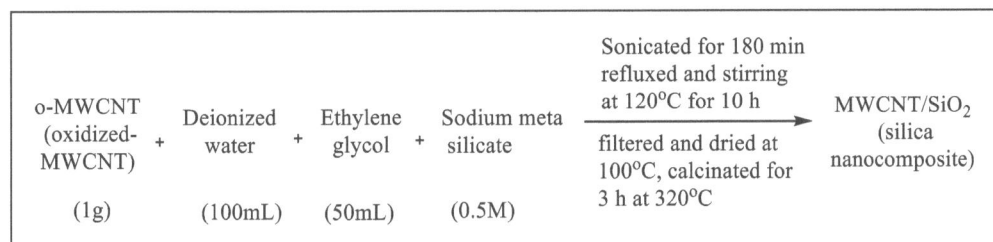

o-MWCNT   Deionized   Ethylene   Sodium meta   Sonicated for 180 min   MWCNT/SiO$_2$
(oxidized- +   water   + glycol   + silicate   refluxed and stirring   (silica
MWCNT)   at 120°C for 10 h   nanocomposite)
                                                       filtered and dried at
(1g)   (100mL)   (50mL)   (0.5M)   100°C, calcinated for
                                                       3 h at 320°C

Scheme 16. Synthesis of MWCNT/SiO$_2$ (silica nanocomposite).

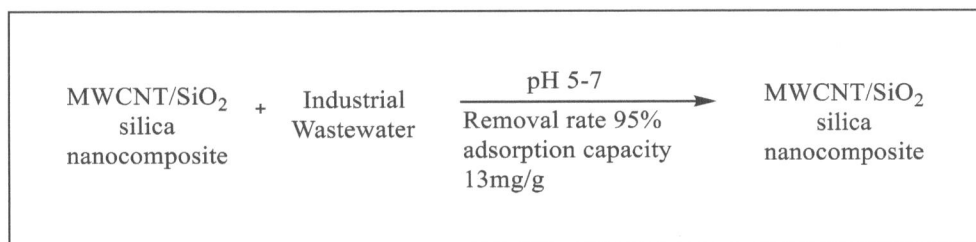

MWCNT/SiO$_2$ +   Industrial   pH 5-7 →   MWCNT/SiO$_2$
silica   Wastewater   Removal rate 95%   silica
nanocomposite   adsorption capacity nanocomposite
                                   13mg/g

Scheme 17. Adsorption of lead (II) metal ions present in industrial wastewater.

To eliminate heavy metal pollutants from water, a nanocomposite was created. Their capacity to absorb lead from aqueous solutions was put to the test. Comparing the nanocomposite to silica nanoparticles (~50%) and carbon nanotubes (~45%), the results indicated that the nanocomposite had a superior adsorption capability (~95%). An activation energy of 15.8 kJ/mol was determined for the adsorption process. More evidence for the nanocomposite's promise for environmental applications came from the results, which also showed that it could be recycled with great efficiency. With five different nanocomposites, the adsorption cycle was carried out.

Gholipour, M. et al. evaluated the efficacy of multi-walled carbon nanotubes in eliminating hexavalent chromium from water [49]. (Scheme **18**).

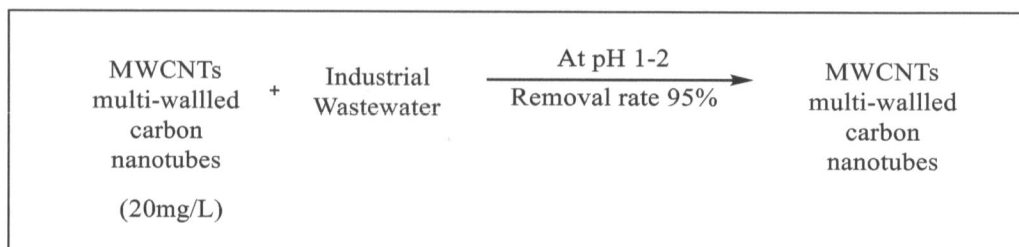

		At pH 1-2	
MWCNTs multi-wallled carbon nanotubes	+ Industrial Wastewater	⟶ Removal rate 95%	MWCNTs multi-wallled carbon nanotubes
(20mg/L)			

Scheme 18. Adsorption of chromium (VI) metal ions.

Higher temperatures, longer contact times, low starting concentrations of chromium ions, and low pH were determined to be the ideal conditions for chromium ion elimination. To comprehend the connection between the concentration of chromium ions in the solution and the quantity of ions that have been adsorbed, equilibrium isotherms were also examined. The experimental data was found to be accurately fitted by the BET model. According to the findings, chromium adsorption onto carbon nanotubes is a spontaneous, physical, endothermic process. The reversibility of the adsorption process and the impact of different variables on the desorption of chromium from the nanotubes are also examined. Multi-walled carbon nanotubes can be regenerated without a significant decrease in their performance, making them suitable for reuse in multiple cycles of adsorption and desorption.

Alguacil, F. J. et al. fabricated MWCNT as an adsorbent. It was specifically used for the heavy metals in nuclear tank waste [50]. (Scheme **19**).

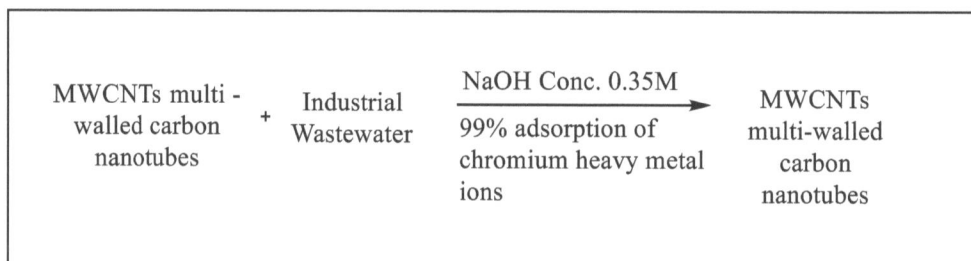

		NaOH Conc. 0.35M	
MWCNTs multi - walled carbon nanotubes	+ Industrial Wastewater	⟶ 99% adsorption of chromium heavy metal ions	MWCNTs multi-walled carbon nanotubes

Scheme 19. Adsorption of heavy metal ions present in the wastewater.

The investigation aims to understand the adsorption behavior of heavy metal on the nanotubes and determine the optimal experimental conditions for efficient adsorption. The dependence of adsorption on NaOH concentration indicates that the presence of alkaline conditions influences the adsorption process. The maximum adsorption of heavy metal is found to be 99% at 0.35M concentration of NaOH. The trend of the adsorption process slowing down with increasing contact times suggests that the initial adsorption sites on the multi-walled carbon nanotubes may have become saturated.

Adsorption has demonstrated potential as a successful technique for reducing heavy metals from aqueous solutions when carbon-based materials, such as carbon nanotubes, are used. The metal may be recovered by simply eluting the adsorbed chromium ion from the nanotubes using acidic solutions.

Atieh, M. A. et al. examined the adsorption of heavy metal ions by activated carbon-supported nanotubes using a pseudo-second-order kinetic model. The effectiveness of employing carbon nanotube-supported activated carbon as an adsorbent for heavy metal ions in polluted water is demonstrated, and this method has promise for improving wastewater treatment and reducing pollution in the environment [51]. (Scheme **20**).

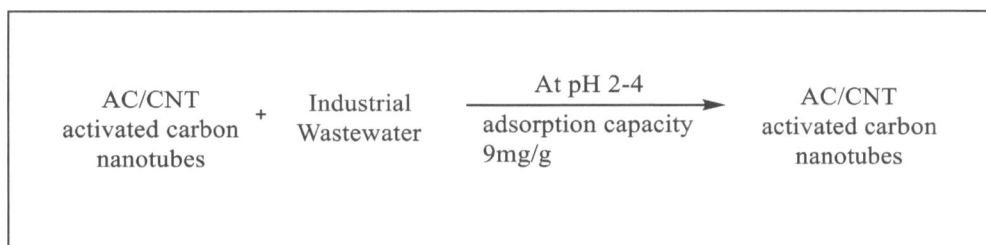

| AC/CNT activated carbon nanotubes | + | Industrial Wastewater | At pH 2-4 → adsorption capacity 9mg/g | AC/CNT activated carbon nanotubes |

Scheme 20. Adsorption of Cr(VI) metal ions.

The high adsorption capacity and efficiency of this adsorbent make it a potential solution for addressing water pollution issues caused by heavy metal ions. Because carbon nanotubes have functional groups and a larger surface area than activated carbon, their presence on the surface of the carbon material boosted the adsorption capacity. With activated carbon covered with carbon nanotubes as the adsorbent, the batch adsorption experiment yielded the greatest adsorption capacity of 9.0 mg/g. The study's conclusions show that activated carbon-supported by carbon nanotubes can remove heavy metal ions from contaminated water, which may lead to its use in environmental remediation procedures.

Parlayici, S. et al. examined the removal of heavy metals from an aqueous solution using powdered activated carbon supported by carbonaceous nano adsorbents. Catalytic chemical vapor deposition has been used to synthesize carbon nanotubes [52]. (Schemes **21 - 23**).

MWCNT Multi-walled carbon nanotubes	+	KMnO$_4$	+ H$_2$SO$_4$	+ HCl	Sonicated for 30 min refluxed for 5 h at 150°C —————————————→ cooled at r.t and dried at 100°C in vaccum	f-MWCNT functionalized carbon nanotubes
(100mg)		(250mg)	(0.5M)	(10ml)		

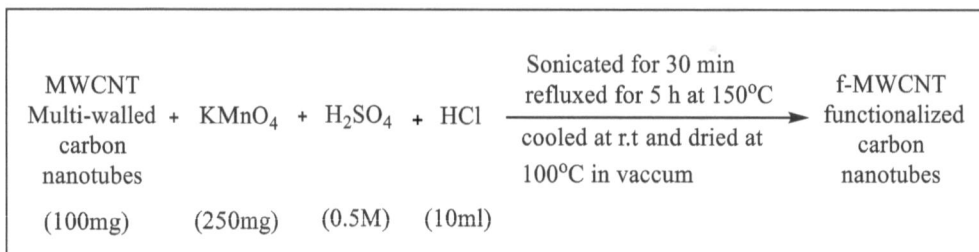

Scheme 21. Functionalization of MWCNTs.

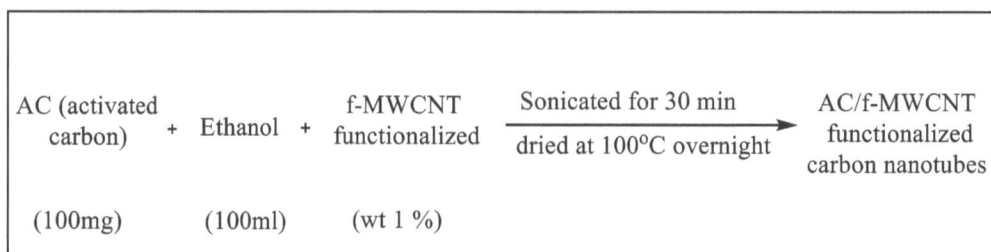

AC (activated carbon)	+ Ethanol +	f-MWCNT functionalized	Sonicated for 30 min —————————————→ dried at 100°C overnight	AC/f-MWCNT functionalized carbon nanotubes
(100mg)	(100ml)	(wt 1 %)		

Scheme 22. Preparation of AC/f-MWCNT composites.

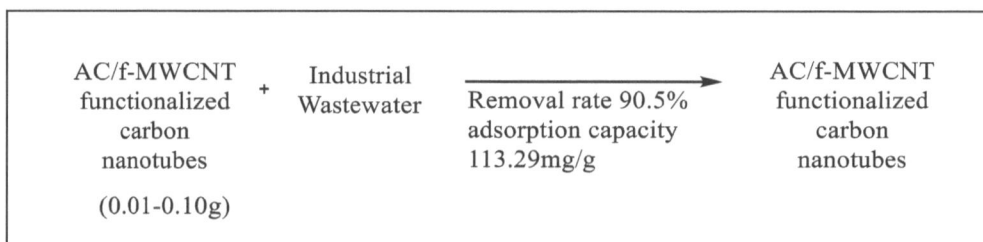

AC/f-MWCNT functionalized carbon nanotubes	+ Industrial Wastewater	—————————————→ Removal rate 90.5% adsorption capacity 113.29mg/g	AC/f-MWCNT functionalized carbon nanotubes
(0.01-0.10g)			

Scheme 23. Adsorption of chromium (VI) metal ions present in the wastewater bodies.

There is a large amount of unoccupied surface area that is accessible for adsorption since multi-walled carbon nanotubes have a narrow size range, usually between 30 and 50 nm in diameter. The enhanced surface area of the composite material amplifies its adsorption capability. The activated carbon functionalized MWCNTs and activated carbon functionalized CNTs' respective adsorption capacities of 113.29 mg/g and 105.48 mg/g indicate that the adsorption process was monolayer in nature. Moreover, even with simple regeneration and repurposing, these adsorbents operate poorly. Given that adsorbents are considered ecologically benign materials, they do not pose any environmental concerns. Their benefits include high adsorption capacity, large internal surface area, low cost, rapid regeneration, and great mechanical strength. For this reason, they might be effectively applied to eliminate heavy metal ions from wastewater.

Balog, R. et al. have investigated how nitrogen-doped MWCNTs, both pure and oxidized, adsorb heavy metal ions. Using *n*-butylamine as the carbon source and nickel nitrate and magnesium oxide as the catalyst, the nitrogen-doped carbon nanotube samples were created by the chemical vapor deposition technique [53] (Schemes **24, 25**).

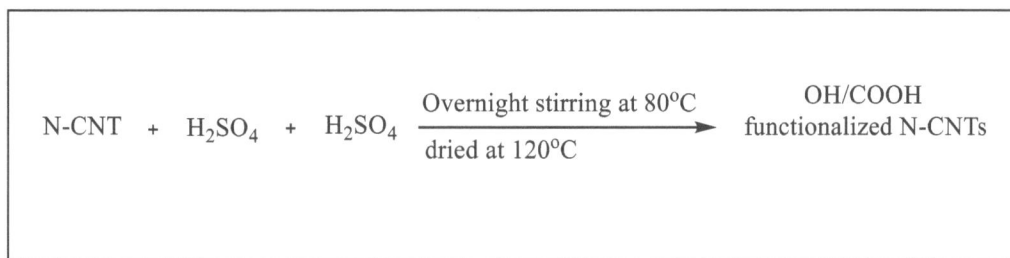

N-CNT + H$_2$SO$_4$ + H$_2$SO$_4$ $\xrightarrow[\text{dried at 120°C}]{\text{Overnight stirring at 80°C}}$ OH/COOH functionalized N-CNTs

Scheme 24. Functionalization of N-CNTs.

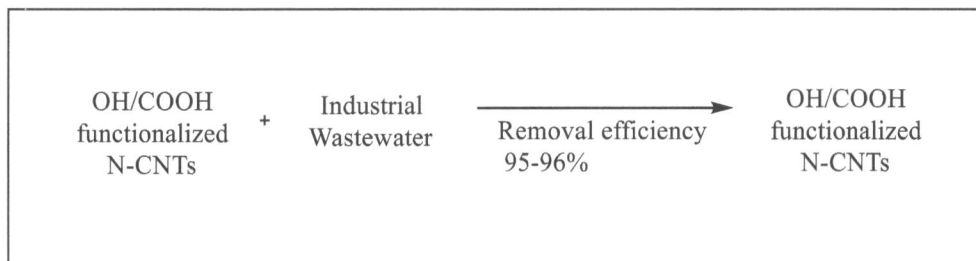

OH/COOH functionalized N-CNTs + Industrial Wastewater $\xrightarrow{}$ Removal efficiency 95-96%   OH/COOH functionalized N-CNTs

Scheme 25. Adsorption of harmful nickel (II) heavy metal ions.

The oxidized nitrogen-doped carbon nanotubes' surface functional groups increase their adsorption capability through an ion exchange process. Furthermore, both pure and oxidized nitrogen-doped carbon nanotubes experience a pH fall in the presence of heavy metal ions in the solution; the oxidized sample has a larger decrease. This implies that proton release from the oxidized nitrogen-doped carbon nanotubes is a step in the adsorption process. Both the prepared and oxidized carbon nanotube adsorption isotherms conform to the Langmuir equation, showing monolayer adsorption. The two samples' adsorption plateau levels are comparable, indicating that the oxidation process has little effect on the maximum adsorption capacity.

Guo, L. et al. have used a novel approach combining mussel-inspired chemistry and the Mannich reaction to prepare composites. CNTs were chosen as the base material for modification [54]. (Schemes **26 - 28**).

Dopamine solution	+	Pristine CNT carbon nanotubes	Stirring at r.t for 8 h centrifugation at 750 rpm for 8 min ⟶ Poured in dialysis bag for 3 days, cooled and dried at 40°C for 2 days in oven	CNT/PDA nanocomposite
(5gL^{-1})		(500mg)		

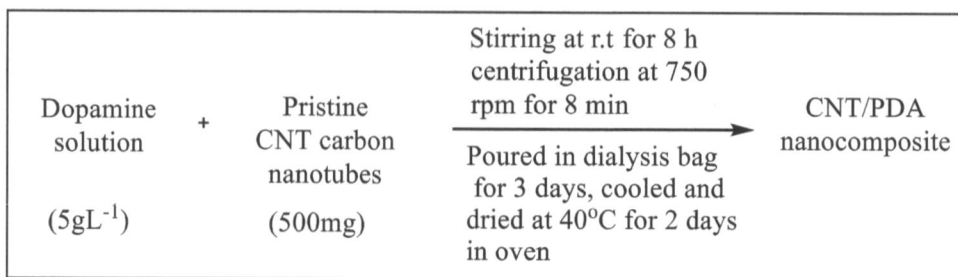

Scheme 26. Synthesis of CNT/PDA nanocomposite.

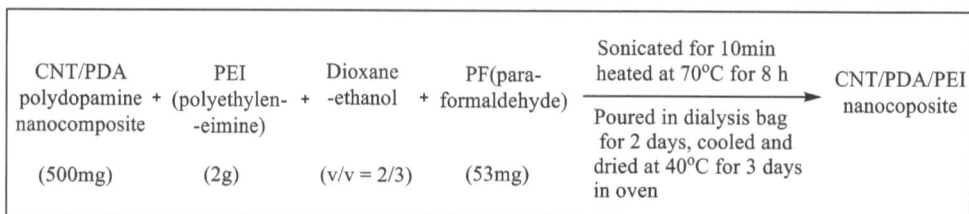

CNT/PDA polydopamine nanocomposite	+	PEI (polyethylen- -eimine)	+	Dioxane -ethanol	+	PF(para- formaldehyde)	Sonicated for 10min heated at 70°C for 8 h ⟶ Poured in dialysis bag for 2 days, cooled and dried at 40°C for 3 days in oven	CNT/PDA/PEI nanocoposite
(500mg)		(2g)		(v/v = 2/3)		(53mg)		

Scheme 27. Synthesis of CNT nanocomposites.

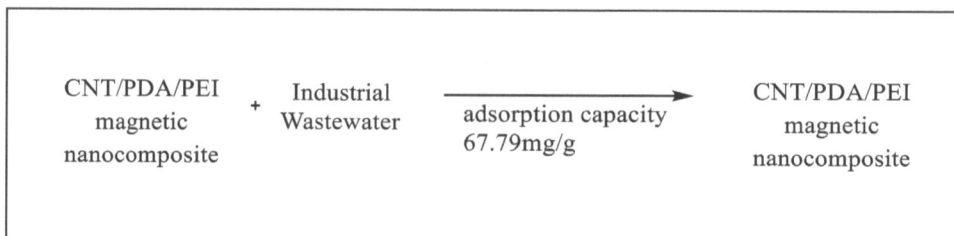

CNT/PDA/PEI magnetic nanocomposite	+	Industrial Wastewater	adsorption capacity 67.79mg/g ⟶	CNT/PDA/PEI magnetic nanocomposite

Scheme 28. Adsorption of copper (II) metals present in the water bodies.

The efficiency of the surface modification was assessed by comparing and studying the adsorption behaviors of clean and modified carbon nanotubes toward metal ions. The preparation method used in this work has shown to be successful in producing adsorbents with a high adsorption capacity. The outcomes showed that the combination of the Mannich process with mussel-inspired chemistry produced functionalized carbon nanotubes, which had good adsorption properties for removing metal ions.

Temnuch, N. et al. synthesized the multi-walled carbon nanotubes with an acid functionalization, treated them with acid, and then decorated them with magnetic oxide nanoparticles through the co-precipitation of iron(II) and iron(III) in the multi-walled carbon nanotubes with an acid functionalization colloidal suspension [55]. (Schemes **29 - 31**).

MWCMTs (Multi-walled carbon nanotubes $+$ Con.HNO$_3$ → Heated for 8 h at 100 °C, filtered and dried at 100 °C for 24 h → MWCNTs-COOH (Functionalized MWCNTs)

(1 g)   (75 ml)

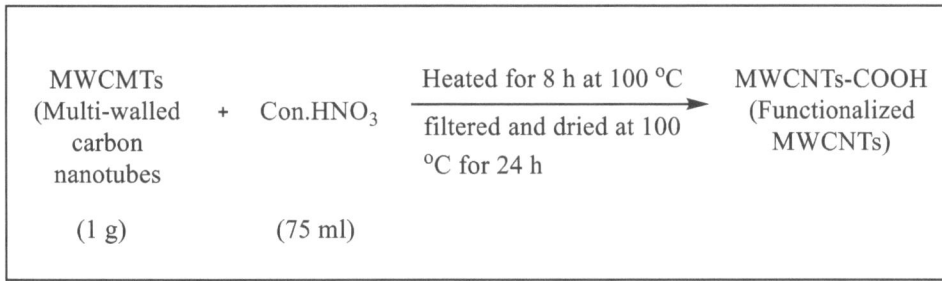

Scheme 29. Functionalization of MWCNTs.

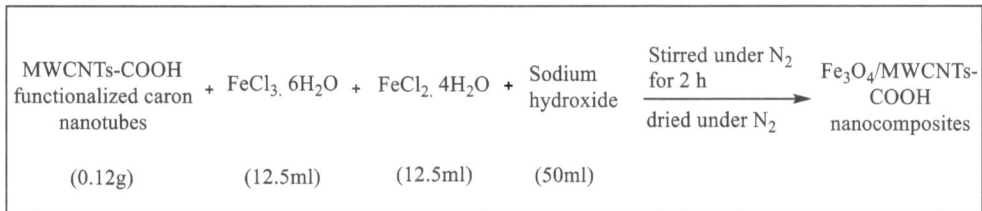

MWCNTs-COOH functionalized caron nanotubes $+$ FeCl$_3$. 6H$_2$O $+$ FeCl$_2$. 4H$_2$O $+$ Sodium hydroxide → Stirred under N$_2$ for 2 h, dried under N$_2$ → Fe$_3$O$_4$/MWCNTs-COOH nanocomposites

(0.12g)   (12.5ml)   (12.5ml)   (50ml)

Scheme 30. Synthesis of magnetic nanoparticles grafted on COOH-MWCNTs.

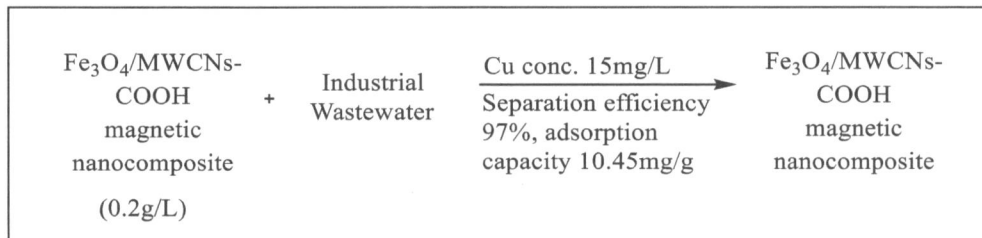

Fe$_3$O$_4$/MWCNs-COOH magnetic nanocomposite $+$ Industrial Wastewater → Cu conc. 15mg/L, Separation efficiency 97%, adsorption capacity 10.45mg/g → Fe$_3$O$_4$/MWCNs-COOH magnetic nanocomposite

(0.2g/L)

Scheme 31. Reduction of copper (II) ions from wastewater.

The adsorption capacity and separation efficiency of the two synthetic adsorbents were used to investigate the removal of copper ions from an aqueous solution. The optimal parameters for reaching the maximal adsorption capacity of 10.45 mg/g were found to be a contact length of 10 minutes, an adsorbent dosage of 0.2 g/L, and an initial concentration of copper ions of 15 mg/L. The elimination of copper ions from aqueous solution was examined by varying the weight percentages of magnetic oxide and 25 weight percent of magnetic nanotubes. The data demonstrated an optimal adsorption capacity of 9.50 mg/g and 97% separation efficiency.

The adsorption capacities of magnetic biochar and functionalized carbon nanotubes have been compared by ***Mubarak, N. M. et al.*** Regarding the extraction of zinc (II) ions from the solutions, a comparative analysis was conducted [56]. (Schemes **32, 33**).

EFB (Empty fruit branch) + FeCl$_3$ + N$_2$ gas $\xrightarrow[\text{pyrolysis in microwave muffel, cooled at r.t}]{\text{Agitated for 4h dried for 24 h at 100 °C}}$ Magnetic biochar

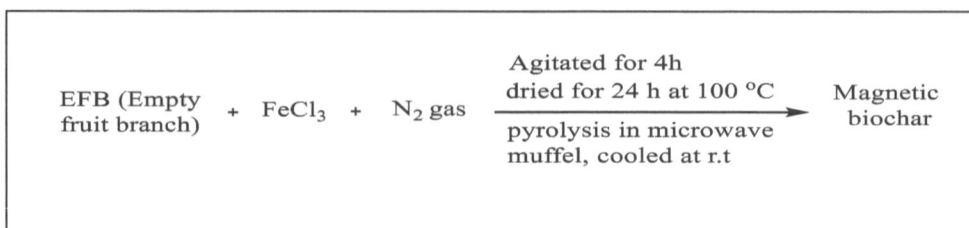

Scheme 32. Preparation of magnetic biochar.

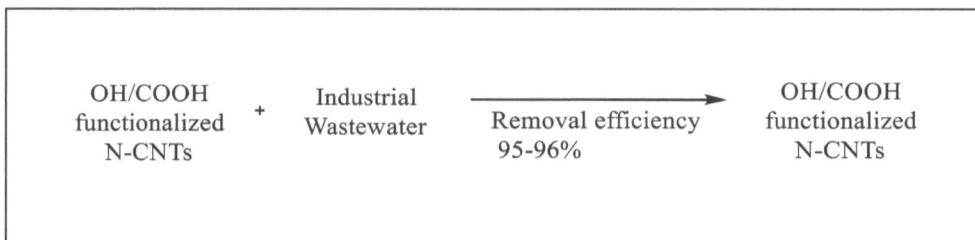

OH/COOH functionalized N-CNTs + Industrial Wastewater $\xrightarrow[\text{95-96\%}]{\text{Removal efficiency}}$ OH/COOH functionalized N-CNTs

Scheme 33. Adsorption of Zinc (II) ions from wastewater.

The results showed that using functionalized carbon nanotubes, the removal efficiency of zinc was 99% for an initial concentration of 1.1 mg/L, while using magnetic biochar, the removal efficiency was 75%. The maximum adsorption capacities were also determined, with functionalized carbon nanotubes having a capacity of 1.05 mg/g and magnetic biochar having a capacity of 1.18 mg/g. The study showed that CNTs have potential as effective adsorbents for the removal of zinc (II) ions. However, functionalized carbon nanotubes demonstrated higher removal efficiency and adsorption capacity compared to magnetic biochar. Both the materials showed a good fit with the Langmuir and Freundlich models.

Luan, H. et al. have created a novel nanocomposite material by incorporating copper nanoparticles CNTs, which were then deposited onto a traditional polymeric membrane. The nanocomposite membranes show great potential for removing Arsenic (III) from contaminated water [57]. (Schemes **34, 35**).

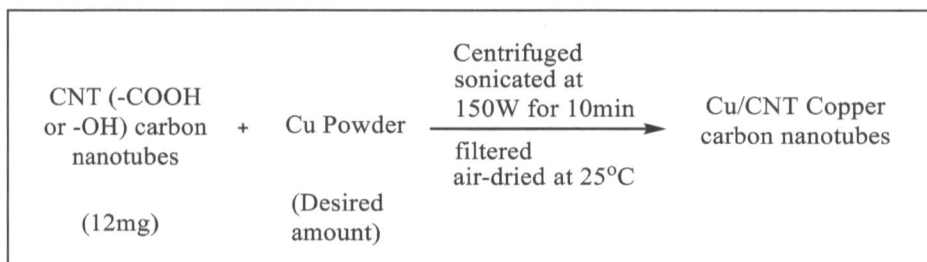

CNT (-COOH or -OH) carbon nanotubes (12mg) + Cu Powder (Desired amount) $\xrightarrow[\substack{\text{filtered} \\ \text{air-dried at 25°C}}]{\substack{\text{Centrifuged} \\ \text{sonicated at} \\ \text{150W for 10min}}}$ Cu/CNT Copper carbon nanotubes

Scheme 34. Preparation of Cu/CNT nanocomposites.

Cu/CNT Copper carbon nanotubes	+ Drinking water $\xrightarrow[\text{Removal capacity 90\%}]{\text{pH range 5-9}}$	Cu/CNT Copper carbon nanotubes

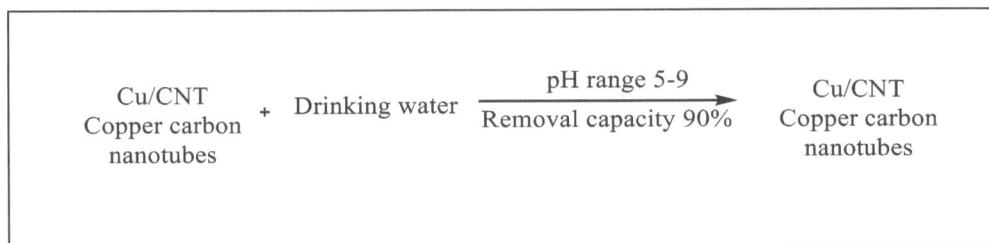

Scheme 35. Adsorption of arsenic (III).

The membrane demonstrated high permeability to pure water (4639 - 4854 L/m^2 h bar) and the ability to remove over 90% of arsenic at low transmembrane pressures below 0.01 bar. This makes it suitable for gravity-driven membrane filtration applications. The nanocomposite membrane also remained effective in the presence of sulfate, bicarbonate, nitrate, natural organic matter, and chloride. The presence of copper facilitated the oxidation in the presence of dissolved oxygen, and the resulting oxidized arsenic species were also adsorbed by the copper nanotube composites. The composite has shown great potential for removing arsenic from contaminated water.

Jabbari, V. et al. synthesized hybrid nanocomposites for the degradation of dye methylene blue [58]. (Schemes **36, 37**).

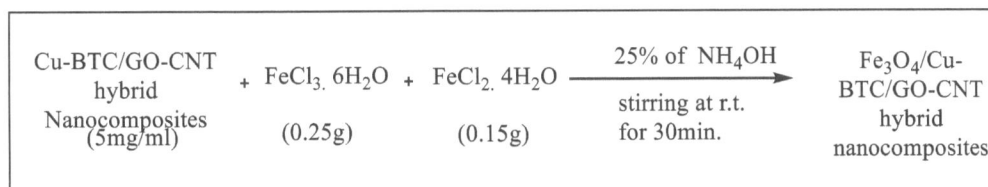

Cu-BTC/GO-CNT hybrid Nanocomposites (5mg/ml)	+ $FeCl_3. 6H_2O$ + $FeCl_2. 4H_2O$ (0.25g) (0.15g)	$\xrightarrow[\substack{\text{stirring at r.t.}\\\text{for 30min.}}]{\text{25\% of } NH_4OH}$	Fe_3O_4/Cu- BTC/GO-CNT hybrid nanocomposites

Scheme 36. Synthesis of Fe_3O_4/Cu-BTC/GO-CNT hybrid nanocomposites.

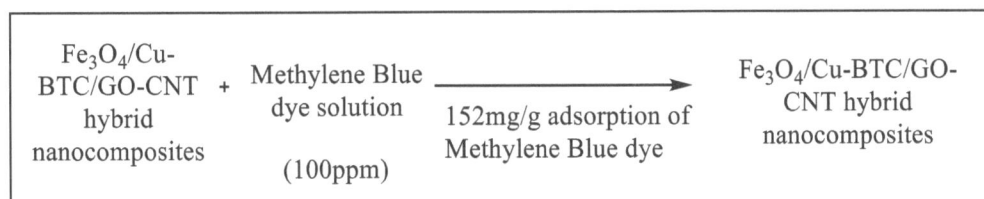

Fe_3O_4/Cu- BTC/GO-CNT hybrid nanocomposites	+ Methylene Blue dye solution (100ppm)	$\xrightarrow[\substack{\text{152mg/g adsorption of}\\\text{Methylene Blue dye}}]{}$	Fe_3O_4/Cu-BTC/GO- CNT hybrid nanocomposites

Scheme 37. Degradation of methylene blue dye.

The materials were created using an easy-to-understand green solvothermal method. The produced hybrid nanomaterials are more capable of adsorbing pollutants than the original materials. Both the distinct characteristics of the nanoscale MOF and the combined impact of covalent bonding between the parent materials are responsible for this increased adsorption capacity. It has been demonstrated that when pollution levels rise, so does the absorbent's ability to absorb it.

Wang, Y. et al. fabricated MOF hybrids for the reduction of organic pollutants from wastewater [59]. (Schemes **38, 39**).

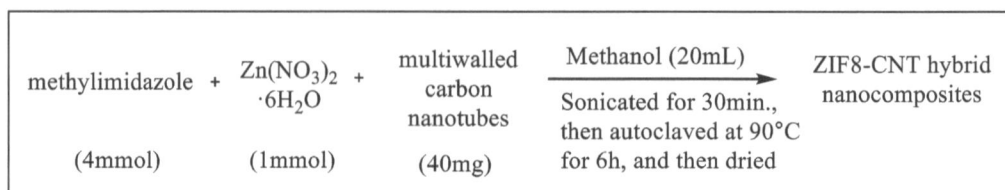

methylimidazole	+	$Zn(NO_3)_2$ ·$6H_2O$	+	multiwalled carbon nanotubes	Methanol (20mL) → Sonicated for 30min., then autoclaved at 90°C for 6h, and then dried	ZIF8-CNT hybrid nanocomposites
(4mmol)		(1mmol)		(40mg)		

Scheme 38. Synthesis of ZIF8-CNT hybrid nanocomposites.

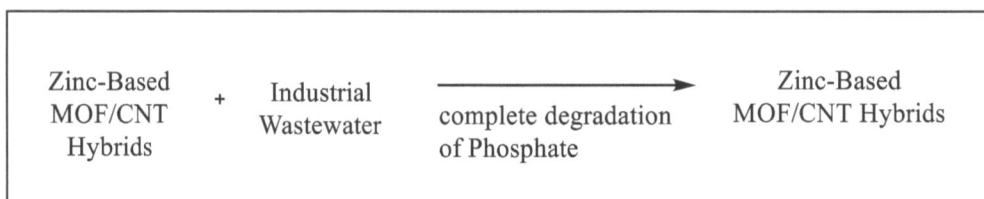

Zinc-Based MOF/CNT Hybrids	+	Industrial Wastewater	→ complete degradation of Phosphate	Zinc-Based MOF/CNT Hybrids

Scheme 39. Removal of phosphate.

The remarkable pollution removal capability of 92-10% is maintained in the actual water. The major π-π interactions that take place during the adsorption phase do not appreciably alter the removal of contaminants affected by Zn-O-P and H-bond interactions. Higher initial pollutant concentrations may result in less efficient clearance, hence an increase in inhibition is anticipated in the presence of acetaminophen.

EFFECTS OF CARBON NANOTUBES ON THE ENVIRONMENT

CNTs have attracted a lot of interest due to their extraordinary qualities and diverse range of uses. However, their influence on the environment is a source of worry. Given their potential toxicity, one of the main environmental problems with CNTs is their use. Research has indicated that some varieties of carbon nanotubes (CNTs) have the potential to cause detrimental consequences on living things, such as inflammation, oxidative stress, and fatal cell death. Regarding the safety of CNTs in different environmental situations, these harmful effects raise doubts.

Further complicating their removal and cleanup is the fact that CNTs are persistent in the environment. Long-term environmental damage can result from

CNT accumulation in soil and water because of their tiny size and great stability. The health of the species living there may be impacted by this buildup, upsetting ecosystems.

Furthermore, energy-intensive procedures and the usage of potentially hazardous chemicals are frequently used in the manufacture of CNTs, which raises the risk of carbon emissions and environmental contamination. Research is now being done to address these issues by figuring out how CNTs cause harm, creating ecologically safe synthesis processes, and investigating safe disposal and remediation techniques. In order to reduce the environmental effect of carbon nanotubes (CNTs), regulatory frameworks for their proper usage and management are also being established.

CONCLUSION

Nowadays, carbon nanotubes (CNTs) are beneficial for treating wastewater because of their unique chemical and structural characteristics. These nanoparticles are perfect for eliminating various pollutants, such as pathogens, heavy metals, and organic pollutants, because of their enormous surface area, strong mechanical strength, and exceptional adsorption properties. The effectiveness and selectivity of CNTs in absorbing particular contaminants can be further improved by functionalization. They aid in the purifying process by acting as both adsorbents and catalysts, which enables the breakdown of dangerous chemicals. Furthermore, the overall effectiveness and sustainability of wastewater treatment technologies have improved by introducing CNTs into filtration membranes and other treatment systems, showing encouraging results. Notwithstanding their potential, CNTs must be integrated into common wastewater treatment applications by addressing cost, scalability, and environmental effects. However, current research and development in this area indicate that CNTs will be essential for expanding water purification technologies and guaranteeing the availability of clean water.

LIST OF ABBREVIATIONS

CNTs	Carbon Nanotubes
MWCNTs	Multi-Walled Carbon Nanotubes
M-MWCNTs	Magnetic Multi-Walled Carbon Nanotubes
SWCNTs	Single-Walled Carbon Nanotubes
NPs	Nanoparticles

ACKNOWLEDGEMENTS

The authors would like to express their sincere thanks to the management of SGT University, Gurugram, Haryana, India, and Aditya Institute of Technology and Management, Tekkali-532201, Andhra Pradesh, India, for providing the facilities to write and submit the book chapter for publication.

REFERENCES

[1]   Deng, Z.; Chen, Y.; Tian, Q.; Guo, T.; Zhang, Y.; Huang, Z.; Hu, H.; Gan, T. Cobalt single atoms supported on CNT/MoS2 heterojunction nanocomposite for highly-efficient reduction of 4-nitrophenol wastewater: Enhanced electron transport. *Separ. Purif. Tech.,* **2024**, *336*, 126284.
[http://dx.doi.org/10.1016/j.seppur.2024.126284]

[2]   Chen, L.; Wu, J.; Huang, H.; Zhang, X.; Tang, W.; He, J.; Zheng, C.; Yang, Y. Peroxymonosulfate activated by CoMn@CNT nanocomposite for moxifloxacin degradation. *Inorg. Chem. Commun.,* **2024**, *169*, 113007.
[http://dx.doi.org/10.1016/j.inoche.2024.113007]

[3]   Jatoi, A.S.; Hashmi, Z.; Usman, T.; Mubarak, N.M.; Ali Mazari, S.; Karri, R.R.; Koduru, J.R.; Dehghani, M.H. *Role of carbon nanomaterials for wastewater treatment—A brief review*; Water Treatment Using Engineered Carbon Nanotubes, **2023**, pp. 29-62.
[http://dx.doi.org/10.1016/B978-0-443-18524-3.00016-7]

[4]   Ma, L.; Dong, X.; Chen, M.; Zhu, L.; Wang, C.; Yang, F.; Dong, Y. Fabrication and water treatment application of carbon nanotubes (CNTs)-based composite membranes: A review. *Membranes (Basel),* **2017**, *7*(1), 16.
[http://dx.doi.org/10.3390/membranes7010016] [PMID: 28335452]

[5]   Garg, A.; Chalak, H.D.; Belarbi, M-O.; Zenkour, A.M.; Sahoo, R. Estimation of carbon nanotubes and their applications as reinforcing composite materials–An engineering review. *Compos. Struct.,* **2021**, *272*, 114234.
[http://dx.doi.org/10.1016/j.compstruct.2021.114234]

[6]   Han, D.; Yan, G.; Wang, C. Influence of multi-walled carbon nanotubes (MWCNTs) content on metal friction and wear in thermally cracked carbon black (CBp) formulation system during mixing. *Polym. Test.,* **2022**, *113*, 107674.
[http://dx.doi.org/10.1016/j.polymertesting.2022.107674]

[7]   Zhao, J.; Liu, H.; Qi, Y.; Wang, R.; Lv, Z.; Yu, Y.; Sun, S. Construction of robust DA@CNTs/CS hydrogel coatings on hydrophobic PVDF membrane by deposition of self-assembly for efficient separation of oil-water emulsion and dye wastewater. *Separ. Purif. Tech.,* **2024**, *333*, 125956.
[http://dx.doi.org/10.1016/j.seppur.2023.125956]

[8]   Qi, Y.; Chen, Y.; Liu, H.; Zhao, J.; Yu, Y.; Wang, R.; Lv, Z.; Sun, S. Construction of highly hydrophilic PAN-co-IA/CNTs@TA composite membrane with large flux and high retention for purification of dye wastewater. *Chem. Eng. J.,* **2024**, *487*, 150752.
[http://dx.doi.org/10.1016/j.cej.2024.150752]

[9]   Mahmoud, K.A.; Mansoor, B.; Mansour, A.; Khraisheh, M. Functional graphene nanosheets: The next generation membranes for water desalination. *Desalination,* **2015**, *356*, 208-225.
[http://dx.doi.org/10.1016/j.desal.2014.10.022]

[10]   Warsi, M.F.; Ihsan, A.; Alzahrani, F.M.A.; Tariq, M.H.; Alrowaili, Z.A.; Al-Buriahi, M.S.; Shahid, M. A cost-effective strategy to synthesize perovskite holmium doped LaFeO$_3$ and its composite with CNTs for wastewater treatment. *Mater. Sci. Eng. B,* **2024**, *301*, 117181.
[http://dx.doi.org/10.1016/j.mseb.2024.117181]

[11]   Baratta, M.; Nezhdanov, A.V.; Mashin, A.I.; Nicoletta, F.P.; De Filpo, G. Carbon nanotubes

buckypapers: A new frontier in wastewater treatment technology. *Sci. Total Environ.,* **2024**, *924*, 171578.
[http://dx.doi.org/10.1016/j.scitotenv.2024.171578] [PMID: 38460681]

[12]   Jhaveri, J.H.; Murthy, Z.V.P. A comprehensive review on anti-fouling nanocomposite membranes for pressure driven membrane separation processes. *Desalination,* **2016**, *379*, 137-154.
[http://dx.doi.org/10.1016/j.desal.2015.11.009]

[13]   Zhang, Z.; Li, W.; Zhao, B.; Yang, X.; Zhao, C.; Wang, W.; Yang, X.; Shen, A.; Ye, M. Novel CNT/MXene composite membranes with superior electrocatalytic efficiency and durability for sustainable wastewater treatment. *Chem. Eng. J.,* **2024**, *495*, 153605.
[http://dx.doi.org/10.1016/j.cej.2024.153605]

[14]   Mocan, T.; Matea, C.T.; Pop, T.; Mosteanu, O.; Buzoianu, A.D.; Suciu, S.; Puia, C.; Zdrehus, C.; Iancu, C.; Mocan, L. Carbon nanotubes as anti-bacterial agents. *Cell. Mol. Life Sci.,* **2017**, *74*(19), 3467-3479.
[http://dx.doi.org/10.1007/s00018-017-2532-y] [PMID: 28536787]

[15]   Chen, S.; Wang, X.; Zhu, G.; Lu, Z.; Zhang, Y.; Zhao, X.; Hou, B. Developing multi-wall carbon nanotubes/Fusion-bonded epoxy powder nanocomposite coatings with superior anti-corrosion and mechanical properties. *Colloids Surf. A Physicochem. Eng. Asp.,* **2021**, *628*, 127309.
[http://dx.doi.org/10.1016/j.colsurfa.2021.127309]

[16]   Xu, Y.; Yu, Y.; Yang, Y.; Sun, T.; Dong, S.; Yang, H.; Liu, Y.; Fan, X.; Song, C. Improved separation performance of carbon nanotube hollow fiber membrane by peroxydisulfate activation. *Separ. Purif. Tech.,* **2021**, *276*, 119328.
[http://dx.doi.org/10.1016/j.seppur.2021.119328]

[17]   Ihsanullah, ; Abbas, A.; Al-Amer, A.M.; Laoui, T.; Al-Marri, M.J.; Nasser, M.S.; Khraisheh, M.; Atieh, M.A. Heavy metal removal from aqueous solution by advanced carbon nanotubes: Critical review of adsorption applications. *Separ. Purif. Tech.,* **2016**, *157*, 141-161.
[http://dx.doi.org/10.1016/j.seppur.2015.11.039]

[18]   de Oliveira, T.C.; Ferreira, F.V.; de Menezes, B.R.C.; da Silva, D.M.; dos Santos, A.S.; Kawachi, E.Y.; Simonetti, E.A.N.; Cividanes, L.S. Engineering the surface of carbon-based nanomaterials for dispersion control in organic solvents or polymer matrices. *Surf. Interfaces,* **2021**, *24*, 101121.
[http://dx.doi.org/10.1016/j.surfin.2021.101121]

[19]   Khandelia, T.; Patel, B.K. Carbon nanotube-based oil-water separation. In: *Advances in Oil-Water Separation*; Das, P.; Manna, S.; Pandey, J.K., Eds.; Elsevier, **2022**; pp. 195-206.
[http://dx.doi.org/10.1016/B978-0-323-89978-9.00019-7]

[20]   Hosseini, H.; Ghaffarzadeh, M. Surface functionalization of carbon nanotubes *via* plasma discharge: A review. *Inorg. Chem. Commun.,* **2022**, *138*, 109276.
[http://dx.doi.org/10.1016/j.inoche.2022.109276]

[21]   Li, S.; Zhang, Y.; Cheng, C.; Wei, H.; Du, S.; Yan, J. Surface-treated carbon nanotubes in cement composites: Dispersion, mechanical properties and microstructure. *Constr. Build. Mater.,* **2021**, *310*, 125262.
[http://dx.doi.org/10.1016/j.conbuildmat.2021.125262]

[22]   Tariq, F.; Rafiq, U.; Siddique, S.; Alothman, Z. A.; Shakir, I.; and Warsi, M. F. Evaluating the synergistic effect of Cu/Zr co-doped Co3O4/CNTs nanocomposite for removal of drugs and dyes from industrial wastewater. *Ceramics International.,* **2025**.
[http://dx.doi.org/10.1016/j.ceramint.2025.01.042]

[23]   Gong, A.; Zhao, Y.; Liang, B.; Li, K. Stepwise hollow Prussian blue/carbon nanotubes composite as a novel electrode material for high-performance desalination. *J. Colloid Interface Sci.,* **2022**, *605*, 432-440.
[http://dx.doi.org/10.1016/j.jcis.2021.07.103] [PMID: 34332416]

[24]   Zhu, L.; Chen, M.; Dong, Y.; Tang, C.Y.; Huang, A.; Li, L. A low-cost mullite-titania composite

ceramic hollow fiber microfiltration membrane for highly efficient separation of oil-in-water emulsion. *Water Res.,* **2016**, *90*, 277-285.
[http://dx.doi.org/10.1016/j.watres.2015.12.035] [PMID: 26748205]

[25]   Ma, T.; Liu, M.; Li, T.; Ren, H.; Zhou, R. Nitrogen-doped carbon nanotubes derived from carbonized polyaniline as a robust peroxydisulfate activator for the oxidation removal of organic pollutants: Singlet oxygen dominated mechanism and structure-activity relationship. *Separ. Purif. Tech.,* **2022**, *293*, 121124.
[http://dx.doi.org/10.1016/j.seppur.2022.121124]

[26]   Peng, J.; He, Y.; Zhou, C.; Su, S.; Lai, B. The carbon nanotubes-based materials and their applications for organic pollutant removal: A critical review. *Chin. Chem. Lett.,* **2021**, *32*(5), 1626-1636.
[http://dx.doi.org/10.1016/j.cclet.2020.10.026]

[27]   Bhuvaneswari, K.; Palanisamy, G.; Sivashanmugan, K.; Pazhanivel, T.; Maiyalagan, T. ZnO nanoparticles decorated multiwall carbon nanotube assisted ZnMgAl layered triple hydroxide hybrid photocatalyst for visible light-driven organic pollutants removal. *J. Environ. Chem. Eng.,* **2021**, *9*(1), 104909.
[http://dx.doi.org/10.1016/j.jece.2020.104909]

[28]   Anjum, H.; Johari, K.; Gnanasundaram, N.; Ganesapillai, M.; Arunagiri, A.; Regupathi, I.; Thanabalan, M. A review on adsorptive removal of oil pollutants (BTEX) from wastewater using carbon nanotubes. *J. Mol. Liq.,* **2019**, *277*, 1005-1025.
[http://dx.doi.org/10.1016/j.molliq.2018.10.105]

[29]   Engel, M.; Chefetz, B. Removal of triazine-based pollutants from water by carbon nanotubes: Impact of dissolved organic matter (DOM) and solution chemistry. *Water Res.,* **2016**, *106*, 146-154.
[http://dx.doi.org/10.1016/j.watres.2016.09.051] [PMID: 27710798]

[30]   Fu, L.; Mao, S.; Chen, F.; Zhao, S.; Su, W.; Lai, G.; Yu, A.; Lin, C.T. Graphene-based electrochemical sensors for antibiotic detection in water, food and soil: A scientometric analysis in CiteSpace (2011–2021). *Chemosphere,* **2022**, *297*, 134127.
[http://dx.doi.org/10.1016/j.chemosphere.2022.134127] [PMID: 35240147]

[31]   Palmas, S.; Vacca, A.; Mais, L. Bibliometric analysis on the papers dedicated to microplastics in wastewater treatments. *Catalysts,* **2021**, *11*(8), 913.
[http://dx.doi.org/10.3390/catal11080913]

[32]   Li, M.; Wang, Y.; Shen, Z.; Chi, M.; Lv, C.; Li, C.; Bai, L.; Thabet, H.K.; El-Bahy, S.M.; Ibrahim, M.M.; Chuah, L.F.; Show, P.L.; Zhao, X. RETRACTED: Investigation on the evolution of hydrothermal biochar. *Chemosphere,* **2022**, *307*(Pt 2), 135774.
[http://dx.doi.org/10.1016/j.chemosphere.2022.135774] [PMID: 35921888]

[33]   Ali, I.; Basheer, A.A.; Mbianda, X.Y.; Burakov, A.; Galunin, E.; Burakova, I.; Mkrtchyan, E.; Tkachev, A.; Grachev, V. Graphene based adsorbents for remediation of noxious pollutants from wastewater. *Environ. Int.,* **2019**, *127*, 160-180.
[http://dx.doi.org/10.1016/j.envint.2019.03.029] [PMID: 30921668]

[34]   Xiang, W.; Zhang, X.; Chen, J.; Zou, W.; He, F.; Hu, X.; Tsang, D.C.W.; Ok, Y.S.; Gao, B. Biochar technology in wastewater treatment: A critical review. *Chemosphere,* **2020**, *252*, 126539.
[http://dx.doi.org/10.1016/j.chemosphere.2020.126539] [PMID: 32220719]

[35]   Zhang, L.C.; Jia, Z.; Lyu, F.; Liang, S.X.; Lu, J. A review of catalytic performance of metallic glasses in wastewater treatment: Recent progress and prospects. *Prog. Mater. Sci.,* **2019**, *105*, 100576.
[http://dx.doi.org/10.1016/j.pmatsci.2019.100576]

[36]   Asif, M.B.; Zhang, Z. Ceramic membrane technology for water and wastewater treatment: A critical review of performance, full-scale applications, membrane fouling and prospects. *Chem. Eng. J.,* **2021**, *418*, 129481.
[http://dx.doi.org/10.1016/j.cej.2021.129481]

[37]   Jatoi, A.S.; Hashmi, Z.; Mubarak, N.M.; Ali Mazari, S.; Karri, R.R.; Koduru, J.R.; Dehghani, M.H.

Industrial wastewater treatment using carbon nanotube membranes—A brief review; Water Treatment Using Engineered Carbon Nanotubes, **2023**, pp. 179-207.
[http://dx.doi.org/10.1016/B978-0-443-18524-3.00001-5]

[38]   Anjaneyulu, B.; Chauhan, V.; Chinmay, ; Afshari, M. Enhancing photocatalytic wastewater treatment: investigating the promising applications of nickel ferrite and its novel nanocomposites. *Environ. Sci. Pollut. Res. Int.,* **2024**, *31*(31), 43453-43475.
[http://dx.doi.org/10.1007/s11356-024-33502-8] [PMID: 38684612]

[39]   Bendi, A.; Bhathiwal, A.S.; Chanchal, ; Chauhan, V.; Tiwari, A.; Raghav, N.; Chinmay, ; Praveen, P.L. Exploration of pyridine-based self-assembled complexes-An overview. *J. Mol. Struct.,* **2024**, *1312*, 138568.
[http://dx.doi.org/10.1016/j.molstruc.2024.138568]

[40]   Bendi, A.; Chauhan, V.; Vashisth, C.; Yogita, ; Chinmay, ; Raghav, N. Revolutionizing industrial wastewater Treatment: MXenes conquer organic pollutants in a paradigm shifting breakthrough towards Sustainability. *Chem. Eng. J.,* **2024**, *490*, 151373.
[http://dx.doi.org/10.1016/j.cej.2024.151373]

[41]   Anjaneyulu, B. Chinmay; Chauhan, V.; Carabineiro, S. A. C.; Afshari, M. Recent advances on zinc ferrite and its derivatives as the forerunner of the nanomaterials in catalytic applications. *J. Inorg. Organomet. Polym. Mater.,* **2023**.
[http://dx.doi.org/10.1007/s10904-023-02952-x]

[42]   Bhathiwal, A.S.; Bendi, A.; Tiwari, A. A study on synthesis of benzodiazepine scaffolds using biologically active chalcones as precursors. *J. Mol. Struct.,* **2022**, *1258*, 132649.
[http://dx.doi.org/10.1016/j.molstruc.2022.132649]

[43]   Wang, Z.; Xu, W.; Jie, F.; Zhao, Z.; Zhou, K.; Liu, H. The selective adsorption performance and mechanism of multiwall magnetic carbon nanotubes for heavy metals in wastewater. *Sci. Rep.,* **2021**, *11*(1), 16878.
[http://dx.doi.org/10.1038/s41598-021-96465-7] [PMID: 34413419]

[44]   Ghasemi, S.; Mohammadnia, E.; Hadavifar, M.; Veisi, H. AUT journal of civil engineering cadmium removal from aqueous solution by magnetized and polydopamine surface functionalized single-walled carbon nanotubes. *Civ. Eng.,* **2020**, *4*(3), 315-322.
[http://dx.doi.org/10.22060/ajce.2020.17477.5635]

[45]   Fayazi, M. Removal of mercury(II) from wastewater using a new and effective composite: sulfur-coated magnetic carbon nanotubes. *Environ. Sci. Pollut. Res. Int.,* **2020**, *27*(11), 12270-12279.
[http://dx.doi.org/10.1007/s11356-020-07843-z] [PMID: 31993910]

[46]   Kończyk, J.; Żarska, S.; Ciesielski, W. Adsorptive removal of Pb(II) ions from aqueous solutions by multi-walled carbon nanotubes functionalised by selenophosphoryl groups: Kinetic, mechanism, and thermodynamic studies. *Colloids Surf. A Physicochem. Eng. Asp.,* **2019**, *575*, 271-282.
[http://dx.doi.org/10.1016/j.colsurfa.2019.04.058]

[47]   Li, Z.; Chen, J.; Ge, Y. Removal of lead ion and oil droplet from aqueous solution by lignin-grafted carbon nanotubes. *Chem. Eng. J.,* **2017**, *308*, 809-817.
[http://dx.doi.org/10.1016/j.cej.2016.09.126]

[48]   Saleh, T.A. Nanocomposite of carbon nanotubes/silica nanoparticles and their use for adsorption of Pb(II): from surface properties to sorption mechanism. *Desalination Water Treat.,* **2016**, *57*(23), 10730-10744.
[http://dx.doi.org/10.1080/19443994.2015.1036784]

[49]   Gholipour, M.; Hashemipour, H. Procenjivanje performansi višeslojnih ugljeničnih nanocevi u adsorpciji i desorpciji šestovalentnog hroma. *Chemical Industry and Chemical Engineering Quarterly,* **2012**, *18*(4 I), 509-523.
[http://dx.doi.org/10.2298/CICEQ111104025G]

[50]   Alguacil, F.J.; López, F.A. On the active adsorption of chromium(III) from alkaline solutions using

multiwalled carbon nanotubes. *Appl. Sci. (Basel),* **2019**, *10*(1), 36.
[http://dx.doi.org/10.3390/app10010036]

[51]   Ali Atieh, M. Removal of chromium (VI) from polluted water using carbon nanotubes supported with activated carbon. In: *In Procedia Environmental Sciences*; Elsevier B.V., **2011**; 4, pp. 281-293.
[http://dx.doi.org/10.1016/j.proenv.2011.03.033]

[52]   Parlayici, S.; Eskizeybek, V.; Avcı, A.; Pehlivan, E. Removal of chromium (VI) using activated carbon-supported-functionalized carbon nanotubes. *J. Nanostructure Chem.,* **2015**, *5*(3), 255-263.
[http://dx.doi.org/10.1007/s40097-015-0156-z]

[53]   Balog, R.; Manilo, M.; Vanyorek, L.; Csoma, Z.; Barany, S. Comparative study of Ni(II) adsorption by pristine and oxidized multi-walled N-doped carbon nanotubes. *RSC Advances,* **2020**, *10*(6), 3184-3191.
[http://dx.doi.org/10.1039/C9RA09755D] [PMID: 35497765]

[54]   Guo, L.; Liu, Y.; Dou, J.; Huang, Q.; Lei, Y.; Chen, J.; Wen, Y.; Li, Y.; Zhang, X.; Wei, Y. Surface modification of carbon nanotubes with polyethyleneimine through "mussel inspired chemistry" and "Mannich reaction" for adsorptive removal of copper ions from aqueous solution. *J. Environ. Chem. Eng.,* **2020**, *8*(3), 103721.
[http://dx.doi.org/10.1016/j.jece.2020.103721]

[55]   Temnuch, N.; Suwattanamala, A.; Inpaeng, S.; Tedsree, K. Magnetite nanoparticles decorated on multi-walled carbon nanotubes for removal of Cu^{2+} from aqueous solution. *Environ. Technol.,* **2021**, *42*(23), 3572-3580.
[http://dx.doi.org/10.1080/09593330.2020.1740328] [PMID: 32149580]

[56]   Mubarak, N.M.; Alicia, R.F.; Abdullah, E.C.; Sahu, J.N.; Haslija, A.B.A.; Tan, J. Statistical optimization and kinetic studies on removal of Zn^{2+} using functionalized carbon nanotubes and magnetic biochar. *J. Environ. Chem. Eng.,* **2013**, *1*(3), 486-495.
[http://dx.doi.org/10.1016/j.jece.2013.06.011]

[57]   Luan, H.; Teychene, B.; Huang, H. Efficient removal of As(III) by Cu nanoparticles intercalated in carbon nanotube membranes for drinking water treatment. *Chem. Eng. J.,* **2019**, *355*, 341-350.
[http://dx.doi.org/10.1016/j.cej.2018.08.104]

[58]   Jabbari, V.; Veleta, J. M.; Zarei-Chaleshtori, M.; Gardea-Torresdey, J.; Villagrán, D. *Green Synthesis of Magnetic MOF/GO and MOF/CNT Hybrid Nanocomposites with High Adsorption Capacity towards Organic Pollutants,* **2016**.

[59]   Wang, Y.; Gao, Z.; Shang, Y.; Qi, Z.; Zhao, W.; Peng, Y. Proportional modulation of zinc-based MOF/carbon nanotube hybrids for simultaneous removal of phosphate and emerging organic contaminants with high efficiency. *Chem. Eng. J.,* **2021**, *417*, 128063.
[http://dx.doi.org/10.1016/j.cej.2020.128063]

CHAPTER 5

Graphene-Based Nanocomposites in Wastewater Treatment

Anirudh Singh Bhathiwal[1] and **Anjaneyulu Bendi**[2,*]

[1] *Department of Chemistry, MMV, Banaras Hindu University, Varanasi, UP, India*

[2] *Innovation and Translational Research Hub (iTRH) & Department of Chemistry, Presidency University, Bangalore, Karnataka, India*

Abstract: The problem of protecting water, which is a limited resource for future generations, has been brought to light by the startling annual rise in its contamination. Human and environmental health is negatively impacted by an excessive reliance on synthetic chemicals. Sustainable development holds the key to sustaining social and economic development in this concerning situation. The remarkable properties of graphene-based nanocomposite materials, such as their high mechanical strength, large surface area, and versatile reactivity toward polar and nonpolar contaminants, which makes them ideal for broad-spectrum contaminant removal, have led to their selection. This chapter encompasses the utility of these materials in wastewater treatment.

Keywords: Adsorption, Dye degradation, Graphene, Graphene oxide, Heavy metals, Organic dye, Water pollution, Wastewater.

INTRODUCTION

To meet both current and future demands, the world is searching for development that is sustainable due to the faster growth of the global population and the depletion of natural resources [1]. Creating, using, recycling, getting rid of, and eliminating chemicals with the fewest possible adverse effects on human health and the environment are global goals in the field of chemistry. The misuse of artificial chemicals has a negative impact on water, among other things [2, 3]. The most crucial natural resource for maintaining life on Earth is, without a doubt, water.

The estimated amount of wastewater generated annually is 380 billion m^3, and by 2050, this amount is predicted to rise by 51% [4]. Fungicides, industrial effluents, heavy metals, dyes, detergents, soaps, and other substances are the main sources

[*] **Corresponding author Anjaneyulu Bendi:** Innovation and Translational Research Hub (iTRH) & Department of Chemistry, Presidency University, Bangalore, Karnataka, India; E-mail: anjaneyulu.bendi@gmail.com

of pollution in water [5 - 11]. The treatment of wastewater is essential to a sustainable future because this can result in a global shortage of water.

Many techniques, including chemical precipitation, ion exchange, neutralization, adsorption, disinfection, and others, have been developed over time for the treatment of wastewater [12 - 16]. These techniques involve the use of chemicals like sodium chlorite, hydrogen peroxide, chlorine, and sodium hypochlorite, among others [17 - 20]. Researching heavy metal removal and dye degradation techniques that are both economical and efficient is humanity's most urgent need.

Graphene is a sp^2 hybridized form of carbon with a honeycomb-like structure and conjugated π-bonds, which account for its mechanical and thermal strength [21, 22]. Due to its excellent adsorption capacities, modified graphene has been utilized in water purification, with graphene nanocomposite showing promising ability due to its high efficiency and long-term durability. The presence of different functional groups in these nanocomposites helps remove heavy metal ions and absorb organic dyes from wastewater.

DEGRADATION OF ORGANIC DYES USING GRAPHENE AND ITS DERIVATIVES

Durmus *et al.* synthesized a nanocomposite by the amalgamation of graphene oxide (GO) nanosheets with zinc oxide (ZO) nanoparticles for the removal of basic fuchsin (BF) dye. A two-dimensional structure was formed using the sol-gel method with a ratio of GO/ZO of 0.54/0.46 (w/w), and zinc oxide nanoparticles had an average size of 25–30 nm (Scheme **1**). The synthesized nanocomposite showed a degradation efficiency of 92.5% in the first reaction cycle and 84.5% in the fifth reaction cycle, which shows that the nanocomposite can be reused many times [23] (Scheme **2**).

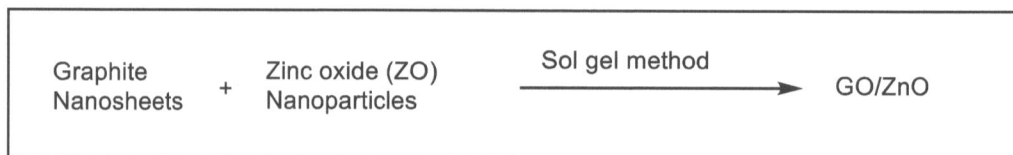

Scheme 1. Graphene oxide and zinc oxide nanocomposite synthesis.

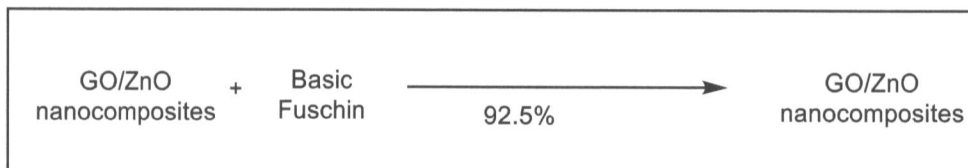

Scheme 2 . Basic Fuschin dye degradation by GO/ZnO nanocomposite.

To make the degradation of methyl orange efficient, ***Wang et al.*** mixed the zinc oxide nanoparticles with porous graphene oxides. The resultant nanocomposite showed excellent photocatalytic activity against the organic dye, removing it 100% in just 150 minutes. The effective charge separation between electron-hole pairs due to the addition of porous graphene with zinc oxide is the primary reason for achieving such high degradation efficiency [24] (Schemes **3** and **4**).

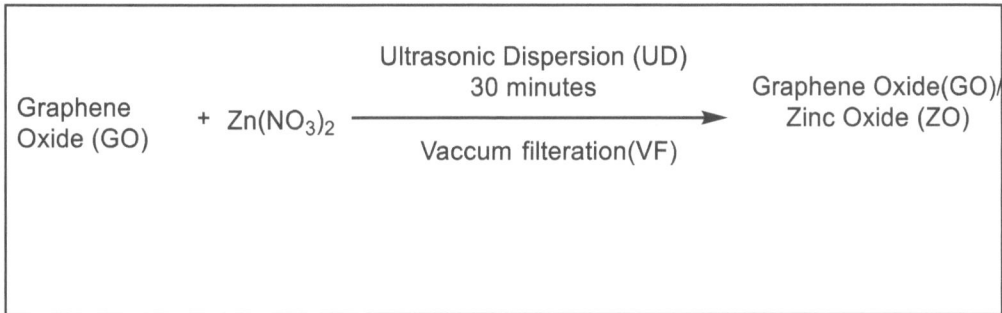

| Graphene Oxide (GO) | + Zn(NO$_3$)$_2$ | Ultrasonic Dispersion (UD) 30 minutes ⟶ Vaccum filteration(VF) | Graphene Oxide(GO)/ Zinc Oxide (ZO) |

Scheme 3. Formation of graphene oxide/zinc oxide nanocomposite.

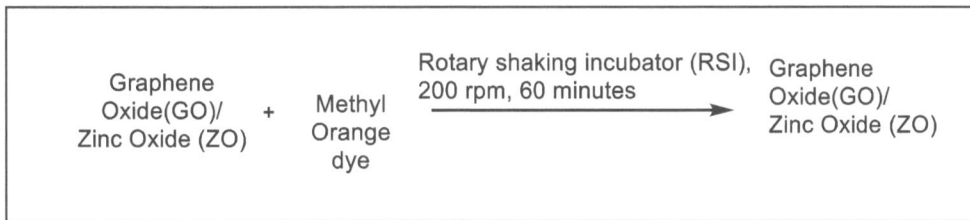

| Graphene Oxide(GO)/ Zinc Oxide (ZO) | + | Methyl Orange dye | Rotary shaking incubator (RSI), 200 rpm, 60 minutes ⟶ | Graphene Oxide(GO)/ Zinc Oxide (ZO) |

Scheme 4. Degradation of methyl orange dye by GO/ZnO nanocomposite.

Khan et al. formed a nanocomposite by combining titanium dioxide (TO) with graphene oxide (GO) in different quantities, *i.e* ., 2%, 4%, 6%, and 8%, having a size of 12.5 nm with a crystalline nature and spherical structure. The results revealed that the synthesized nanocomposite with 8% GO had a larger surface area as well as a narrow band gap, which was the reason for its high and effective catalytic activity against methyl orange and ciprofloxacin [25] (Schemes **5-7**).

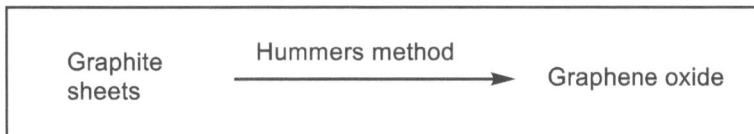

| Graphite sheets | Hummers method ⟶ | Graphene oxide |

Scheme 5. Formation of graphene oxide using Hummer's method.

Graphene oxide (GO) + $(NH_4)_2TiF_6$ $\xrightarrow{60°C, H_3BO_3, 2 h}$ TiO_2/GO composites

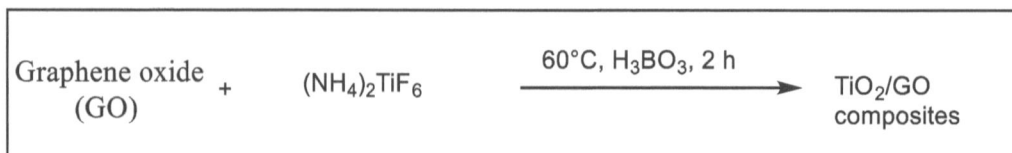

Scheme 6. Formation of TiO_2/GO composites.

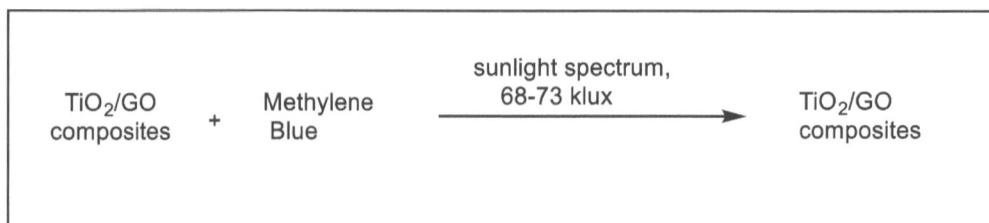

TiO_2/GO composites + Methylene Blue $\xrightarrow{\text{sunlight spectrum, 68-73 klux}}$ TiO_2/GO composites

Scheme 7. Degradation of methylene blue using TiO_2/GO composites.

Li et al. developed a novel composite of $CoFe_2O_4$/BiOBr/graphene for the degradation of rhodamine B dye and reduction of Cr (IV) in the presence of visible light. The synthesized composites proved to be highly efficient with degradation efficiency and a metal removal rate of 100% up to five cycles of use. It was shown that the holes that were generated during the photocatalyst process were stabilized by highly conductive graphene molecules [26] (Schemes **8-10**).

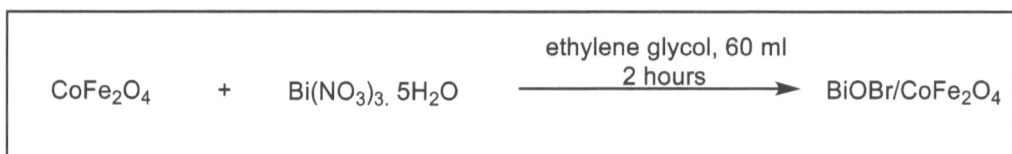

$CoFe_2O_4$ + $Bi(NO_3)_3. 5H_2O$ $\xrightarrow{\text{ethylene glycol, 60 ml} \atop \text{2 hours}}$ $BiOBr/CoFe_2O_4$

Scheme 8. Synthesis of $CoFe_2O_4$/BiOBr.

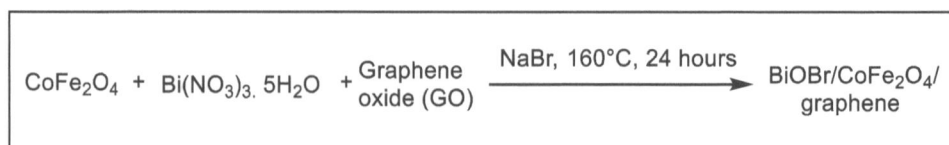

$CoFe_2O_4$ + $Bi(NO_3)_3. 5H_2O$ + Graphene oxide (GO) $\xrightarrow{\text{NaBr, 160°C, 24 hours}}$ $BiOBr/CoFe_2O_4$/ graphene

Scheme 9. Development of novel composite of $CoFe_2O_4$/BiOBr/graphene.

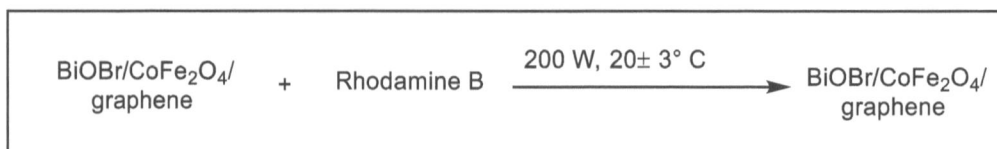

$BiOBr/CoFe_2O_4$/ graphene + Rhodamine B $\xrightarrow{200 W, 20\pm 3° C}$ $BiOBr/CoFe_2O_4$/ graphene

Scheme 10. Application of $CoFe_2O_4$/BiOBr/graphene composites in the degradation of rhodamine B dye.

Al-Rawashdeh et al. synthesized graphene oxide and zinc oxide nanocomposite and embedded them with silver and copper metal nanoparticles to study their effect on the degradation of methyl orange dye. The results showed that the graphene oxide/zinc oxide composite showed excellent photocatalytic activity with a value of 84% when exposed to sunlight for 90 minutes. The addition of copper metal to the nanocomposite decreased the catalytic activity, but when silver metal was added, 100% of the dye was degraded in less time, *i.e* ., 40 minutes [27] (Schemes **11-14**).

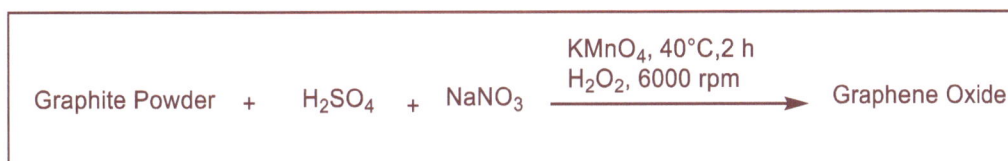

Graphite Powder $+$ H_2SO_4 $+$ $NaNO_3$ $\xrightarrow{\substack{KMnO_4,\ 40°C,2\ h \\ H_2O_2,\ 6000\ rpm}}$ Graphene Oxide

Scheme 11. Graphene oxide synthesized by modified Hummers process.

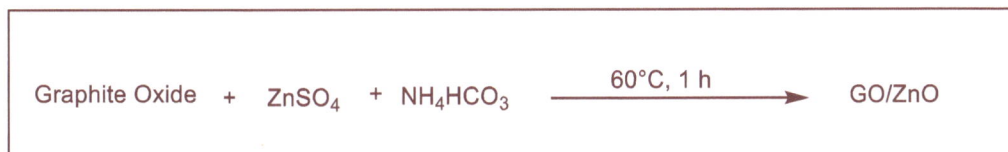

Graphite Oxide $+$ $ZnSO_4$ $+$ NH_4HCO_3 $\xrightarrow{60°C,\ 1\ h}$ GO/ZnO

Scheme 12. Formation of GO/ZnO nanocomposite.

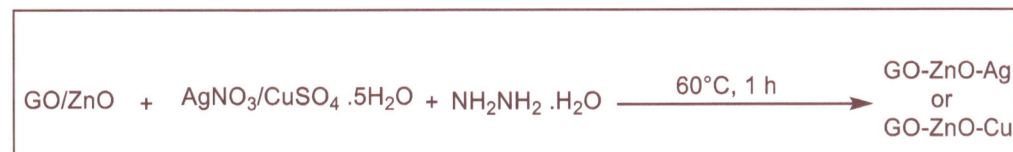

GO/ZnO $+$ $AgNO_3/CuSO_4\ .5H_2O$ $+$ $NH_2NH_2\ .H_2O$ $\xrightarrow{60°C,\ 1\ h}$ GO-ZnO-Ag or GO-ZnO-Cu

Scheme 13. Synthesis of GO-ZnO-Ag or GO-ZnO-Cu.

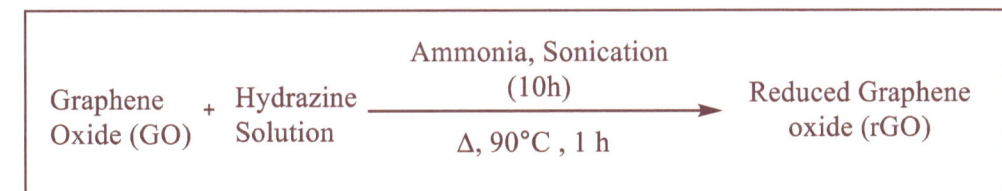

Graphene Oxide (GO) $+$ Hydrazine Solution $\xrightarrow[\Delta,\ 90°C\ ,\ 1\ h]{\substack{Ammonia,\ Sonication \\ (10h)}}$ Reduced Graphene oxide (rGO)

Scheme 14. Synthesis of reduced graphene oxide.

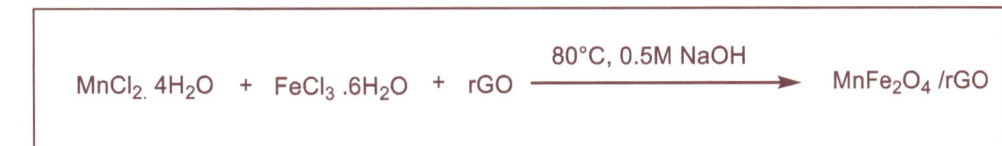

$MnCl_2.\ 4H_2O$ $+$ $FeCl_3\ .6H_2O$ $+$ rGO $\xrightarrow{80°C,\ 0.5M\ NaOH}$ $MnFe_2O_4\ /rGO$

Scheme 15. Formation of manganese ferrite/reduced graphene oxide nanocomposite.

$MnFe_2O_4/rGO$ +   Methylene Blue (MB) $\xrightarrow{\hspace{3cm}}$ $MnFe_2O_4/rGO$

(10mg/L)

60 minutes
97% removal efficiency of
MB

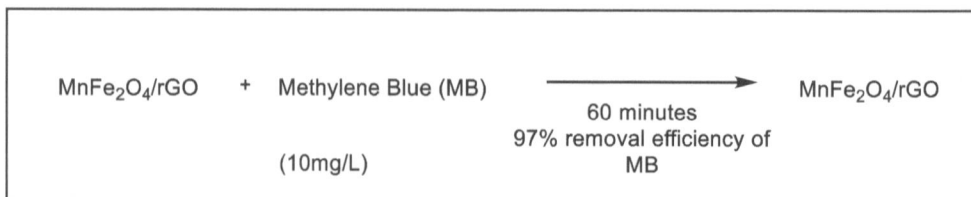

Scheme 16. Methylene blue degradation by manganese ferrite/reduced graphene oxide nanocomposite.

Mandal et al. studied and compared the photocatalytic activity of manganese ferrite nanoparticles ($MnFe_2O_4$) and manganese ferrite/reduced graphene oxide nanocomposite against methylene blue dye (MB). The results showed that the nanocomposite was more effective in the degradation of dye, achieving a value of 97% in 60 minutes, whereas manganese ferrite was able to degrade only 84% of dye in 290 minutes [28] (Schemes **15** and **16**).

El-Aziz, M. E. A. et al. fabricated bio-nanocomposites to achieve effective elimination of heavy metals and degradation of basic red dye. The size of the synthesized nanocomposites was determined to be around 104.79 nm. The sorbents' kinetic study results adhered to the pseudo-second-order mode. The sorption capacity of 0.15g of the created bio-nanocomposites was 100 mg/L lead ions and 25 mg/L basic-red 46. According to the findings, red dye adsorption was highest in grafted chitosan and bio-nanocomposites at 79.54% and heavy metal adsorption at 79.98%. The results showed that Freundlich adsorption was followed for heavy metals and Langmuir adsorption for red dye in the isothermal investigation [29] (Schemes **17** and **18**).

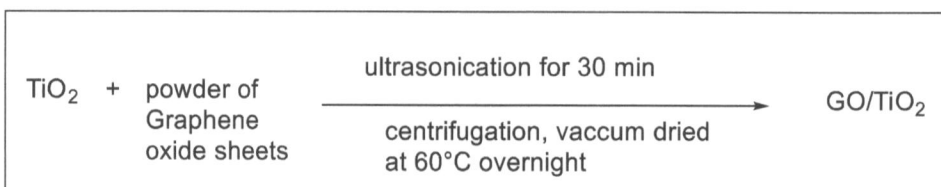

TiO_2 +   powder of Graphene oxide sheets $\xrightarrow[\text{at 60°C overnight}]{\text{ultrasonication for 30 min}}$ GO/TiO_2

centrifugation, vaccum dried

Scheme 17. Synthesis of GO/TiO_2 nanocomposites.

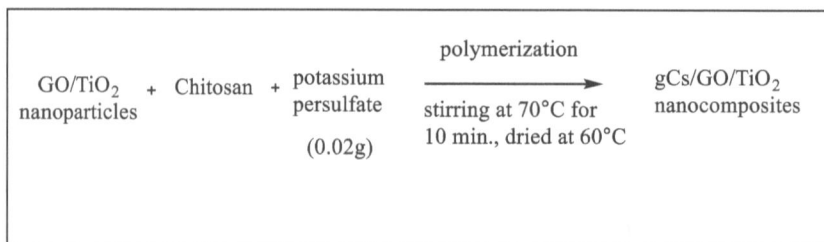

GO/TiO_2 nanoparticles   +   Chitosan   +   potassium persulfate

(0.02g)

polymerization

$\xrightarrow[\text{stirring at 70°C for}]{}$ $gCs/GO/TiO_2$ nanocomposites

10 min., dried at 60°C

Scheme 18. Synthesis of $gCs/gO/TiO_2$ bio-nanocomposites.

Zhang et al. concentrated on creating a novel $CQDs$-GO-Ag_2S nanocomposite to remove methylene blue dye from aqueous solutions. The composite's larger specific surface area was responsible for its astounding 99.0 percent removal rate, with 83.6% achieved within 5 minutes. The functional groups and surface morphology were described, emphasizing the effectiveness of the composite. It was discovered that pH levels had an impact on the adsorption capacity [30] (Schemes **19** and **20**).

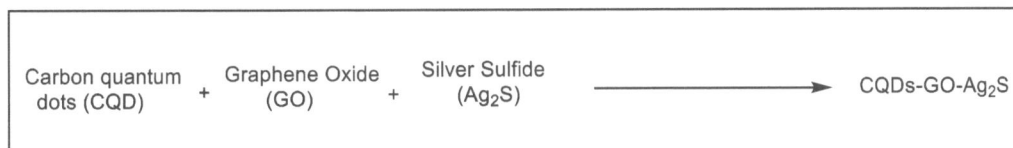

Carbon quantum dots (CQD) + Graphene Oxide (GO) + Silver Sulfide (Ag_2S) ⟶ $CQDs$-GO-Ag_2S

Scheme 19. Formation of novel $CQDs$-GO-Ag_2S nanocomposites.

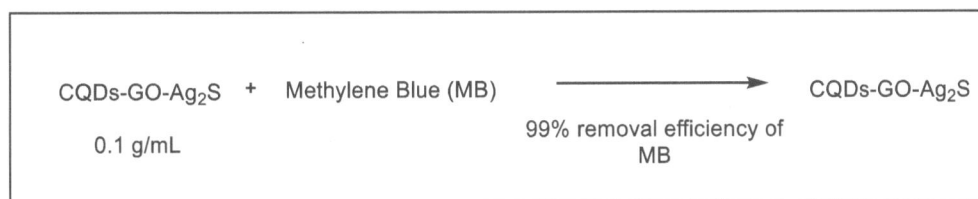

$CQDs$-GO-Ag_2S + Methylene Blue (MB) ⟶ $CQDs$-GO-Ag_2S

0.1 g/mL   99% removal efficiency of MB

Scheme 20. Methylene blue degradation using $CQDs$-GO-Ag_2S nanocomposites.

Pham Thi et al. showed that GO–AgNPs–TNTs nanocomposites were produced by the application of gamma-ray irradiation, effectively assembling silver nanoparticles over the graphene oxide (GO) sheets and TiO_2 nanotubes (TNTs). The synthesis of graphene oxide sheets (GO) and silver nanoparticles (AgNPs) was achieved using previously known methods. The products were vacuum-dried and mixed in a PEG solution. The mixtures were gamma irradiated using the COBALT-60/B, and then they were labeled as GAT-5,10,15, 20, and 25 (different doses of the nanocomposites).

Graphene and silver nanoparticles combined with titanium dioxide (TiO_2) showed improved photocatalytic activity for pollutant removal. The nanocomposites demonstrated notable photocatalytic efficiency, with 81.21% being the highest decolorization efficiency value. The GAT-10 sample exhibited the highest decolorization efficiency value among all nanocomposite materials, according to the results, which also showed higher values for the individual components [31] (Schemes **21** and **22**).

REMOVAL OF HEAVY METALS USING GRAPHENE AND ITS DERIVATIVES

Shahzad *et al.* formed novel graphene-based nanocomposites by amalgamating ethylenediaminetetraacetic acid (EDTA) and magnetic chitosan (CS). The EDTA-MCS/GO nanocomposite was synthesized using a modified Hummers' method, which involved the dispersion of $FeCl_3.6H_2O$ in the presence of HCl, followed by the addition of GO and CS, along with a distilled water stir. Then, the solutions of $FeCl_3$, Na_2SO_3, and NH_3 were added, and a blackish precipitate was obtained. After that, the mixture was combined with dried ground MCS/GO and Na_2EDTA, separated, filtered, and oven-dried.

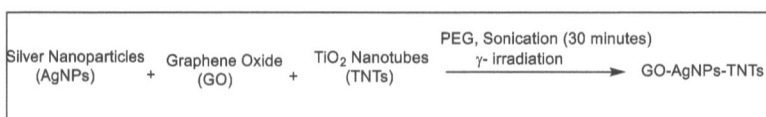

Scheme 21. Synthesis of GO–AgNPs–TNTs nanocomposites.

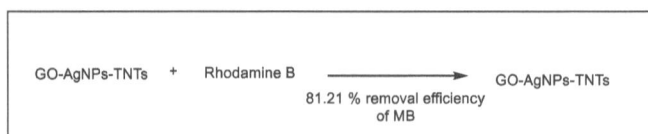

Scheme 22. Removal of MB using GO-AgNPs-TNTs.

The nanocomposites were examined for removing heavy metals in aqueous solutions, and from the results, it was concluded that they acted as excellent removal agents for various heavy metals, with adsorption capacity values of 42.7, 206.5, and 207.2 mg/g for As^{+3}, Pb^{+2}, and Cu^{+2} respectively [32] (Scheme **23**).

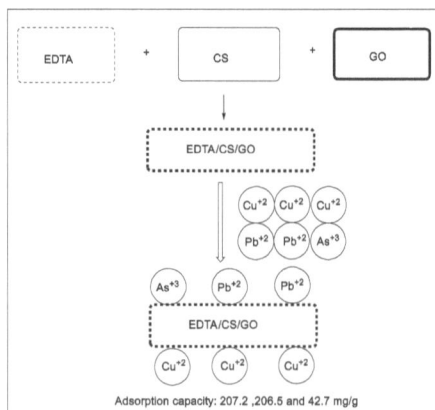

Scheme 23. Synthesis of novel graphene-based nanocomposites by the amalgamation of ethylenediaminetetraacetic acid (EDTA) and magnetic chitosan (CS).

Khatamian *et al.* developed several nanocomposites based on graphene and zeolites and tested them for the adsorption of arsenic ions. The results showed that the composites improved various properties, such as the specific surface area and electrical conductivity of reduced graphene oxide. The nanocomposite $Fe_3O_4/RGO/Cu$-ZEA was most effective, removing approximately 15 pbb of arsenic ions from a 100 pbb solution of H_2AsO_4 [33] (Scheme **24**).

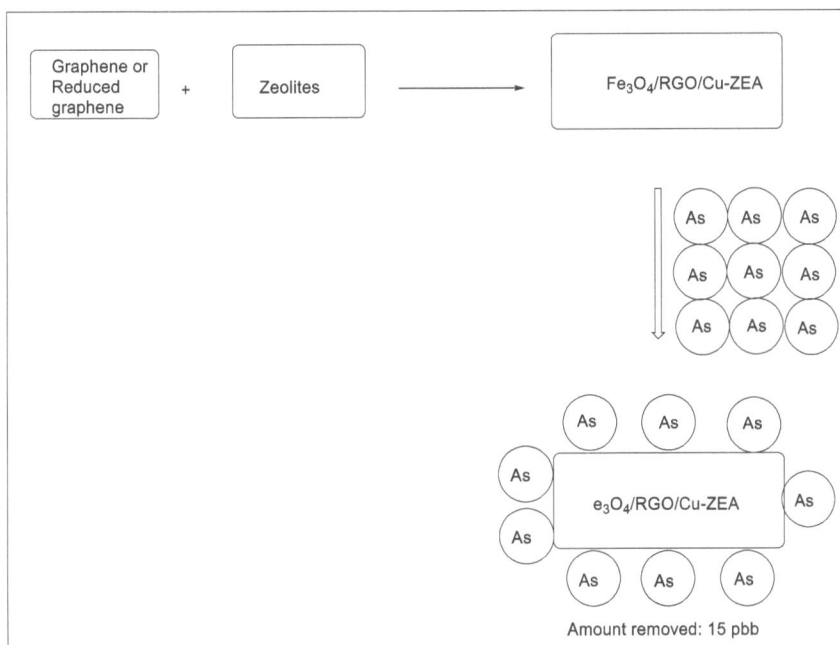

Scheme 24. Formation of several nanocomposites based on graphene and zeolites.

Galunin et al. studied the effect of graphene oxide and newly constructed nanocomposites of polyquinone/graphene (PQ/G) and polyamine cumulene/graphene (PAC/G) on the extraction of different heavy metals from aqueous solutions. The comparative study among the three showed that graphene oxide was most effective in the removal of ions of Cu^{+2}, Zn^{+2}, and Cr^{+3} metals with adsorption capacities of 60, 26, and 5.5 mg/g, respectively, whereas the nanocomposites were only able to remove copper ions with adsorption capacities of 40 mg/g for PQ/G and 26 mg/g for PAC/G [34] (Scheme **25**).

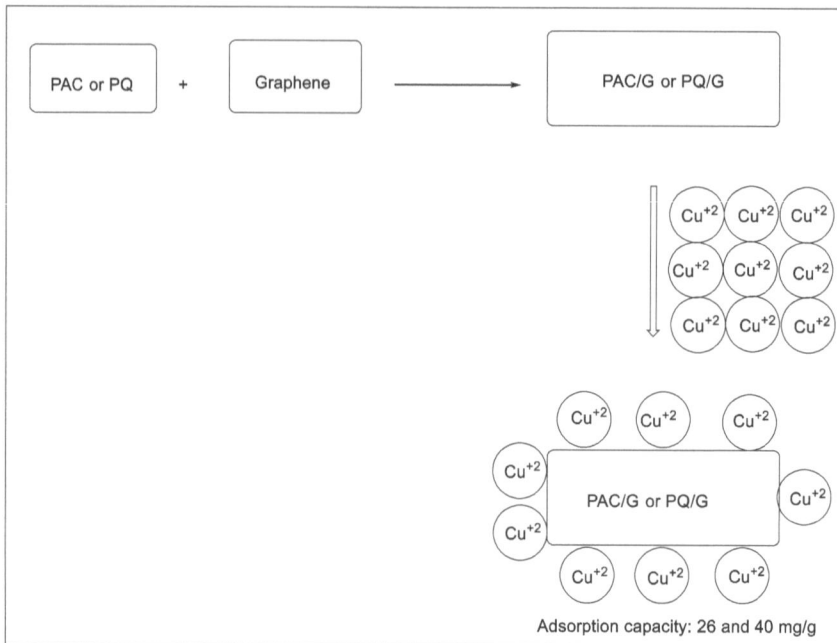

Scheme 25. Formation of novel nanocomposites based on graphene.

Shahrin et al. noted the development of a mixed matrix membrane (MMM) by forming a hybrid nanocomposite with the amalgamation of graphene oxide with manganese ferrite, followed by putting the nanocomposite on a polymeric membrane matrix. Then, it was evaluated for adsorbing the arsenic ion (As^{+5}) from aqueous solutions. The outcomes demonstrated that at pH 4, the best adsorption capacity of 75.5 mg/g was obtained [35] (Scheme **26**).

Zhang et al. synthesized a nanocomposite by the amalgamation of graphene oxide with nickel oxide. The formation of graphene oxide was achieved using the modified Hummers method by mixing GO with $NiCl_2 \cdot 6H_2O$ and deionized water. After that, a hydrazine hydrate solution was added, and the mixture was stirred for a few minutes to form the target nanocomposite.

Then, they were evaluated for the removal of chromium heavy metal from an aqueous solution. The maximum adsorption capacity of 198 mg/g was observed when the pH was 4 and the temperature was 25 °C. The results also showed that changing pH affected the values of adsorption, whereas there was no effect with the variation of temperature [36] (Scheme **27**).

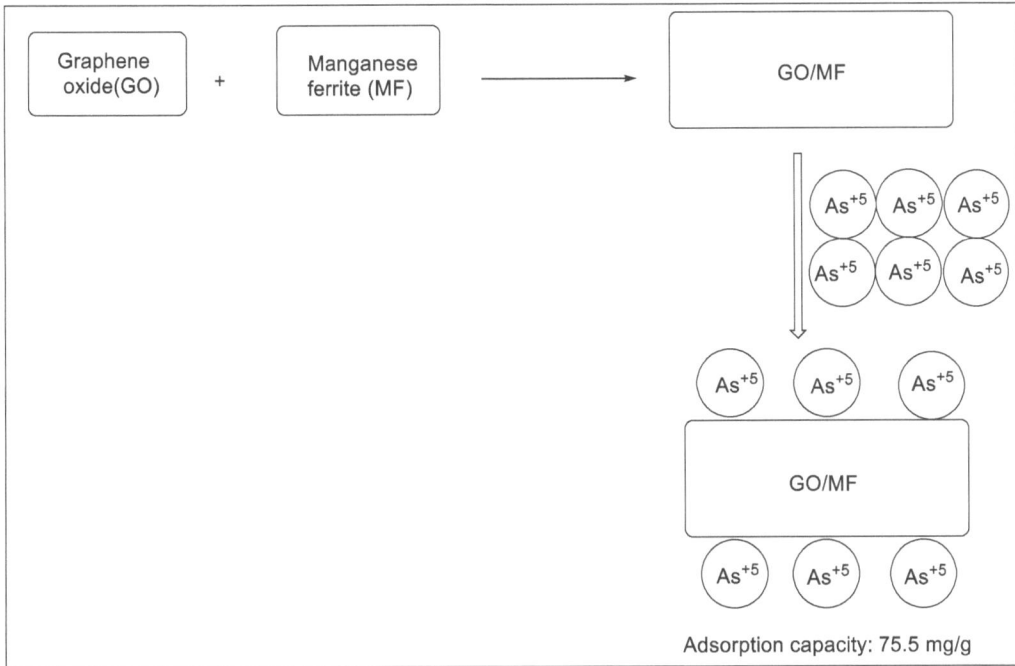

Scheme 26. Synthesis of MMM (mixed matrix membrane).

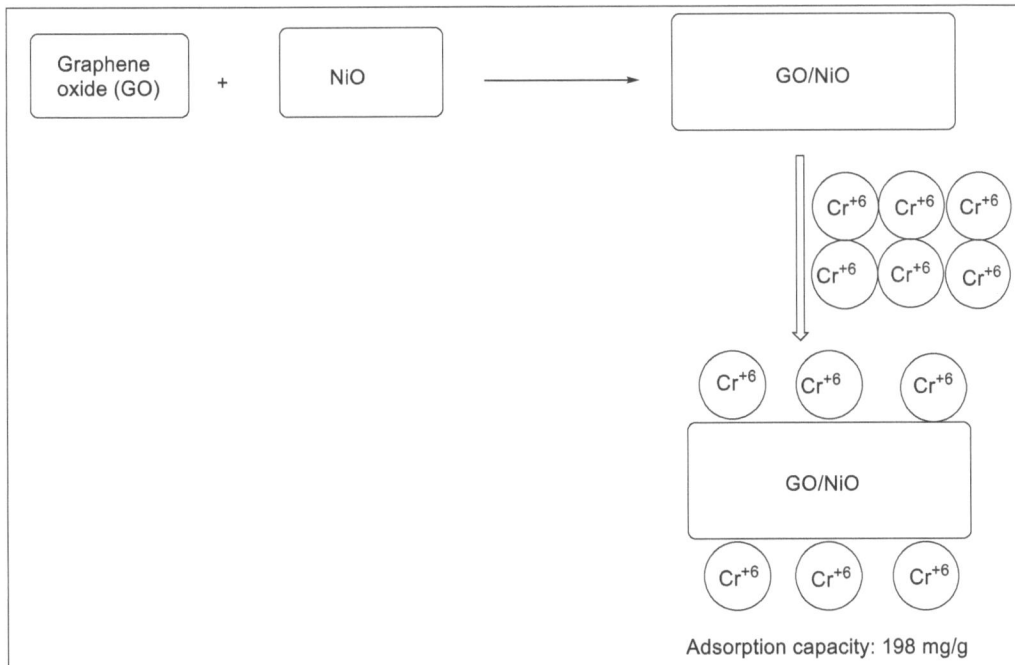

Scheme 27. Application of GO/NiO composite in adsorption of heavy metal chromium (VI).

Wu et al. noted the effect of copper oxide and magnetite nanocomposite arranged on graphene oxide sheets on the removal of heavy metal arsenic from water. $GO/Fe_3O_4@CuO$ composites were made, and GO was synthesized using a modified Hummers technique. Nine percent GO with 91 percent Fe+Cu was the ideal mass ratio to maintain. For As (III) and As (V) adsorption, the atomic ratio of Fe/Cu was optimized to be 3:1. After adding GO to deionized water and stirring, Fe_3Cl_3, $FeSO_4$, NaOH, and $CuCl_2$ were added to form the composite.

The results showed that the synthesized nanocomposite was able to adsorb As^{+3} and As^{+5} ions with values of adsorption capacities of 70.36 and 62.60 mg/g. The results also showed that with the change in pH, the adsorption capacity of As^{+5} decreased as the pH increased, whereas for the As^{+3} ion, no considerable change was observed [37] (Scheme **28**).

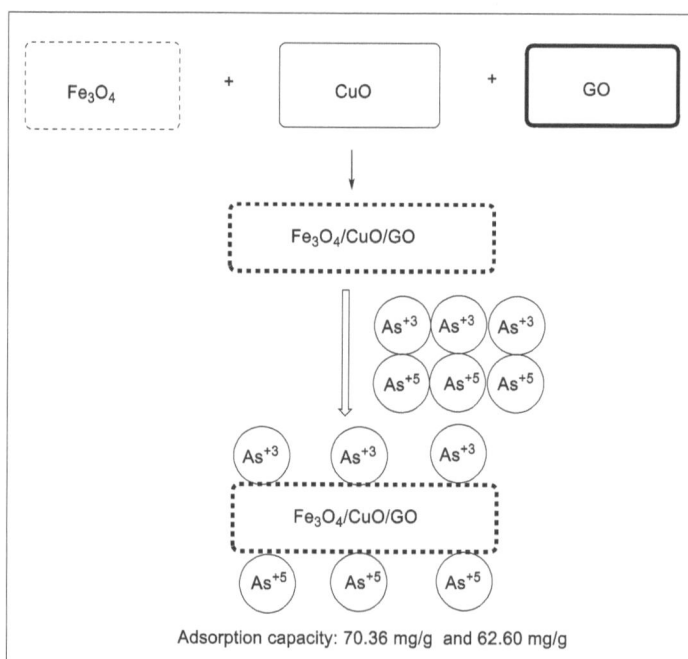

Scheme 28. Adsorption of Arsenic ions using Fe_3O_4/CuO/GO composites.

Wei et al. used attapulgite (ATP) to improve the adsorption capacity of graphene oxide for the adsorption of lead ions from aqueous solutions. The chemical reaction between GO and ATP involved dissolving ATP in deionized water, adding sodium hexametaphosphate, stirring under magnetic stirring, and allowing it to stand stationary for a few hours. The suspension solution was diluted, and GO and ATP were sonicated individually and mixed. The reaction was stirred for 10 hours at 45 °C [38] (Scheme **29**).

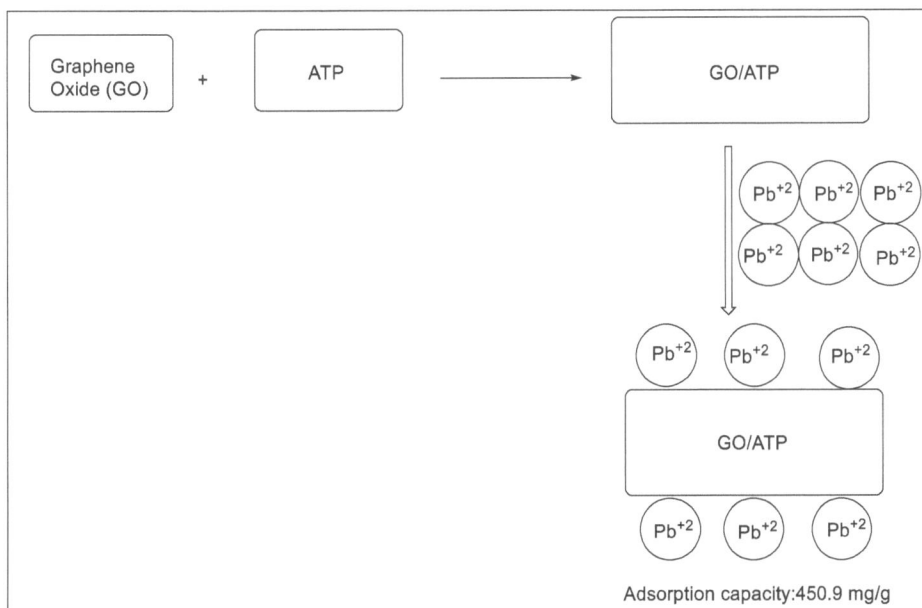

Scheme 29. Application of GO/ATP in removing lead ions.

Cao et al. generated nanosheets of graphene and examined them to remove copper ions from aqueous solutions through adsorption. The nanosheets were synthesized by oxidizing multi-layer graphene nanosheets. The process involved adding concentrated H_2SO_4 and H_3PO_4 to a beaker, then stirring it for a few minutes, followed by adding KMnO4. The mixture was heated to 98 °C and then deionized water and hydrogen peroxide were added. The mixture was centrifuged, washed, and lyophilized to form the target compound. The results showed that nanosheets were excellent at removing copper ions from water, with a removal percentage of 96% [39] (Scheme **30**).

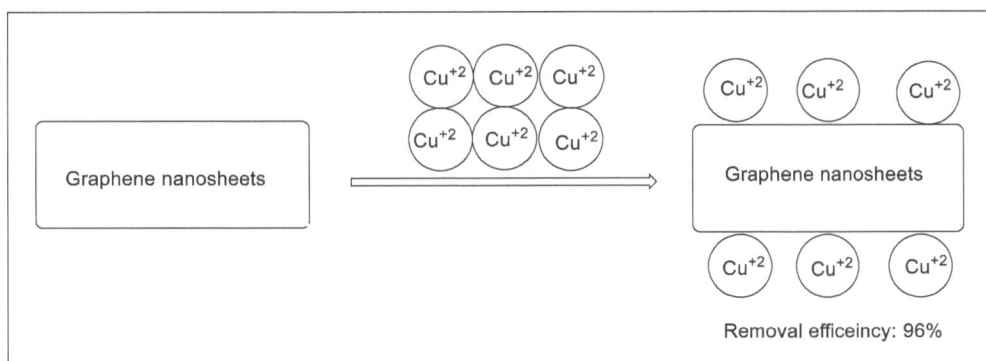

Scheme 30. Extraction of copper ions using graphene nanosheets.

Wang et al. developed a nanocomposite by amalgamation of magnetite with graphene and examined it for removing ions of chromium from aqueous solutions. After dissolving HA-K in DI water and adding $FeCl_3 \cdot 6H_2O$, Fe_3O_4/graphene was obtained. After stirring and drying the mixture, powdered sodium chloride was added. The sample was dehydrated and cleaned with DI water after being heated for four hours in a tube furnace with an Ar atmosphere.

The results showed that the nanocomposite was effective in the removal of chromium ions, and it followed pseudo-second-order kinetics for the adsorption of the heavy metal. A maximum adsorption capacity of 280.6 mg/g was observed when the nanocomposite was prepared at 600 °C [40] (Scheme 31).

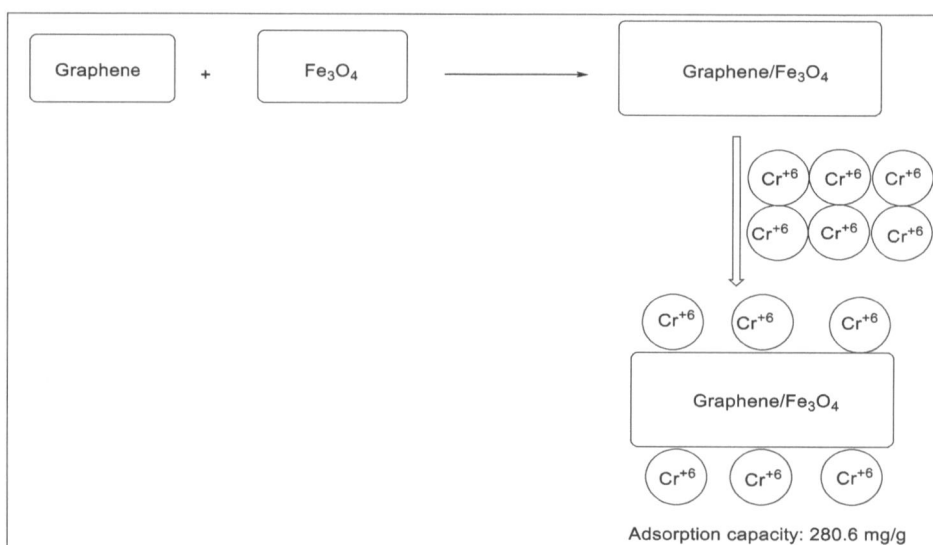

Scheme 31. Synthesis of graphene/Fe_3O_4 nanocomposites.

Ain et al. noted the effect of magnetic graphene oxide modified with iron particles on the extraction of toxic heavy metals such as Pb^{+2}, Cr^{+3}, Cu^{+2}, Zn^{+2}, and Ni^{+2} from aqueous solutions. The graphene oxide was synthesized by mixing graphite powder, sodium nitrate, and sulfuric acid for a couple of hours. To the mixture, potassium permanganate was added, and the resultant solution was kept for a few days, allowing the product to be formed.

The synthesized graphene oxides were checked to see if heavy metals had been removed, and the results showed that the extraction of metals took place through an endothermic and spontaneous process that followed the Langmuir adsorption isotherm. They showed maximum adsorption capacity values of 200, 24.330,

62.893, 63.694, and 51.020 mg/g for Pb^{+2}, Cr^{+3}, Cu^{+2}, Zn^{+2}, and Ni^{+2}, respectively [41] (Scheme **32**).

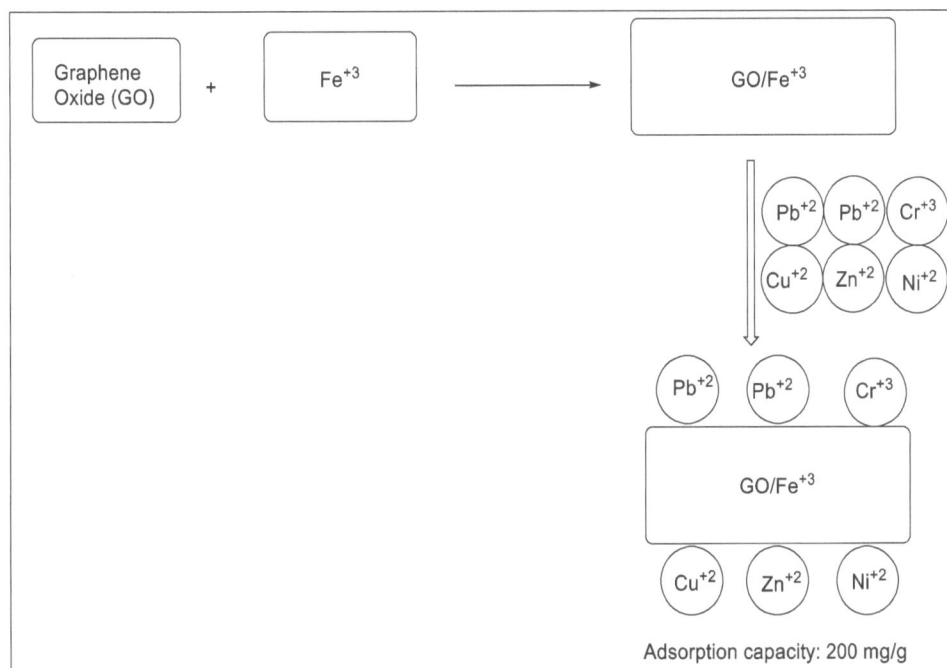

Scheme 32. Effect of magnetic graphene oxide modified with iron particles.

Jibin *et al.* reported the removal of the heavy metal chromium from aqueous solutions using graphene oxide (GO) and graphene oxide covered with silicon dioxide. The study involved synthesizing graphene oxide and modifying amine on nano-silica using 3-aminopropyl triethoxysilane in an ethanol/water mixture. The resulting nanoparticles were self-assembled and prepared. The core shell was then washed and dried. Then, the amine-modified silica and graphene oxide were dispersed in distilled water in separate beakers. Finally, both solutions were allowed to mix for a few hours, followed by drying the precipitate at 80 °C.

From the results obtained, it can be concluded that both showed excellent properties, but with the presence of a silicon dioxide coating, graphene oxide performed better, with a removal percentage of 92.28%. It was also shown that the adsorption values depended on the pH of the solution and the initial concentration of the solution [42] (Scheme **33**).

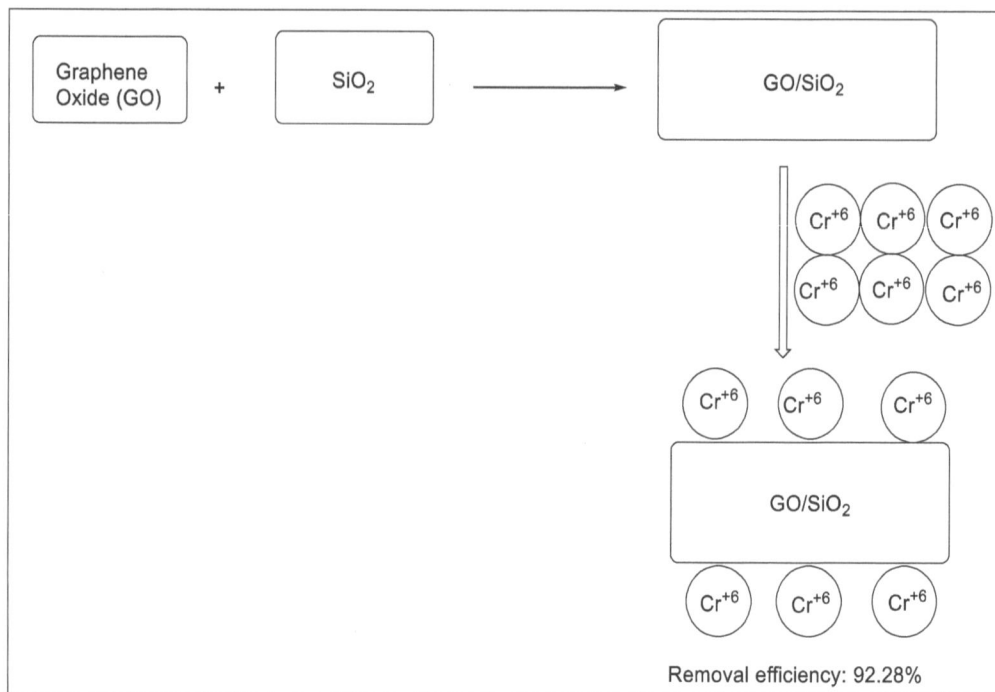

Scheme 33. Graphene oxide (GO) and graphene oxide covered in silicon dioxide: Their respective effects on heavy metal elimination (chromium).

CONCLUSION

Graphene-based nanomaterials have proven to be effective chemicals for removing various organic dyes and heavy metals from wastewater. In the present study, we reviewed various nanoparticles, particularly composites, for the purification of wastewater. We observed that the nanocomposite of graphene oxide and zinc oxide was able to remove methyl orange dye completely, and graphene nanosheets showed up to 96% removal efficiency of copper heavy metal. This study provides a comprehensive overview of graphene and its derivatives for wastewater treatment applications, summarizing recent advancements and trends over the past 5–6 years. It highlights the various properties of graphene-based materials that make them suitable for removing contaminants and pollutants from wastewater. By compiling and analyzing recent data, this study offers valuable insights.

LIST OF ABBREVIATIONS

Ag₂S   Silver Sulfide

AgNPs   Silver Nanoparticles

ATP   Attapulgite

CoFe$_2$O$_4$	Cobalt Iron Oxide
CQD	Carbon Quantum Dots
CuO	Copper Oxide
GO	Graphene Oxide
MB	Methylene Blue Dye
MF	Manganese Ferrite
MMM	Mixed Matrix Membrane
MnFe$_2$O$_4$	Manganese Ferrite
NiO	Nickel Oxide
PAC	Polyamine Cumulene
PEG	Polyethylene Glycol
PQ	Polyquinone
TNT	Titanium Dioxide Nanoparticles
TO	Titanium Dioxide
ZO	Zinc Oxide

ACKNOWLEDGEMENTS

The authors would like to express their sincere thanks to the management of BHU, Varanasi, Uttar Pradesh, India, and Presidency University, Bangalore, Karnataka, India, for providing the facilities to write and submit the book chapter for publication.

REFERENCES

[1]   Blum, C.; Bunke, D.; Hungsberg, M.; Roelofs, E.; Joas, A.; Joas, R.; Blepp, M.; Stolzenberg, H.C. The concept of sustainable chemistry: Key drivers for the transition towards sustainable development. *Sustain. Chem. Pharm.,* **2017**, *5*, 94-104.
[http://dx.doi.org/10.1016/j.scp.2017.01.001]

[2]   Akhtar, N.; Syakir Ishak, M. I.; Bhawani, S. A.; Umar, K. Various natural and anthropogenic factors responsible for water quality degradation: A review. *Water (Switzerland).,* **2021**.
[http://dx.doi.org/10.3390/w13192660]

[3]   Zahoor, I.; Mushtaq, A. Water pollution from agriculure activities: A Critical Global Review. *IJCBS,* **2023**, *23*(1), 164-176.

[4]   Qadir, M.; Drechsel, P.; Jiménez Cisneros, B.; Kim, Y.; Pramanik, A.; Mehta, P.; Olaniyan, O. Global and regional potential of wastewater as a water, nutrient and energy source. *Nat. Resour. Forum,* **2020**, *44*(1), 40-51.
[http://dx.doi.org/10.1111/1477-8947.12187]

[5]   Yan, X.; Xia, Y.; Ti, C.; Shan, J.; Wu, Y.; Yan, X. Thirty years of experience in water pollution control in Taihu Lake: A review. *Sci. Total Environ.,* **2024**, *914*, 169821.
[http://dx.doi.org/10.1016/j.scitotenv.2023.169821] [PMID: 38190921]

[6]   Sahoo, S.K.; Goswami, S.S. Theoretical framework for assessing the economic and environmental impact of water pollution: A detailed study on sustainable development of India. *Journal of Future*

Sustainability, **2024**, *4*(1), 23-34.
[http://dx.doi.org/10.5267/j.jfs.2024.1.003]

[7]   Sankhla, M.S. Water contamination through pesticide & their toxic effect on human health. *Int. J. Res. Appl. Sci. Eng. Technol.,* **2018**, *6*(1), 967-970.
[http://dx.doi.org/10.22214/ijraset.2018.1146]

[8]   Verma, R.; Dwivedi, P. Heavy metal water pollution-a case study. *Science and Technology,* **2013**, *5*(5), 98-99.

[9]   Farhan Hanafi, M.; Sapawe, N. A Review on the water problem associate with organic pollutants derived from phenol, methyl orange, and remazol brilliant blue dyes. In: *Materials Today: Proceedings*; Elsevier Ltd, **2020**; Vol. 31, pp. A141-A150.
[http://dx.doi.org/10.1016/j.matpr.2021.01.258]

[10]   Goel, G.; Kaur, S. A study on chemical contamination of water due to household laundry detergents. *J. Hum. Ecol.,* **2012**, *38*(1), 65-69.
[http://dx.doi.org/10.1080/09709274.2012.11906475]

[11]   Tekade, P. V; Mohabansi, N. P.; Patil, V. B. *Study of physico-chemical properties of effluents from soap industry in wardha.,* **2011**, *4*(2), 461-465.

[12]   Harper, T. R.; Kingham, N. W. *Removal of arsenic from wastewater using chemical precipitation methods.,*
[http://dx.doi.org/10.2175/WER.64.3.2]

[13]   Kansara, N.; Bhati, L.; Narang, M.; Vaishnavi, R. *Critical review wastewater treatment by ion exchange method: a review of past and recent researches.,*

[14]   Institute of Electrical and Electronics Engineers *ICIAS2012 : 2012 4th international conference on intelligent and advanced systems : a conference of world engineering, science & technology congress (ESTCON),* Kuala Lumpur Convention Centre : Conference Proceedings.; IEEE,**2012**, pp. 12-14.

[15]   Rashid, R.; Shafiq, I.; Akhter, P. Muhammad, & Iqbal, J.; Hussain, M. A state-of-the-art review on wastewater treatment techniques: the effectiveness of adsorption method. *Environ Sci Pollut Res Int.,* **2021**, *28*(8), 9050-9066.
[http://dx.doi.org/10.1007/s11356-021-12395-x] [PMID: 33483933]

[16]   Amin, M. M.; Hashemi, H.; Boyini, A. M. A review on wastewater disinfection. *Environ Sci Pollut Res Int.,* **2021**, *28*(8), 9050-9066.
[http://dx.doi.org/10.1007/s11356-021-12395-x] [PMID: 33483933]

[17]   Liu, Y.; Wang, J.; Zheng, Y.; Wang, A. Adsorption of methylene blue by kapok fiber treated by sodium chlorite optimized with response surface methodology. *Chem. Eng. J.,* **2012**, *184*, 248-255.
[http://dx.doi.org/10.1016/j.cej.2012.01.049]

[18]   Bonetta, S.; Pignata, C.; Bonetta, S.; Amagliani, G.; Brandi, G.; Gilli, G.; Carraro, E. Comparison of UV, peracetic acid and sodium hypochlorite treatment in the disinfection of urban wastewater. *Pathogens,* **2021**, *10*(2), 182.
[http://dx.doi.org/10.3390/pathogens10020182] [PMID: 33572069]

[19]   Karahan, B.N.; Akdag, Y.; Fakioglu, M.; Korkut, S.; Guven, H.; Ersahin, M.E.; Ozgun, H. Coupling ozonation with hydrogen peroxide and chemically enhanced primary treatment for advanced treatment of grey water. *J. Environ. Chem. Eng.,* **2023**, *11*(3), 110116.
[http://dx.doi.org/10.1016/j.jece.2023.110116]

[20]   Sedlak, D.L.; von Gunten, U. Chemistry. The chlorine dilemma. *Science,* **2011**, *331*(6013), 42-43.
[http://dx.doi.org/10.1126/science.1196397] [PMID: 21212347]

[21]   Yang, G.; Li, L.; Lee, W. B.; Ng, M. C. Structure of graphene and its disorders: A review. *Science and Technology of Advanced Materials.,* **2018**, , 613-648.
[http://dx.doi.org/10.1080/14686996.2018.1494493]

[22] Erickson, K.; Erni, R.; Lee, Z.; Alem, N.; Gannett, W.; Zettl, A. Determination of the local chemical structure of graphene oxide and reduced graphene oxide. *Adv. Mater.,* **2010**, *22*(40), 4467-4472.
[http://dx.doi.org/10.1002/adma.201000732] [PMID: 20717985]

[23] Durmus, Z.; Kurt, B.Z.; Durmus, A. Synthesis and characterization of graphene oxide/zinc oxide (go/zno) nanocomposite and its utilization for photocatalytic degradation of basic fuchsin dye. *ChemistrySelect,* **2019**, *4*(1), 271-278.
[http://dx.doi.org/10.1002/slct.201803635]

[24] Wang, L.; Li, Z.; Chen, J.; Huang, Y.; Zhang, H.; Qiu, H. Enhanced photocatalytic degradation of methyl orange by porous graphene/ZnO nanocomposite. *Environ. Pollut.,* **2019**, *249*, 801-811.
[http://dx.doi.org/10.1016/j.envpol.2019.03.071] [PMID: 30953942]

[25] Khan, S.A.; Arshad, Z.; Shahid, S.; Arshad, I.; Rizwan, K.; Sher, M.; Fatima, U. Synthesis of TiO2/Graphene oxide nanocomposites for their enhanced photocatalytic activity against methylene blue dye and ciprofloxacin. *Compos., Part B Eng.,* **2019**, *175*, 107120.
[http://dx.doi.org/10.1016/j.compositesb.2019.107120]

[26] Li, M.; Song, C.; Wu, Y.; Wang, M.; Pan, Z.; Sun, Y.; Meng, L.; Han, S.; Xu, L.; Gan, L. Novel Z-scheme visible-light photocatalyst based on $CoFe_2O_4$/BiOBr/Graphene composites for organic dye degradation and Cr(VI) reduction. *Appl. Surf. Sci.,* **2019**, *478*, 744-753.
[http://dx.doi.org/10.1016/j.apsusc.2019.02.017]

[27] Al-Rawashdeh, N.A.F.; Allabadi, O.; Aljarrah, M.T. Photocatalytic activity of graphene oxide/zinc oxide nanocomposites with embedded metal nanoparticles for the degradation of organic dyes. *ACS Omega,* **2020**, *5*(43), 28046-28055.
[http://dx.doi.org/10.1021/acsomega.0c03608] [PMID: 33163787]

[28] Mandal, B.; Panda, J.; Paul, P.K.; Sarkar, R.; Tudu, B. $MnFe_2O_4$ decorated reduced graphene oxide heterostructures: Nanophotocatalyst for methylene blue dye degradation. *Vacuum,* **2020**, *173*, 109150.
[http://dx.doi.org/10.1016/j.vacuum.2019.109150]

[29] El-Aziz, M.E.A.; Youssef, A.M.; Kamal, K.H.; Kelnar, I.; Kamel, S. Preparation and performance of bionanocomposites based on grafted chitosan, GO and TiO_2-NPs for removal of lead ions and basic-red 46. *Carbohydr. Polym.,* **2023**, *305*, 120571.
[http://dx.doi.org/10.1016/j.carbpol.2023.120571] [PMID: 36737211]

[30] Zhang, D.; Xu, J.; Liang, L.; Li, H.; Du, M.; Liu, X.; Wang, K. Synthesis of a novel CQDs-GO-Ag_2S Composite and study on the adsorption of methylene blue. *ChemistrySelect,* **2020**, *5*(8), 2501-2507.
[http://dx.doi.org/10.1002/slct.202000102]

[31] Thi, T.L.P.; Vu, D.K.N.; Thi, P.A.N.; Nguyen, D.K.V. Silver nanoparticles-assembled graphene oxide sheets on TiO2 nanotubes: synthesis, characterization, and photocatalytic investigation. *Appl. Nanosci.,* **2020**, *10*(10), 3735-3743.
[http://dx.doi.org/10.1007/s13204-020-01445-4]

[32] Shahzad, A.; Miran, W.; Rasool, K.; Nawaz, M.; Jang, J.; Lim, S.R.; Lee, D.S. Heavy metals removal by EDTA-functionalized chitosan graphene oxide nanocomposites. *RSC Advances,* **2017**, *7*(16), 9764-9771.
[http://dx.doi.org/10.1039/C6RA28406J]

[33] Khatamian, M.; Khodakarampoor, N.; Saket-Oskoui, M. Efficient removal of arsenic using graphene-zeolite based composites. *J. Colloid Interface Sci.,* **2017**, *498*, 433-441.
[http://dx.doi.org/10.1016/j.jcis.2017.03.052] [PMID: 28349886]

[34] Galunin, E.; Burakova, I.; Neskoromnaya, E.; Babkin, A.; Melezhik, A.; Burakov, A.; Tkachev, A. Adsorption of heavy metals from aqueous media on graphene-based nanomaterials. *In AIP Conference Proceedings,* **2018**, *2041*.
[http://dx.doi.org/10.1063/1.5079338]

[35] Shahrin, S.; Lau, W.J.; Goh, P.S.; Ismail, A.F.; Jaafar, J. Adsorptive mixed matrix membrane

incorporating graphene oxide-manganese ferrite (GMF) hybrid nanomaterial for efficient As(V) ions removal. *Compos., Part B Eng.,* **2019**, *175*, 107150.
[http://dx.doi.org/10.1016/j.compositesb.2019.107150]

[36]   Zhang, K.; Li, H.; Xu, X.; Yu, H. Synthesis of reduced graphene oxide/NiO nanocomposites for the removal of Cr(VI) from aqueous water by adsorption. *Microporous Mesoporous Mater.,* **2018**, *255*, 7-14.
[http://dx.doi.org/10.1016/j.micromeso.2017.07.037]

[37]   Wu, K.; Jing, C.; Zhang, J.; Liu, T.; Yang, S.; Wang, W. Magnetic Fe3O4@CuO nanocomposite assembled on graphene oxide sheets for the enhanced removal of arsenic(III/V) from water. *Appl. Surf. Sci.,* **2019**, *466*, 746-756.
[http://dx.doi.org/10.1016/j.apsusc.2018.10.091]

[38]   Wei, B.; Cheng, X.; Wang, G.; Li, H.; Song, X.; Dai, L. Graphene oxide adsorption enhanced by attapulgite to remove Pb (II) from aqueous solution. *Appl. Sci. (Basel),* **2019**, *9*(7), 1390.
[http://dx.doi.org/10.3390/app9071390]

[39]   Cao, M. Functionalized graphene nanosheets as absorbent for copper (II) removal from water. *Ecotoxicology and Environmental Safety.,* **2019**, , 28-36.
[http://dx.doi.org/10.1016/j.ecoenv.2019.02.011]

[40]   Wang, X.; Lu, J.; Cao, B.; Liu, X.; Lin, Z.; Yang, C.; Wu, R.; Su, X.; Wang, X. Facile synthesis of recycling Fe3O4/graphene adsorbents with potassium humate for Cr(VI) removal. *Colloids Surf. A Physicochem. Eng. Asp.,* **2019**, *560*, 384-392.
[http://dx.doi.org/10.1016/j.colsurfa.2018.10.036]

[41]   Ain, Q.U.; Farooq, M.U.; Jalees, M.I. Application of magnetic graphene oxide for water purification: Heavy metals removal and disinfection. *J. Water Process Eng.,* **2020**, *33*, 101044.
[http://dx.doi.org/10.1016/j.jwpe.2019.101044]

[42]   Jibin, K.; Augustine, S.; Velayudhan, P.; George, J.; Krishnageham Sidharthan, S.; Paulose, S.; Thomas, S. Unleashing the power of graphene-based nanomaterials for chromium(VI) ion elimination from water. *Crystals (Basel),* **2023**, *13*(7), 1047.
[http://dx.doi.org/10.3390/cryst13071047]

CHAPTER 6

Carbonaceous Nanoparticles in Wastewater Treatment

Aditi Tiwari[1], M. Radha Sirija[2] and **G. B. Dharma Rao[3,*]**

[1] *Intertek India, Gurugram, Haryana-122001, India*

[2] *Department of Chemistry, Vardhaman College of Engineering, Hyderabad-501218, India*

[3] *Department of Chemistry, Kommuri Pratap Reddy Institute of Technology, Hyderabad-500088, Telangana, India*

Abstract: Nanoparticles are minute particles with particle sizes ranging from 1nm to 100nm, possessing distinctive physical and chemical properties due to their large surface area and nanosized nature. Nanoparticles, due to their nano size, have unique optical properties, such as adsorption in the visible region, reactivity, and toughness, due to which they tend to be a fit candidate for a variety of commercial and domestic applications. Carbon-based nanomaterials, when combined with nanotechnology are widely used to provide clean and affordable water. Carbon nanomaterials possess properties like positive correlation with specific structural character, alternating dimension factor, mechanical strength, electrical conductivity, *etc.*, and hence are used in the application of wastewater treatment. This paper focuses on the use of carbon-based nanomaterials, including fullerenes, carbon quantum dots, and nanodiamonds, for dye degradation and heavy metal removal from wastewater.

Keywords: Carbonaceous nanoparticles, Dye degradation, Fullerenes, Heavy metal removal, Nanodiamond, Quantum dots, Wastewater treatment.

INTRODUCTION

The rapid increase in the global population, coupled with ongoing industrialization, has transformed pollution into a significant and pressing concern. Conservationists have become more worried about effectively managing water due to water pollution. They utilize various methods of treatment, including adsorption, photocatalysis, lignin-based biosorbents, advanced oxidation methods (AOPs), electrochemical treatment, and biological treatments [1 - 5]. Wastewater emerging from industries contains toxins that can be inorganic pollutants (*e.g.*, rare earth metals, heavy metals [Pb, Mn, Ni, Hg, Cd, Co, Cr, Zn, Cu, Sb]) and

---

* **Corresponding author G. B. Dharma Rao:** Department of Chemistry, Kommuri Pratap Reddy Institute of Technology, Hyderabad-500088, Telangana, India; E-mail: gbdharmarao@gmail.com

Anjaneyulu Bendi (Ed.)

organic pollutants (*e.g.*, phenols, dyes, herbicides, pharmaceuticals, and polycyclic aromatic hydrocarbons (PAH)), which often tend to be carcinogenic and persist longer in environments [6 - 10]. Among all the other wastewater treatment procedures, the adsorption approach has been an encouraging choice due to its high efficiency, ease of large-scale implementation, easy handling and operation, and cost-effectiveness [11, 12].

ROLE OF FULLERENES, NANODIAMONDS, AND CARBON QUANTUM DOTS IN WASTEWATER TREATMENT

Fabrics that are produced synthetically or from natural sources can be colored using dyes. Anthraquinones (red), indigoids (blue), and flavonoids (yellow) are the compounds found in dyes. From 1856 to 1970, mauveine was the first synthetic dye to be used [13]. Dyes such as lactone, indophenol, indamine, azodyes, anthraquinone, phthalocyanine, indigo, xanthene, thiazine, *etc.*, are categorized according to their chemical structures which were found in wastewater (Fig. **1**). The most poisonous and carcinogenic of them is azodye, which lowers dissolved oxygen and water transparency [14]. Because dyes contain chromophores and auxochromes in their chemical composition, colored water blocks light penetration and is harmful to aquatic life. Dyes from wastewater can be degraded by a variety of techniques (Fig. **2**).

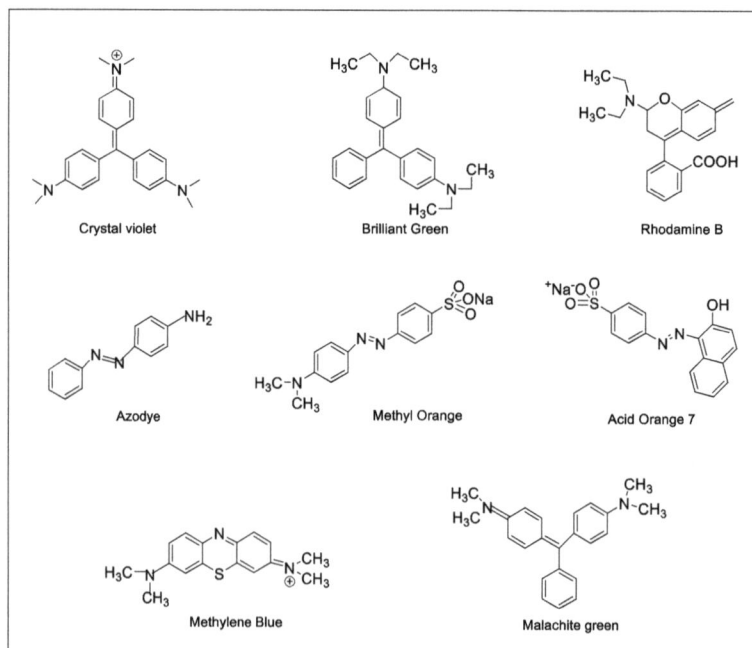

Fig. (1). Structures of some commonly found dyes in wastewater.

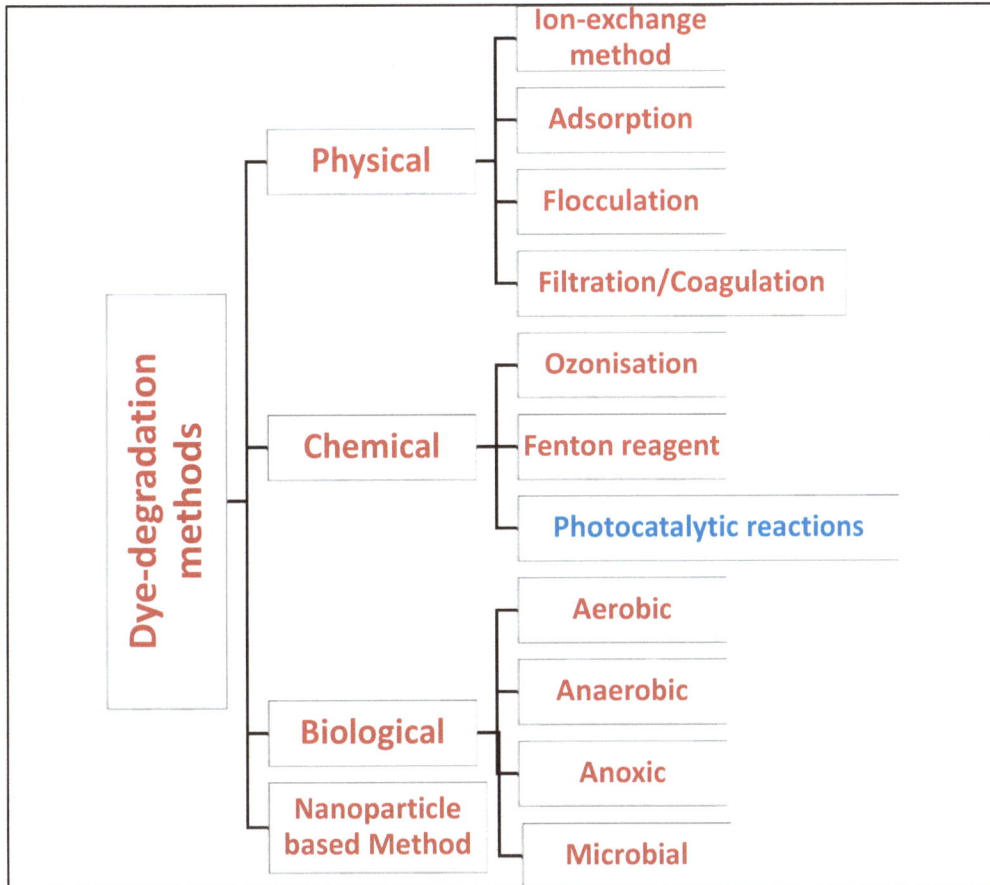

Fig. (2). Methods available for dye degradation.

The global population expansion and increasing industrialization necessitate large amounts of clean water. However, water is becoming more and more contaminated and toxic due to industrial waste.

To turn tainted water into consumable water, multiple methods are used. "Photocatalysis" is among the mentioned. "Carbon-based photocatalysts", which include CNs, NDs, graphene, GCNs, and CQDs [15 - 17], are among the many available photocatalysts (Fig. **3**). Due to their large surface area, high light absorption, excellent stability, ease of use, good porosity, environmental friendliness, ease of preparation, and good band gap, carbon-based photocatalysts are highly preferred. When combined with other semiconductors, the carbon atom functions as an electron sink, essential for improving photocatalytic efficiency [18 - 21].

Fig. (3). Type of carbon-based nanoparticles.

NANODIAMONDS, QUANTUM DOTS, AND FULLERENES FOR WASTEWATER TREATMENT

Fullerenes-based Nanophotocatalysts

C_{60}/SiO_2 powder functions as a photocatalyst for the decomposition of MO in the presence of AA (MO+AA). Under light irradiation at >420nm, MO is degraded to DMPD and SA (Scheme **1**) [22].

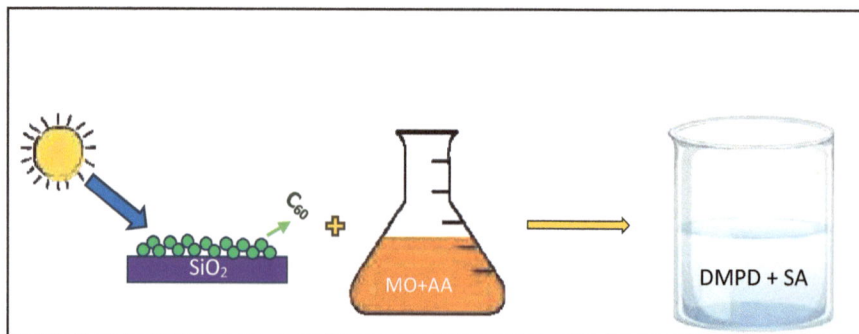

Scheme 1. C_{60}/SiO_2 for the degradation of MO.

Under ultraviolet irradiation at 365 nm, the photocatalytic function of C_{60}nanowhiskers/NiS$_2$nanocomposites was examined, and the results showed efficient destruction of brilliant green and methylene blue. The heated C_{60} nanowhiskers/NiS$_2$ nanocomposites, having a needle and rod-like morphology, are more efficient in degrading methylene blue rather than brilliant green solution (Scheme **2**) [23].

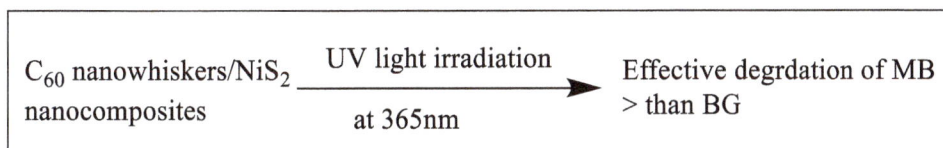

Scheme 2. C_{60} nanoparticles for degradation of methyl orange and brilliant blue.

Pure ZnO, heated ZnO, pure C_{60}, heated C_{60},and C_{60}/ZnO nanocomposites were synthesized for the photodegradation of organic dyes such as MB, MO, and RhB under UV light irradiation (254nm). C_{60}/ZnO nanoparticles work more effectively as photocatalysts than the heated ZnO, heated C_{60}, pure ZnO, and pure C_{60}. Heated ZnO C_{60}works more effectively as a photocatalyst than the unheated ones due to changes in morphology. ZnO particles change from needle to rod-like particles, whereas C_{60} changed from crystalline to amorphous due to heat treatment for 2h at 700°C (Scheme **3**) [24].

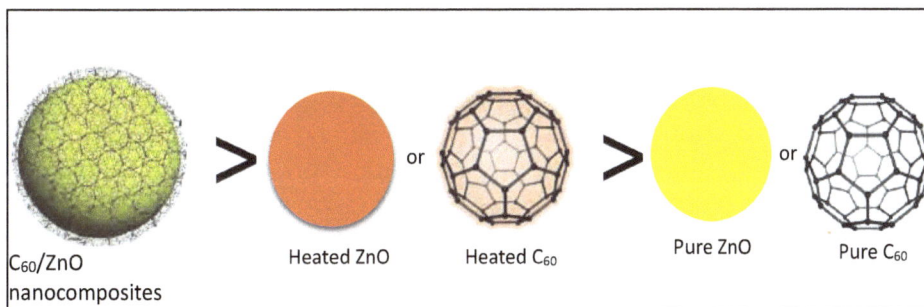

Scheme 3. Effective degradation of MB, MO, and RB.

The synthesis of C_{60}/Au NPs supported by TiO$_2$is addressed as follows. The stabilized gold particles bind to the TiO$_2$ nanoparticles in an effective manner to form the C_{60}/AuNPs/TiO$_2$ nanocomposites. These nanocomposites are employed in the photocatalytic degradation of MO and catalytic breakdown of 4-nitrophenol (4-NP) in the presence of NaBH$_4$. C_{60}/AuNPs/TiO$_2$ shows twice the photocatalytic activity and 132 times the catalytic reduction than the pure TiO$_2$. The stability is shown up to 10 cycles of photocatalytic degradation (Scheme **4**) [25].

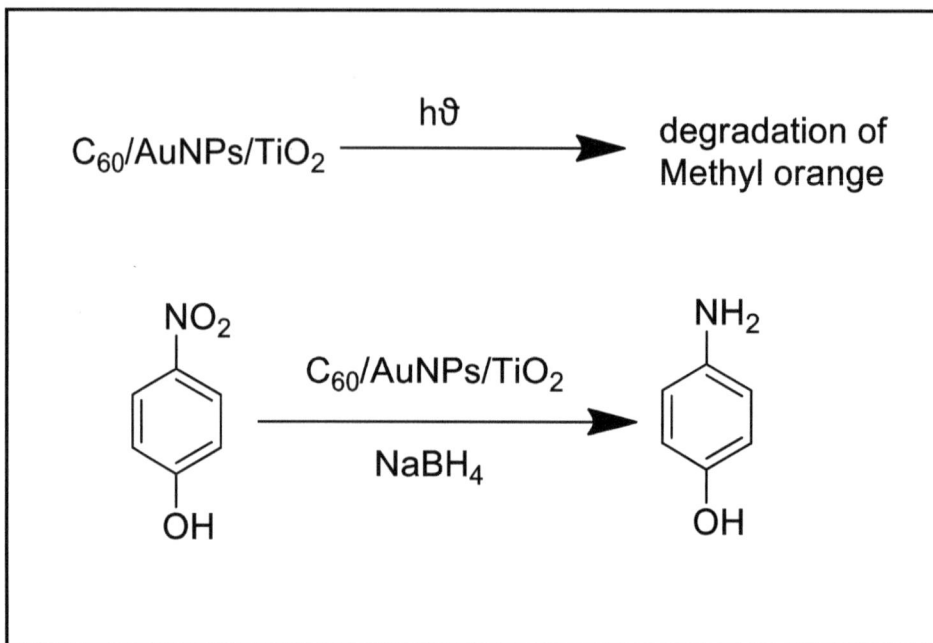

Scheme 4. C_{60}/AuNPs/TiO$_2$ nanocomposites for the degradation of MO.

Underlight irradiation at 254nm, WO$_3$/graphene and WO$_3$/C$_{60}$ are used as photocatalysts to degrade organic dyes such as methylene blue, brilliant green, and rhodamine G. MB> RhB >BG was the order in which the organic dyes were most effectively degraded by WO$_3$/graphene and WO$_3$/C$_{60}$ fullerene nanocomposites when used as a photocatalyst. To degrade methylene blue solution, WO$_3$/C$_{60}$ nanocomposites showed more photocatalytic activity than WO$_3$/graphene nanocomposites. WO$_3$/graphene nanocomposites, on the other hand, degraded RhB and BG more effectively than WO$_3$/C$_{60}$ nanocomposites (Scheme **5**) [26].

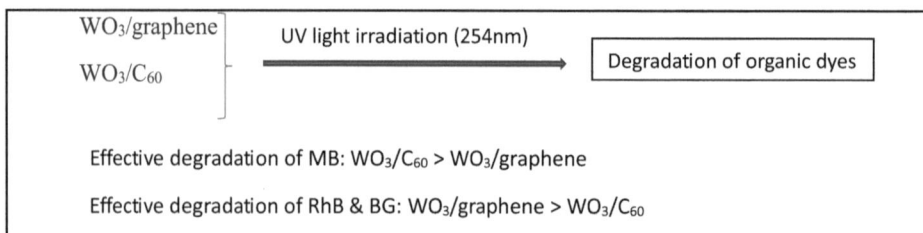

Scheme 5. Graphene and fullerene nanocomposites for dye degradation.

Zichao *et al.* [24] prepared C_{60}/CdS/TiO2 nanocomposites, and under light irradiation, they enhanced the production of H_2. The presence of fullerene over Cds/TiO_2 increased the photogenerated e^- transfer to the C_{60} clusters for the production of H_2 as well as for improved photostability, photocatalytic activity, light absorbing capacity, and suppression of photocorrosion of CdS (Scheme **6**) [27].

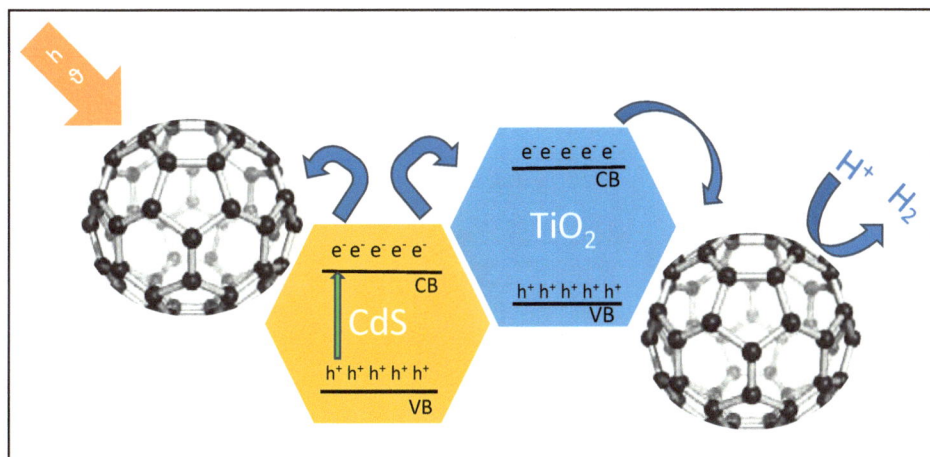

Scheme 6. C_{60}/CdS/TiO2 nanocomposites for dye degradation.

Kezhan *et al.* proposed that the C_{60}@a-TiO_2 photocatalyst can effectively show photocatalytic degradation of MB under UV-A light irradiation. The light absorbing ability, charge separation, and photocatalytic activity of a-TiO_2 have been improved with the combination of fullerene than a-TiO_2 alone (Scheme **7**) [28].

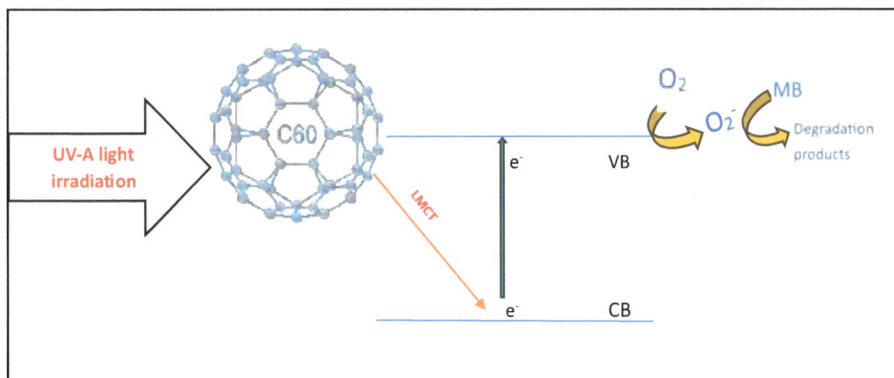

Scheme 7. C_{60}@a-TiO_2 photocatalyst for MB degradation.

Shengbao *et al.* prepared C_{60}–Bi_2WO_6 composites that act as photocatalysts for the degradation of MB and RB under both simulated solar and visible light irradiation. There is an increase in photocatalytic activity with C_{60}/Bi_2WO_6, 5.0 and 1.5 times for MB and RhB under light irradiation (420nm) and 4.6 and 2.1 times for MB and RhB under xenon lamp irradiation (290nm) than with Bi_2WO_6. Due to the presence of a conjugated π-system of C_{60}, enhanced electron migration is observed (Table **1**) [29].

Table 1. C_{60}–Bi_2WO_6 for the degradation of MB and RB.

	Light Source	**Organic Dyes**	**Increase in the % of Photocatalytic Degradation**
$C_{60}/BiWO_6$ nanocomposites	Visible light (420nm)	Methylene blue	5.0
		Rhodamine B	1.5
	Xenon lamp (290nm)	Methylene blue	4.6
		Rhodamine B	2.1

Shahnaz *et al.* synthesized a magnetic $BiVO_4/Fe_3O_4/C_{60}$ nanocomposite under light irradiation (>420nm). Methylene blue undergoes photocatalytic degradation (84%) and shows significant catalytic reduction activity (97%) of nitrophenols (2-NP,4-NP) and nitroanilines (2-NA,4NA) in the presence of $NaBH_4$. In comparison to $BiVO_4$, the $BiVO_4/Fe_3O_4/C_{60}$ composite showed better photocatalytic and catalytic activity and stability (Scheme **8**) [30].

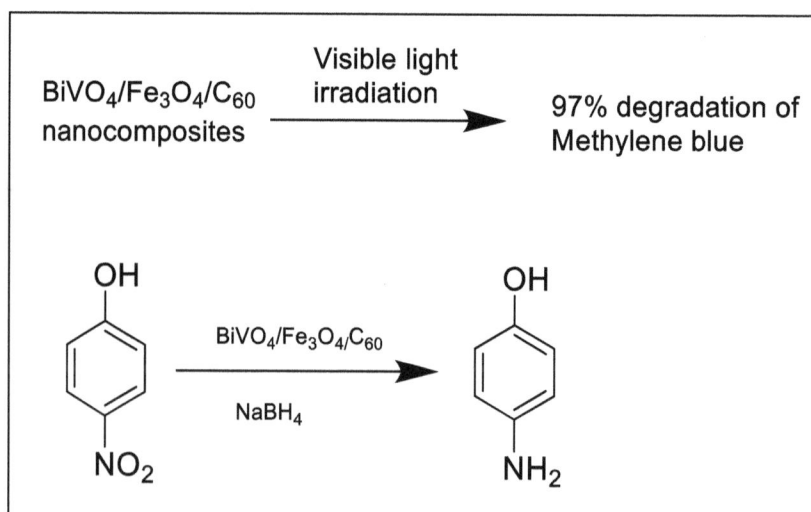

Scheme 8. $BiVO_4/Fe_3O_4/C_{60}$ nanocomposite for methylene blue degradation.

Qiang *et al.* [28] fabricated a C_{60}/CdS composite used for the degradation of RhB and photocatalytic evolution of H_2. Under light irradiation ($\geq$420nm), the photocatalytic experiment was carried out. A suitable layer of 0.4wt% C_{60} covered on CdS works as an acceptor for photogenerated electrons, eases the separation of e^-/h^+ pairs, and increases the photocatalytic activity. The outcomes demonstrated that C_{60}/CdS has higher stability than CdS and does not undergo photocorrosion of CdS, as that happens in pure CdS (Scheme **9**) [31].

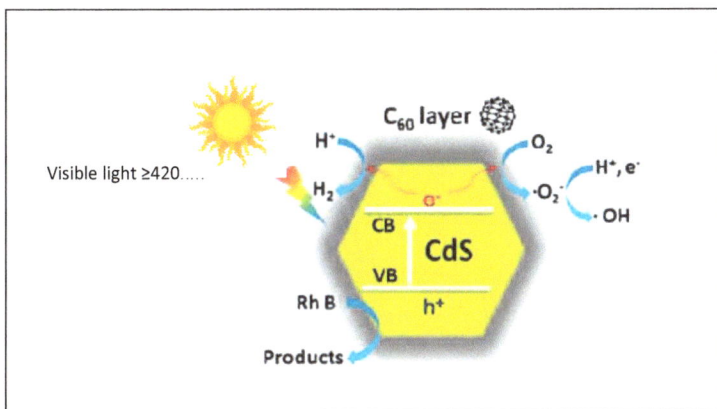

Scheme 9. C_{60}/CdS composite used for the degradation of RhB.

Under UV irradiation (254 nm), CeO_2-C_{60} nanocomposites generate the electron-hole pairs. It consequently causes the development of a significant amount of superoxide anion radicals (O_2^-) and hydroxy radicals (OH; these radicals act as oxidizing agents used for the deterioration of organic colors. Kinetic studies show the efficiency of photocatalytic activity: MB> RB > BG> MO (Scheme **10**) [32].

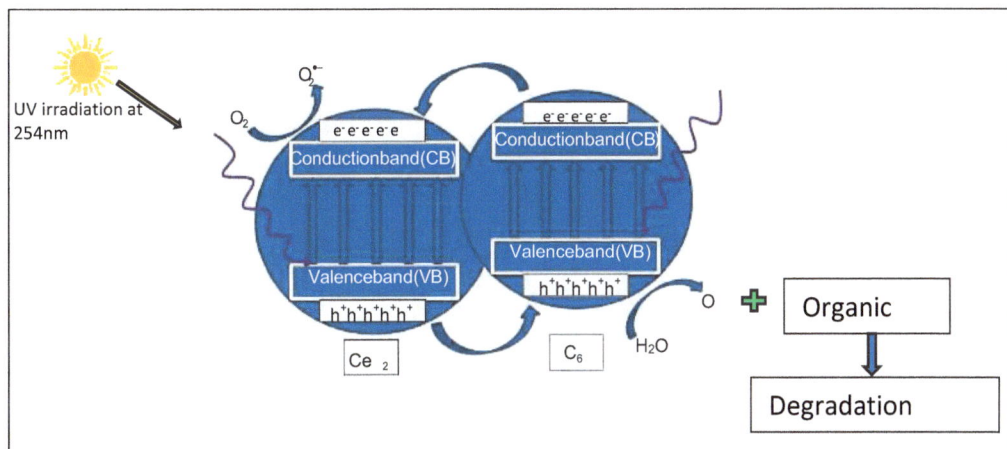

Scheme 10. CeO_2-C_{60} nanocomposites for dye degradation.

M-fullerene/TiO$_2$ composites (M= Pd, Pt, Y) were prepared by the Sol-gel method under light irradiation and MB underwent photodegradation. The combination of the metal added as a dopant and the enhanced photo-absorption action of fullerene resulted in an increase in photocatalytic activity. Among the three different M-fullerene/TiO$_2$ composites to degrade methylene blue, Pd-fullerene/TiO$_2$ shows a high adsorption capacity because of high specific surface area (82.5 m^2/g), high dispersion, and frittage at metal particles at elevated temperatures. Pt-fullerene/TiO$_2$ shows the least adsorption due to less specific surface area (47m^2/g) (Scheme **11**) [33].

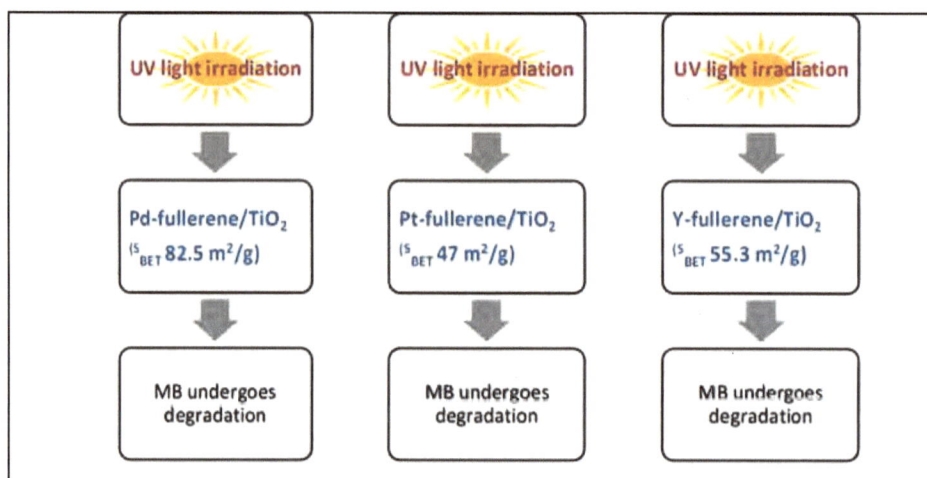

Scheme 11. M-fullerene/TiO$_2$ composites for dye degradation.

M.T. Gabdullin *et al.* utilized fullerenes as sorbents for adsorbing heavy metal ions. Upon testing various shungite and graphite-based sorbents, they concluded that fullerene-containing sorbents efficiently adsorbed heavy metal ions. Their findings have set up new scopes for advancing research on carbon-based nanomaterials for their sorption properties, impacting nanotechnology, ecology, ecotoxicology, and chemical technology (Scheme **12**) [34].

Boayan *et al.* prepared SnO/ND composites for MO degradation under light irradiation. The degradation findings show that the ability of SnO/ND is higher than that of pure SnO. SnO/9.2% ND shows 98% absorption of light, which is 17.9 times greater than with pure SnO (Scheme **13**) [35].

TiO$_2$@ND composite was used as a photocatalyst for the degradation of BPA under light irradiation. The morphology of TiO$_2$@ND is polycrystalline and shows an effective degradation of 92% when the composition is TiO$_2$-3@ND for

100 min of light irradiation. The BPA degradation capacity also depends on the P^H of the electrolyte and is shown better in neutral and acidic media (Scheme **14**) [36].

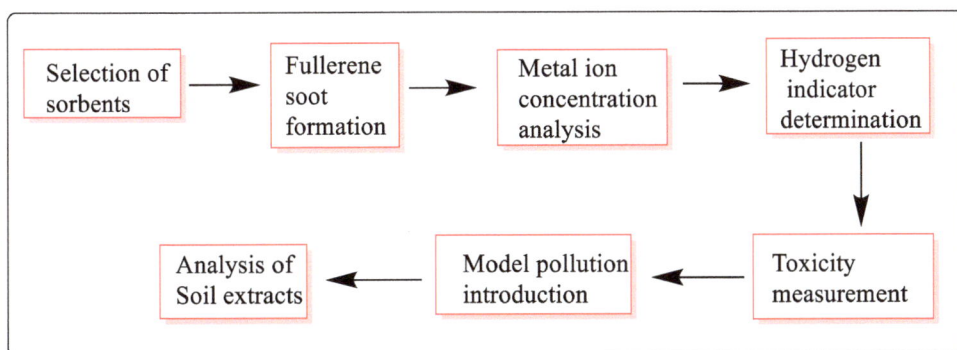

Scheme 12. Research approach of Gabdullin *et al*. Nanodiamond-based nanophotocatalyst.

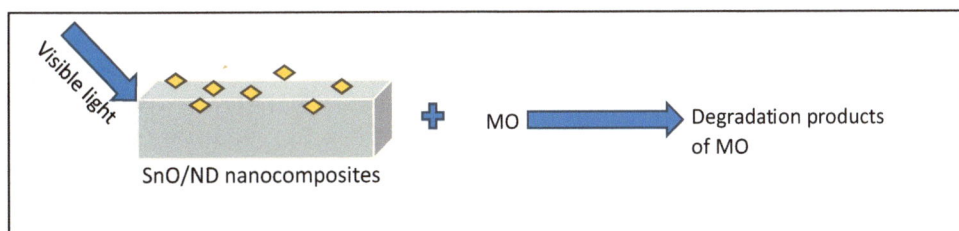

Scheme 13. SnO ND for methyl orange degradation.

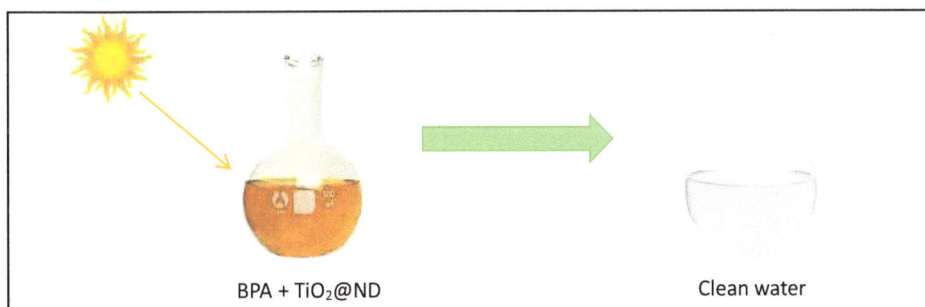

Scheme 14. TiO$_2$-ND for the degradation of BPA.

QUANTUM DOTS-BASED NANOPHOTOCATALYSTS

Prem *et al*. created a combination of CQDs and TiO$_2$ nanofibers for the degradation of MB under light irradiation (natural environment). CQDs/TiO$_2$ composite shows antibacterial and catalytic properties (Scheme **15**) [37].

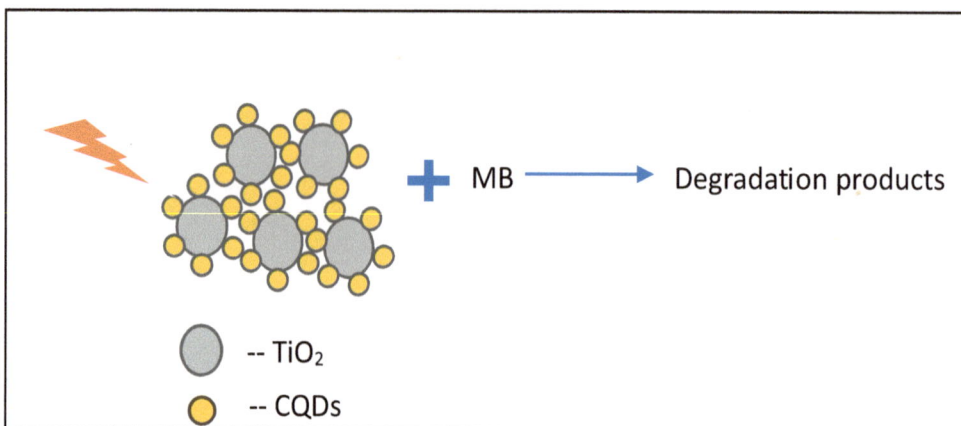

Scheme 15. CQDs with TiO$_2$ nanofibres for MB degradation.

Natercia *et al.*, synthesized N–TiO$_2$(p25)/CQD photocatalysts. N-CQDs act as electron acceptors and accept electrons from CB of TiO2(p25) and separate the charge carriers for a long time, improving the degradation rate of MB under UV and visible spectroscopy. With N–TiO$_2$(p25)/CQDs, 91% degradation of methylene blue occurs, whereas with TiO$_2$(p25), 68% degradation of methylene blue is observed within 1h of irradiation. The rise in the degradation rate is due to the hydrothermal technique used for the preparation of N–TiO$_2$(p25)/CQDs. (Table **2**) [38].

Table 2. Reaction conditions for methylene blue degradation.

Photocatalyst	Light Irradiation Time	Degradation Rate
TiO$_2$(p25)	1 hour	68%
N–TiO$_2$(p25)/CQDs		91%

Ruidi *et al.* prepared C–TiO2/CQD photocatalysts for RB dye degradation under visible light (420nm). Due to the metal-free and carbon-doped TiO$_2$ supported by CQDs, the absorption of light and degradation rate increases. (Scheme **16**) [39].

Shelja *et al.* prepared N-ZnO-CQD (Nitrogen-doped ZnO supported by C-dots) nanoflowers used for malachite green degradation under visible light. When the light irradiation exposure increases to 160 min, 85% degradation for malachite green (MG) is observed, whereas 75% degradation is observed for N-ZnO and 65% degradation for pure ZnO (Scheme **17**) [40].

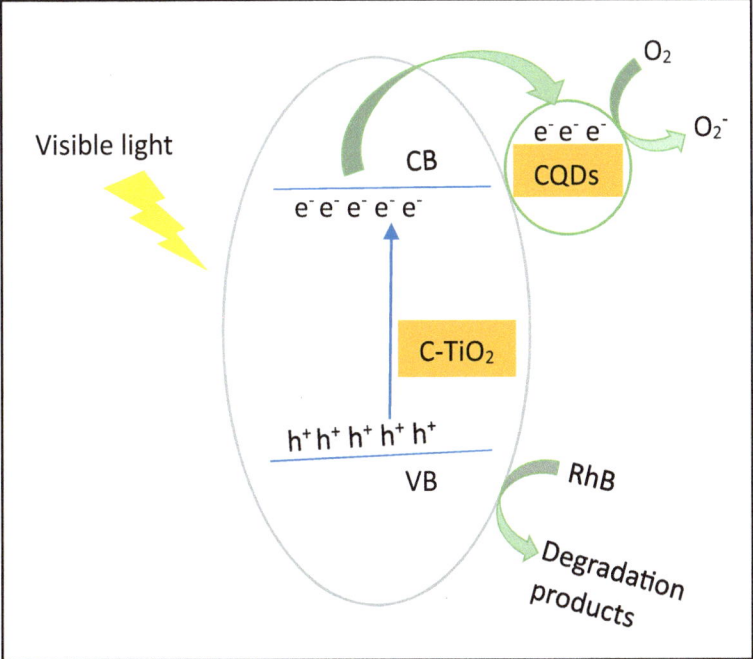

Scheme 16. TiO_2-based quantum dots for RB degradation.

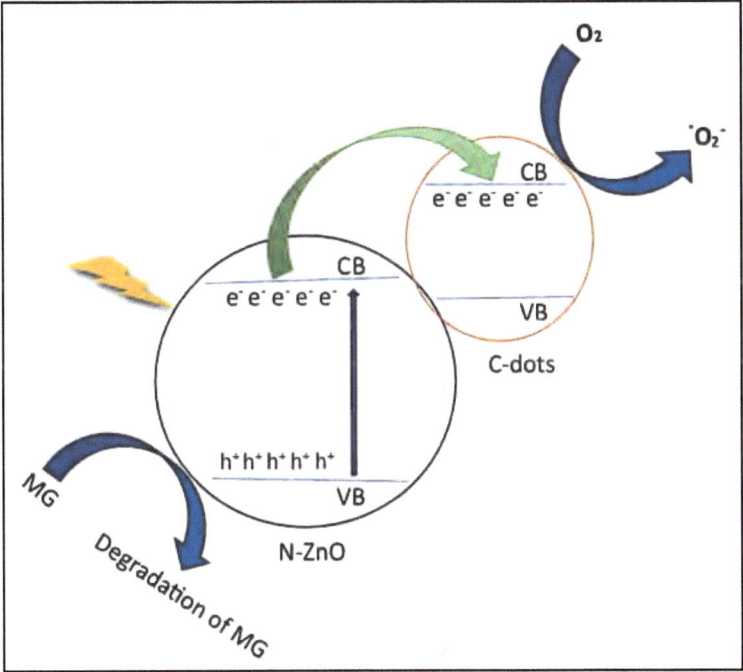

Scheme 17. N-ZnO-CQDs for the degradation of MG.

Hollow ZnO doped with carbon dots forms into ZnO/CQDs photocatalyst. Compared with pure ZnO, ZnO/C-dots can be the best photocatalyst. Under UV-visible light irradiation for 30 min, methylene blue underwent 96% degradation by using ZnO/C-dots, but with pure ZnO, only 63% degradation was observed. There are added advantages of using ZnO/C-dots, such as being inexpensive, stability up to five cycles, low recombination rate of charged carriers, and high absorbance with increased time of light irradiation (Scheme **18**) [41].

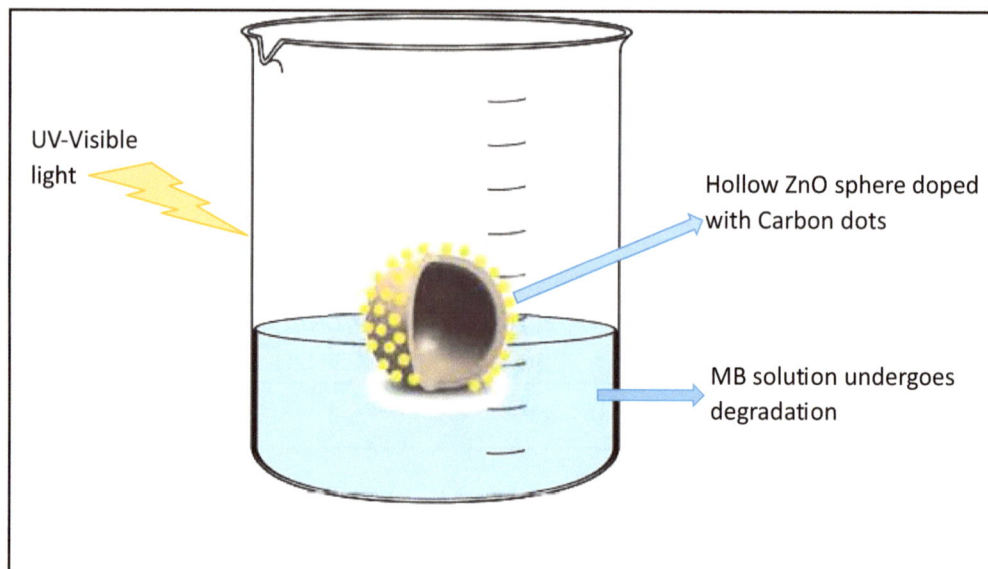

Scheme 18. ZnO-based CDSs for the degradation of MB.

CONCLUSION

In summary, nanotechnology with carbon-based nanomaterials is broadly employed in providing uncontaminated and affordable water due to the characteristics of positive correlation with specific structural character, alternating dimension factor, mechanical strength, and electrical conductivity. So, on the basis of the above literature precedents, carbon-based nanomaterials are experienced in the relevance of wastewater treatment. This chapter chiefly pays attention to the employment of carbon-based nanomaterials, as well as carbon quantum dots, fullerenes, and nanodiamonds, for the degradation of organic dye and eradication of heavy metal from polluted wastewater. This chapter can confirm that it might be tremendously encouraging for industrial/academic chemists or research scholars in the development of novel technologies/procedures in favor of the massive synthesis of new-fangled carbon-based nanomaterials for the eradication and degradation of heavy metals and organic dye impurities from contaminated wastewater.

LIST OF ABBREVIATIONS

AOPs   Advanced Oxidation Process

BPA   Bisphenol A

CQDs   Carbon Quantum Dots

DMPD   N,N-Dimethyl-p-phenylenediamine

MB   Methylene Blue

MG   Malachite Green

MO   Methyl Orange

ND   Nanodiamond

PAHs   Polycyclic Aromatic Hydrocarbons

RB   Rhodamine B

SA   Sulfanilic Acid

ACKNOWLEDGEMENTS

The author, Aditi, is thankful to Intertek India, Gurugram, Haryana, India. The author, Sirija, expresses her deep gratitude to the management of VIE, Hyderabad, India, and the author, G. B. Dharma Rao, conveys his thankfulness to the management of KPRIT (Autonomous), Hyderabad, India, for providing the amenities to write and submit this chapter for publication.

REFERENCES

[1]   Solayman, H.M.; Hossen, M.A.; Abd Aziz, A.; Yahya, N.Y.; Leong, K.H.; Sim, L.C.; Monir, M.U.; Zoh, K-D. Performance evaluation of dye wastewater treatment technologies: A review. *J. Environ. Chem. Eng.,* **2023,** *11*(3), 109610.
[http://dx.doi.org/10.1016/j.jece.2023.109610]

[2]   Babu Ponnusami, A.; Sinha, S.; Ashokan, H.; V Paul, M.; Hariharan, S.P.; Arun, J.; Gopinath, K.P.; Hoang Le, Q.; Pugazhendhi, A. Advanced oxidation process (AOP) combined biological process for wastewater treatment: A review on advancements, feasibility and practicability of combined techniques. *Environ. Res.,* **2023,** *237*(Pt 1), 116944.
[http://dx.doi.org/10.1016/j.envres.2023.116944] [PMID: 37611785]

[3]   Adeoye, J.B.; Tan, Y.H.; Lau, S.Y.; Tan, Y.Y.; Chiong, T.; Mubarak, N.M.; Khalid, M. Advanced oxidation and biological integrated processes for pharmaceutical wastewater treatment: A review. *J. Environ. Manage.,* **2024,** *353*, 120170.
[http://dx.doi.org/10.1016/j.jenvman.2024.120170] [PMID: 38308991]

[4]   Zhang, H.; Xue, K.; Wang, B.; Ren, W.; Sun, D.; Shao, C.; Sun, R. Advances in lignin-based biosorbents for sustainable wastewater treatment. *Bioresour. Technol.,* **2024,** *395*, 130347.
[http://dx.doi.org/10.1016/j.biortech.2024.130347] [PMID: 38242243]

[5]   Mishra, S.; Sundaram, B. A review of the photocatalysis process used for wastewater treatment. *Mater. Today Proc.,* **2023.**

[6]   Widhiastuti, F.; Rajendram, W.; Pramanik, B.K. Understanding the risk of using herbicides for tree root removal into wastewater treatment plant performance. *Chemosphere,* **2023,** *337*, 139345.
[http://dx.doi.org/10.1016/j.chemosphere.2023.139345] [PMID: 37379978]

[7]   Daulay, A.; Nasution, L.H.; Astuti, W.; Mufakhir, F.R.; Sumardi, S.; Prasetia, H. Studies for extraction and separation of rare earth elements by adsorption from wastewater: A review. *Min. Metall. Explor.,* **2024**, *41*(3), 1401-1419.
[http://dx.doi.org/10.1007/s42461-024-00974-8]

[8]   Bacha, A-U-R.; Nabi, I.; Chen, Y.; Li, Z.; Iqbal, A.; Liu, W.; Afridi, M.N.; Arifeen, A.; Jin, W.; Yang, L. Environmental application of perovskite material for organic pollutant-enriched wastewater treatment. *Coord. Chem. Rev.,* **2023**, *495*, 215378.
[http://dx.doi.org/10.1016/j.ccr.2023.215378]

[9]   Zhou, Y.; Liu, Q.; Lu, J.; He, J.; Liu, Y.; Zhou, Y. Accelerated photoelectron transmission by carboxymethyl β-cyclodextrin for organic contaminants removal: An alternative to noble metal catalyst. *J. Hazard. Mater.,* **2020**, *393*, 122414.
[http://dx.doi.org/10.1016/j.jhazmat.2020.122414] [PMID: 32143160]

[10]   Mogashane, Tumelo M.; Maree, Johannes P.; Mokoena, Lebohang Adsorption of polycyclic aromatic hydrocarbons from wastewater using iron oxide nanomaterials recovered from acid mine water: A review. *Minerals 14.8,* **2024**, 826.
[http://dx.doi.org/10.3390/min14080826]

[11]   Zhou, Y.; Hu, Y.; Huang, W.; Cheng, G.; Cui, C.; Lu, J. A novel amphoteric β-cyclodextrin-based adsorbent for simultaneous removal of cationic/anionic dyes and bisphenol A. *Chem. Eng. J.,* **2018**, *341*, 47-57.
[http://dx.doi.org/10.1016/j.cej.2018.01.155]

[12]   Yang, Q.; Wang, B.; Chen, Y.; Xie, Y.; Li, J. An anionic In(III)-based metal-organic framework with Lewis basic sites for the selective adsorption and separation of organic cationic dyes. *Chin. Chem. Lett.,* **2019**, *30*(1), 234-238.
[http://dx.doi.org/10.1016/j.cclet.2018.03.023]

[13]   Samn V.; Forster and Robert, M.C. The significance of the introduction of synthetic dyes in the mid-19th century on the democratisation of western fashion. *Journal of the International Colour Association,* **2013**, *11*, 1-17.

[14]   Ngo, A.C.R.; Tischler, D. Microbial degradation of azo dyes: approaches and prospects for a hazard-free conversion by microorganisms. *Int. J. Environ. Res. Public Health,* **2022**, *19*(8), 4740.
[http://dx.doi.org/10.3390/ijerph19084740] [PMID: 35457607]

[15]   Ge, J.; Zhang, Y.; Park, S.J. Recent advances in carbonaceous photocatalysts with enhanced photocatalytic performances: a mini review. *Materials (Basel),* **2019**, *12*(12), 1916.
[http://dx.doi.org/10.3390/ma12121916] [PMID: 31200594]

[16]   Leary, R.; Westwood, A. Carbonaceous nanomaterials for the enhancement of TiO2 photocatalysis. *Carbon,* **2011**, *49*(3), 741-772.
[http://dx.doi.org/10.1016/j.carbon.2010.10.010]

[17]   Khalid, N.R.; Ahmed, E.; Niaz, N.A.; Nabi, G.; Ahmad, M.; Tahir, M.B.; Rafique, M.; Rizwan, M.; Khan, Y. Highly visible light responsive metal loaded N/TiO 2 nanoparticles for photocatalytic conversion of CO 2 into methane. *Ceram. Int.,* **2017**, *43*(9), 6771-6777.
[http://dx.doi.org/10.1016/j.ceramint.2017.02.093]

[18]   Djellabi, R.; Yang, B.; Wang, Y.; Cui, X.; Zhao, X. Carbonaceous biomass-titania composites with Ti O C bonding bridge for efficient photocatalytic reduction of Cr(VI) under narrow visible light. *Chem. Eng. J.,* **2019**, *366*, 172-180.
[http://dx.doi.org/10.1016/j.cej.2019.02.035]

[19]   Bazli, L.; Siavashi, M.; Shiravi, A. A review of carbon nanotube/TiO2 composite prepared *via* sol-gel method. *Journal of Composites and Compounds,* **2019**, *1*(1), 1-12.
[http://dx.doi.org/10.29252/jcc.1.1.1]

[20]   Ramesh Reddy, N.; Bhargav, U.; Mamatha Kumari, M.; Cheralathan, K.K.; Sakar, M. Review on the

interface engineering in the carbonaceous titania for the improved photocatalytic hydrogen production. *Int. J. Hydrogen Energy,* **2020**, *45*(13), 7584-7615.
[http://dx.doi.org/10.1016/j.ijhydene.2019.09.041]

[21]   Mushtaq, M.; Tan, I.M.; Nadeem, M.; Devi, C.; Lee, S.Y.C.; Sagir, M. A convenient route for the alkoxylation of biodiesel and its influence on cold flow properties. *Int. J. Green Energy,* **2014**, *11*(3), 267-279.
[http://dx.doi.org/10.1080/15435075.2013.772519]

[22]   Wakimoto, R.; Kitamura, T.; Ito, F.; Usami, H.; Moriwaki, H. Decomposition of methyl orange using C60 fullerene adsorbed on silica gel as a photocatalyst *via* visible-light induced electron transfer. *Appl. Catal. B,* **2015**, *166-167*, 544-550.
[http://dx.doi.org/10.1016/j.apcatb.2014.12.010]

[23]   Kim, K.H.; Ko, W.B. Preparation of C60 Nanowhiskers/NiS2 Nanocomposites and Photocatalytic Degradation of Organic Dyes. *Asian J. Chem.,* **2015**, *27*(5), 1811-1814.
[http://dx.doi.org/10.14233/ajchem.2015.17932]

[24]   Hong, S.K.; Lee, J.H.; Ko, W.B. Synthesis of [60]fullerene-ZnO nanocomposite under electric furnace and photocatalytic degradation of organic dyes. *J. Nanosci. Nanotechnol.,* **2011**, *11*(7), 6049-6056.
[http://dx.doi.org/10.1166/jnn.2011.4374] [PMID: 22121656]

[25]   Islam, M.T.; Jing, H.; Yang, T.; Zubia, E.; Goos, A.G.; Bernal, R.A.; Botez, C.E.; Narayan, M.; Chan, C.K.; Noveron, J.C. Fullerene stabilized gold nanoparticles supported on titanium dioxide for enhanced photocatalytic degradation of methyl orange and catalytic reduction of 4-nitrophenol. *J. Environ. Chem. Eng.,* **2018**, *6*(4), 3827-3836.
[http://dx.doi.org/10.1016/j.jece.2018.05.032]

[26]   Kim, K.H.; Ko, J.W.; Ko, W.B. Preparation and kinetics of nanocomposites using wo3 with carbon nanomaterials for photocatalytic degradation of organic dyes. *Asian J. Chem.,* **2016**, *28*(1), 194-198.
[http://dx.doi.org/10.14233/ajchem.2016.19338]

[27]   Lian, Z.; Xu, P.; Wang, W.; Zhang, D.; Xiao, S.; Li, X.; Li, G. C60-decorated CdS/TiO2 mesoporous architectures with enhanced photostability and photocatalytic activity for H2 evolution. *ACS Appl. Mater. Interfaces,* **2015**, *7*(8), 4533-4540.
[http://dx.doi.org/10.1021/am5088665] [PMID: 25658952]

[28]   Qi, K.; Selvaraj, R.; Al Fahdi, T.; Al-Kindy, S.; Kim, Y.; Wang, G-C.; Tai, C-W.; Sillanpää, M. Enhanced photocatalytic activity of anatase-TiO2 nanoparticles by fullerene modification: A theoretical and experimental study. *Appl. Surf. Sci.,* **2016**, *387*, 750-758.
[http://dx.doi.org/10.1016/j.apsusc.2016.06.134]

[29]   Zhu, S.; Xu, T.; Fu, H.; Zhao, J.; Zhu, Y. Synergetic effect of Bi2WO6 photocatalyst with C60 and enhanced photoactivity under visible irradiation. *Environ. Sci. Technol.,* **2007**, *41*(17), 6234-6239.
[http://dx.doi.org/10.1021/es070953y] [PMID: 17937308]

[30]   Sepahvand, S.; Farhadi, S. Preparation and characterization of fullerene (C$_{60}$)-modified BiVO$_4$/Fe$_3$O$_4$ nanocomposite by hydrothermal method and study of its visible light photocatalytic and catalytic activity. *Fuller. Nanotub. Carbon Nanostruct.,* **2018**, *26*(7), 417-432.
[http://dx.doi.org/10.1080/1536383X.2018.1446423]

[31]   Cai, Q.; Hu, Z.; Zhang, Q.; Li, B.; Shen, Z. Fullerene (C60)/CdS nanocomposite with enhanced photocatalytic activity and stability. *Appl. Surf. Sci.,* **2017**, *403*, 151-158.
[http://dx.doi.org/10.1016/j.apsusc.2017.01.135]

[32]   Li, J.; Ko, J.W.; Ko, W.B. Preparation and characterization of CeO2-C60 nanocomposites and their application to photocatalytic degradation of organic dyes. *Asian J. Chem.,* **2016**, *28*(9), 2020-2024.
[http://dx.doi.org/10.14233/ajchem.2016.19880]

[33]   Taneja, Y.; Dube, D.; Singh, R. Recent advances in elemental doping and simulation techniques: improving structural, photophysical and electronic properties of titanium oxide. *J. Mater. Chem. C.,* **2024**, 37, 14774-14808.

[http://dx.doi.org/10.1039/D4TC02031F]

[34]   Gabdullin, M.T.; Khamitova, K.K.; Ismailov, D.V.; Sultangazina, M.N.; Kerimbekov, D.S.; Yegemova, S.S.; Schur, D.V. Nanostructured materials for the sorption or heavy metal ions. *IOP Conf. Ser. Mater. Sci. Eng.,* **2019**.
[http://dx.doi.org/10.1088/1757-899X/511/1/012044]

[35]   Liang, B.; Zhang, W.; Zhang, Y.; Zhang, R.; Zhang, L. Nanodiamond incorporated in SnO composites with enhanced visible-light photocatalytic activity. *Diamond Related Materials,* **2018**, *89*, 108-113.
[http://dx.doi.org/10.1016/j.diamond.2018.08.014]

[36]   Hunge, Y.M.; Yadav, A.A. Photocatalytic degradation of bisphenol A using titanium dioxide@nanodiamond composites under UV light illumination *Journal of Colloid and Interface Science,* **2021**, *582*(B), 1058-1066.
[http://dx.doi.org/10.1016/j.jcis.2020.08.102]

[37]   Saud, P.S.; Pant, B.; Alam, A-M.; Ghouri, Z.K.; Park, M.; Kim, H-Y. Carbon quantum dots anchored TiO2 nanofibers: Effective photocatalyst for waste water treatment. *Ceram. Int.,* **2015**, *41*(9), 11953-11959.
[http://dx.doi.org/10.1016/j.ceramint.2015.06.007]

[38]   Martins, N.C.T.; Ângelo, J.; Girão, A.V.; Trindade, T.; Andrade, L.; Mendes, A. N-doped carbon quantum dots/TiO2 composite with improved photocatalytic activity. *Appl. Catal. B,* **2016**, *193*, 67-74.
[http://dx.doi.org/10.1016/j.apcatb.2016.04.016]

[39]   Liu, R.; Li, H.; Duan, L.; Shen, H.; Zhang, Y.; Zhao, X. *In situ* synthesis and enhanced visible light photocatalytic activity of C-TiO 2 microspheres/carbon quantum dots. *Ceram. Int.,* **2017**, *43*(12), 8648-8654.
[http://dx.doi.org/10.1016/j.ceramint.2017.03.184]

[40]   Sharma, S.; Mehta, S.K.; Kansal, S.K. N doped ZnO/C-dots nanoflowers as visible light driven photocatalyst for the degradation of malachite green dye in aqueous phase. *J. Alloys Compd.,* **2017**, *699*, 323-333.
[http://dx.doi.org/10.1016/j.jallcom.2016.12.408]

[41]   Velumani, A.; Sengodan, P.; Arumugam, P.; Rajendran, R.; Santhanam, S.; Palanisamy, M. Carbon quantum dots supported ZnO sphere based photocatalyst for dye degradation application. *Curr. Appl. Phys.,* **2020**, *20*(10), 1176-1184.
[http://dx.doi.org/10.1016/j.cap.2020.07.016]

CHAPTER 7

Nanocomposites in Wastewater Treatment

G. B. Dharma Rao[1,*], Anirudh Singh Bhathiwal[2], Chinmay Mittal[2] and **Vishaka Chauhan[2]**

[1] *Department of Chemistry, Kommuri Pratap Reddy Institute of Technology, Hyderabad-500088, Telangana, India*

[2] *Department of Chemistry, Faculty of Science, SGT University, Gurugram- 122505, India*

Abstract: In recent decades, nanocomposites and their derivatives have been extensively examined to eradicate heavy metals and organic dyes from polluted water. Several methodologies have been introduced to remove heavy metals and dyes. Nanocomposites have been gaining attention, and many nanocomposites have been synthesized to eradicate heavy metals and organic dyes from wastewater because of their outstanding characteristics ensuing from the nanometer end product. In this chapter, metal-oxide-based nanocomposites and their relevance for the eradication of heavy metal ions and organic dyes from polluted water are described.

Keywords: Heavy metals, Nanocomposites, Organic dyes, Wastewater, Water pollution.

INTRODUCTION

The most important water contaminants implicated in various industrial activities are heavy metals and organic dyes, as they cause ecological imbalance and infections in humans and animals due to their accumulation in the organisms. Water effluents pose a massive difficulty in environmental supervision owing to the extensive collection of causes and complex compound composition of toxic waste sources, which are frequently classified as organic and inorganic wastewater [1]; however, in definite water pollution, it is repeatedly a combination of organic and inorganic complex composition, and heavy metals and organic dye components are the most widespread and prominent toxic waste in water pollution [2]. The extensive employment of metallic resources has led to a diversity of heavy metal contaminants like arsenic, lead, chromium, copper, and cadmium, which are a few of the main heavy metal pollutants regularly controlled in the water resources [3]. The pollution of worldly and aquatic environments

* **Corresponding author G.B. Dharmarao:** Department of Chemistry, Kommuri Pratap Reddy Institute of Technology, Hyderabad-500088, Telangana, India; E-mail: gbdharmarao@gmail.com

Anjaneyulu Bendi (Ed.)

with lethal heavy metals is the most important environmental apprehension that has consequences for civic health. The severe contamination of the environment, human existence, and health troubles caused by heavy metals and organic dyes cannot be disregarded [4]. Therefore, eradicating heavy metal ions and organic dyes from wastewater remains a trendy and attractive research topic. Numerous wastewater treatment procedures with the intention of removing heavy metals and organic dyes have been developed [5] and enhanced throughout time like ion exchange [6], precipitation [7], advanced oxidation process (AOP) [8], lignin-based biosorbents [9], electrolysis [10], photocatalysis [11], phytoremediation [12], and biosorption [13] due to their rapidity, effortlessness of procedure, elevated efficiency, and economical price. In recent years, nanocomposites have experienced a significant responsibility in various industries [14]. Due to their vital function, they are being employed in removing water pollutants from wastewater. In this chapter, we addressed the information regarding eradicating heavy metals and organic dyes from wastewater using nanocomposites.

REMOVAL OF HEAVY METALS FROM WASTEWATER

Umejuru E. C. *et al.* described the applications of nanocomposite material containing zeolite adsorbents for the remediation of wastewater [15]. This research group utilized various materials like carbon, polymers, and metal nanoparticles integrated with zeolite for the elimination of lethal heavy metals and organic dyes from wastewater. The adsorption effectiveness of Pb (II) and As (V) by zeolite/ZnO NCs prepared *via* the co-precipitation method was 93% and 89%, respectively, at pH 4. On the other hand, the removal ability of Pb(II) using zeolite Y/faujasite nanocomposite having a surface area of 19.17 m^2/g was 83.26 mg/g at pH 5.5, and zeolite Y/faujasite coated by cobalt ferrite nanoparticles showed greater adsorption ability of 602.4 mg/g for Pb(II) due to high surface area of 434.4 m^2/g. The adsorption ability of Cr(VI) and Cu (II) from wastewater using fly ash-based zeolite (FZA) incorporated by nano zerovalent iron (nZVI) and nickel (nZVI/Ni@FZA) with a surface area of 154.11 m^2/g was 48.31 and 147.06 mg/g, respectively. Titanium dioxide (TiO_2) incorporated zeolite-4A was used for the removal of Fe(III) and Mn(II) heavy metals, and the adsorption ability was found to be 150.1 and 94.1 mg/g, respectively. Zeolite/Polyvinyl alcohol/sodium alginate incorporated nanocomposites illustrate superior adsorption capacity of Ni(II), Pb (II), Zn, Cu(II), Mn, and Cd, which was 93.1%, 99.5%, 95.6%, 97.2%, 92.4%, and 99.2%, respectively. Moreover, the different zeolite-incorporated nanocomposites (Fig. **1**) have been confirmed to be powerful in the elimination of heavy metals from wastewater.

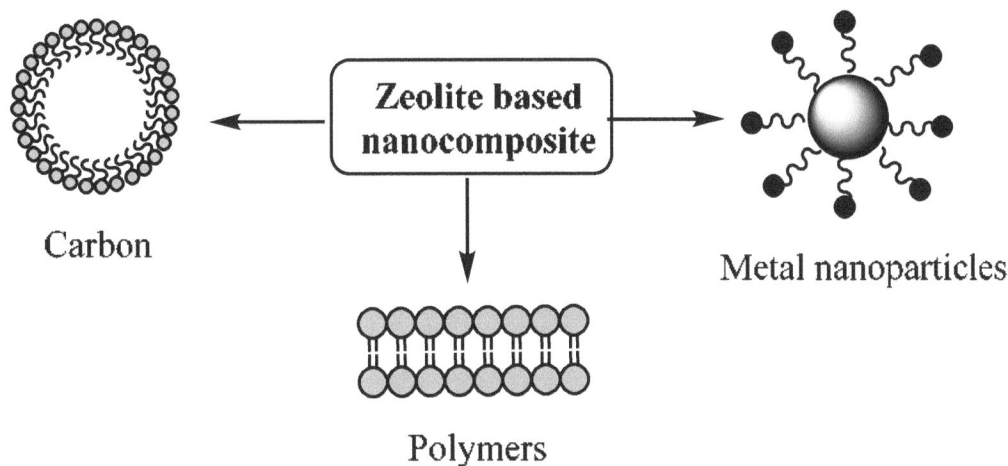

Fig. (1). Zeolite-based nanocomposite using different materials.

Adsorption of reactive orange 16(RO), Congo red (CR), and RO5 in wastewater is carried out by using a nanocomposite of magnetic zeolite/hydroxyapatite (MZeo-HAP). The adsorption of the organic dye system exposed that the nanocomposite material containing functional groups like aromatic, amide, and hydroxyl must contribute to the electrostatic interaction by means of the dyes owing to the positively charged functional groups on CR, RO5, and RO16 with a negatively charged molecule of MZeo-HAP. The sorption capacity of CR, RO5, and RO16 from wastewater is 104.05, 92.45, and 88.31 mg/g, respectively, at pH 2. According to a thermodynamic study, the adsorption capacity of dyes increases with increasing temperature.

Al-Salman *et al.* addressed the preparation of chitosan/graphene nanocomposites [16] for the adsorption of heavy metal ions from wastewater by a solution procedure using various weight percentages of 0.5%, 1.0%, 2.0%, and 5.0%. The preparation of nanographene involves three steps, as shown in (Scheme **1**).

1. *Graphene oxide preparation*
2. *Graphene acylation preparation*
3. *Nanographene preparation*

For the preparation of chitosan grain nanocomposite, first, chitosan solution was added to the functionalized nanophase, followed by magnetic agitation for three days. On the other hand, cross-linked grain nanocomposite was obtained by the addition of formaldehyde and polyethylene glycol, followed by the addition of sodium hydroxide with ethyl acetate to produce granular nanocomposite. The prepared functionalized graphene nanocomposites with an adsorbent phase (25 mg) were placed in 10 ml of cadmium ion solution at a pH of 7. The concentration of cadmium ion in every sample was measured equal to 20 ppm, 30 ppm, 50 ppm, and 100 ppm. For total adsorption of cadmium ion, the sample was kept in a vibrating stirrer at a speed of 200 rpm for 120 minutes. Furthermore, the adsorption capacity is enhanced by enhancing the concentration of cadmium ions. This is due to raising the cadmium ion concentration; the amine active sites, which are not capable of fine connection with cadmium ions, are exposed to other metal ions by escalating the cadmium ion concentration, and therefore, the adsorbent ability is notably advanced.

Step 1: Graphene oxide preparation

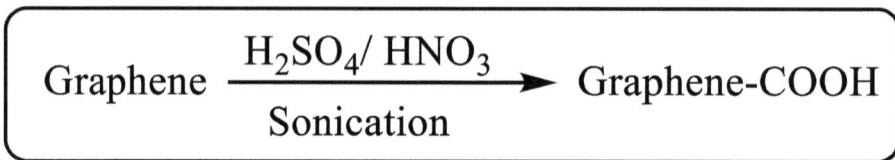

$$Graphene \xrightarrow[\text{Sonication}]{H_2SO_4/\ HNO_3} Graphene\text{-}COOH$$

Step 2: Graphene acylation preparation

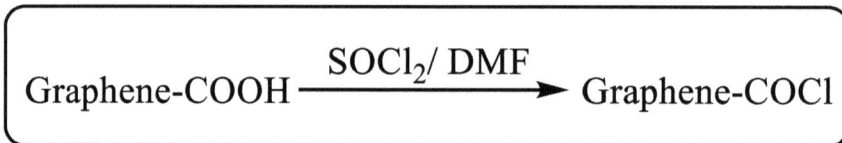

$$Graphene\text{-}COOH \xrightarrow{SOCl_2/\ DMF} Graphene\text{-}COCl$$

Step 3: Nanographene preparation

$$Graphene\text{-}COCl \xrightarrow{TETA} Graphene\text{-}C_6H_{18}N_4$$

Scheme 1. Preparation of chitosan/graphene nanocomposites.

S. Anuma *et al.* addressed the synthesis of polypyrrole functionalized cobalt oxide graphene (COPYGO) nanocomposite materials [17]. First, they prepared the Co_3O_4 nanoparticles by the hydrothermal process using cobalt sulfate heptahydrate, glycine, and hydrazine hydrate at pH 9, followed by the addition of pyrrole and FeCl3 to obtain cobalt oxide-polypyrrole (COPY). Finally, they prepared the cobalt oxide-polypyrrole/GO composite (COPYGO) using graphene oxide in ethanol and COPY, followed by autoclaving at 120°C for 3h, as described in (Scheme 2). The synthesized COPYGO was characterized by FTIR, XRD, FE-SEM, TEM, HR-TEM, Raman Spectroscopy, BET, XPS, and *vs*M. The specific surface area of COPYGO was $133m^2g^{-1}$, and it was up to 195°C thermally stable.

At pH 7.2, 6.1, 5.0, and 5.5, the efficiency of COPYGO nanocomposite in adsorbing Pb(II) and Cd(II) heavy metals from wastewater was 93.08% and 95.28%, respectively. On the other hand, the synthesized COPYGO was used for the removal of heavy metals (Pb^{2+} and Cd^{2+}) from wastewater, and it was observed that the adsorption capacity increased with an increase in pH.

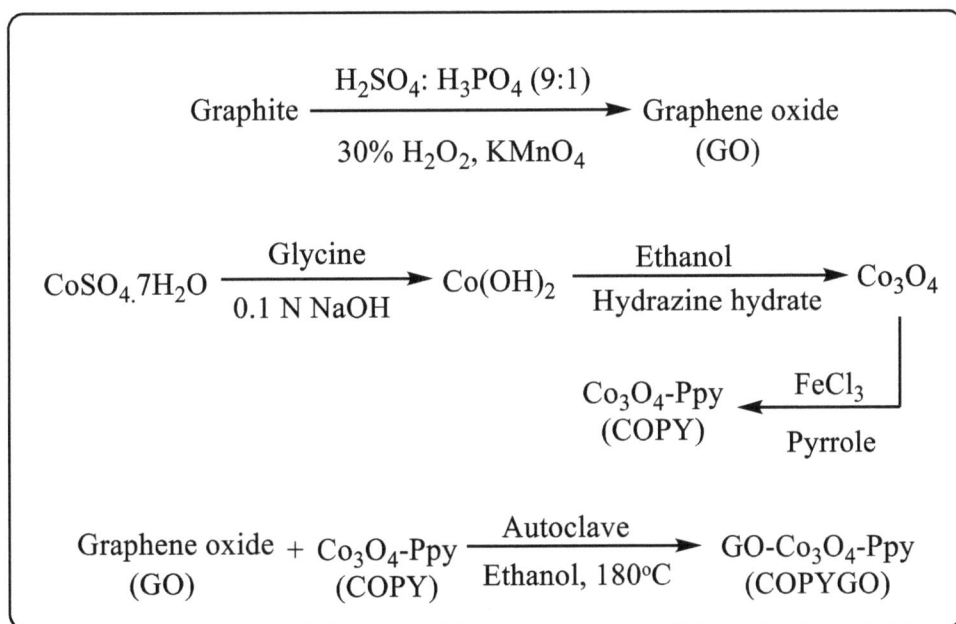

$$Graphite \xrightarrow[\text{30\% } H_2O_2, KMnO_4]{H_2SO_4: H_3PO_4 \text{ (9:1)}} \text{Graphene oxide (GO)}$$

$$CoSO_4.7H_2O \xrightarrow[\text{0.1 N NaOH}]{Glycine} Co(OH)_2 \xrightarrow[\text{Hydrazine hydrate}]{Ethanol} Co_3O_4$$

$$Co_3O_4\text{-Ppy (COPY)} \xleftarrow[\text{Pyrrole}]{FeCl_3} $$

$$\text{Graphene oxide (GO)} + Co_3O_4\text{-Ppy (COPY)} \xrightarrow[\text{Ethanol, 180°C}]{Autoclave} GO\text{-}Co_3O_4\text{-Ppy (COPYGO)}$$

Scheme 2. Synthesis of COPYGO nanocomposite material.

Michele Ferri *et al.* described the design, characterization, synthesis, and applications of environmentally benign hydroxyapatite/carbon (HAP/CMC) composite for the removal of toxic pollutants from wastewater [18]. The HAP/CMC was prepared by co-precipitation at room temperature. First, the mesoporous carbon (CMC) in $(NH_4)_2HPO_4$ solution under ultrasonication was used to obtain the slurry, followed by the addition of NH_4OH to maintain the pH at 9.7. Finally, $Ca(NO_3)_2.4H_2O$ solution is added dropwise through a peristaltic pump to get HAP, followed by stirring, washing, and filtration under vacuum to obtain HAP/CMC. By varying the concentration of CMC, the following four different composites were synthesized under the name HAP/CMC4, HAP/ CMC8, HAP/CMC12, and HAP/CMC16. The acidic and basic nature of the HAP/CMC surface was studied by using the pulsed-injection titration method, and it was noticed that the composite possessed a dual character. The synthesized HAP/CMC composites were characterized by XRPD, N2 adsorption, TEM, and HAADF-

STEM/EDX. The adsorption capacity of Cu(II) and Ni(II) on HAP/CMC was found to be 95.4% and 28.8% from wastewater. The sorption capacity of HAP/CMC8 was greater than the sorption capacity of HAP/CMC4, HAP/CMC12, and HAP/CMC16.

W. Zhang *et al.* reported that alginate-based porous nanocomposite hydrogels (GO@PAN-PPy/SA) are used for the sorption of heavy metals from water [19]. The graphite oxide solution was prepared by the addition of graphite oxide with 1 M HCl solution under ultrasonication, followed by the addition of pyridine (Py) and aniline (An) homogeneously. For the above reaction mixture, add ammonium persulfate (APS) to obtain GO@PAN-PPy-1, GO@PAN-PPy-2, and GO@PAN-PPy-3 by varying the concentration of Py to An as 1:1, 1:2, and 1:3, respectively. Finally, alginate-based porous nanocomposite was prepared by the addition of sodium alginate (SA) through ionic cross-linkage using $CaCO_3$ and $CaCl_2$, as shown in (Scheme 3).

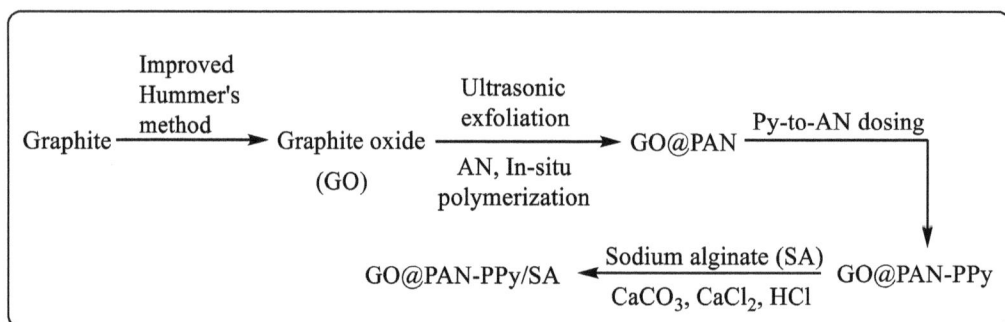

Graphite $\xrightarrow{\text{Improved Hummer's method}}$ Graphite oxide (GO) $\xrightarrow[\text{AN, In-situ polymerization}]{\text{Ultrasonic exfoliation}}$ GO@PAN $\xrightarrow{\text{Py-to-AN dosing}}$ GO@PAN-PPy $\xrightarrow[\text{CaCO}_3\text{, CaCl}_2\text{, HCl}]{\text{Sodium alginate (SA)}}$ GO@PAN-PPy/SA

Scheme 3. Synthesis of alginate-based porous nanocomposite hydrogels.

The prepared GO@PAN-PPy/SA hydrogel was used to remove the Cr(VI) and Cu(II) with adsorption capability of -133.7 mg/g and -87.2 mg/g, respectively. Among all the prepared alginate-based porous nanocomposite hydrogels, GO@PAN-PPy-2 exhibits more sorption capacity of Cr and Cu at pH 3.

D. A. El-Nagar *et al.* addressed the removal of heavy metal using nano-hydroxyapatite (n-HAP), nano-bentonite (n-Bo), and bentonite-hydroxyapatite nanocomposite (B-HAP NC) from water [20]. In the first step, they prepared the n-HAP using $CaCO_3$ with H_3PO_4, followed by the irradiation of ultrasound to obtain a white precipitate. In the second step, they prepared the n-Bo sheets by the reaction between PEG 6000 and bentonite, followed by ultrasound irradiation to obtain a white precipitate. In the final step, they described the synthesis of B-HAP NC by the reaction between n-HAP and n-Bo in a 1:1equivalent ratio under irradiation of ultrasound. The greatest adsorption ability of Pb^{+2} by n-HAP, n-Bo, and B-HAP NC was 5.83, 2.05, and 5.99 mg/g, respectively. On the other hand, the sorption capability of Ni^{+2} was 3.72, 1.26, and 1.01 mg/g, respectively. From the above observations, it was observed that B-HAP NC shows a great sorption capacity of Pb^{+2} than other materials, and Ni^{+2} was adsorped by n-HAP.

Heba Isawi addressed the synthesis of zeolite/polyvinyl alcohol/sodium alginate nanocomposite beads (Zeo/PVA/SA NCs) for the removal of heavy metals from wastewater [21]. First, the zeolite (Zeo) sample was converted into Zeo nanoparticles using a 24-blade FRITSCH Rotar Mortar (Model-Pulverizette114) for 2-3 h at 400°C. The Zeo NPs with various concentrations were reacted with poly vinyl alcohol (PVA) and sodium alginate (SA) in deionized water, with glutaraldehyde (GA) as a cross-linking agent. The resulting solid mass was left for solidification and washed with deionized water until it reached a pH of 7, as shown in (Scheme **4**). The synthesized Zeo/PVA/SA NC was employed for the removal of heavy metals present in wastewater, and the adsorption capacity of Li^{2+}, Cd^{2+}, Cu^{2+}, Ni^{2+}, Zn^{2+}, Mn^{2+}, Sr^{2+}, and Pb^{2+} was found to be 74.5%, 99.2, 97.2, 93.1, 95.6, 92.4, 98.8, and 99.5, respectively, at the pH of 6.0. On the other hand, the maximum elimination was attained for Al^{3+} and Fe^{3+} ions with 94.9% and 96.5%, respectively, at 25°C at a pH of 5.0.

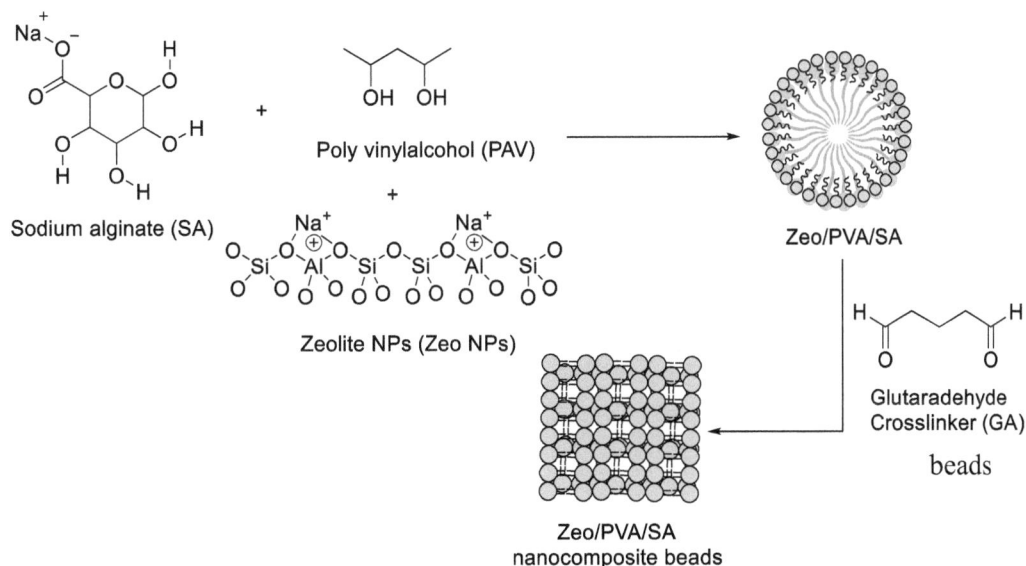

Scheme 4. Synthesis of Zeolite/Polyvinyl alcohol/sodium alginate nanocomposite beads.

REMOVAL OF ORGANIC DYES FROM WASTEWATER

Naushad, M. *et al.* synthesized nanocomposites for the degradation of dyes present in wastewater (Schemes **5-7**) [22].

Simple co-polymerization was used to create the synthesized composites. It was shown that 75% of dyes may be broken down with nanocomposites and hydrogen peroxide under ideal circumstances. The photodegradation process involves hydroxyl radicals as one of the primary active species, as shown by the results of the scavenging tests. Furthermore, after five reuse successions, the nanocomposites were shown to be effective photocatalysts. Creating a nanocomposite hydrogel might be a likely step toward creating a unique, effective photocatalyst for wastewater treatment.

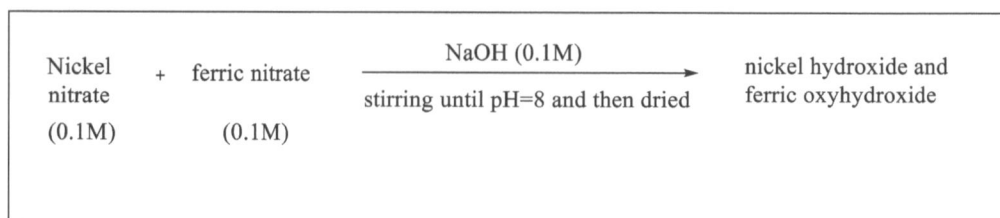

Nickel nitrate (0.1M)	+	ferric nitrate (0.1M)	$\xrightarrow[\text{stirring until pH=8 and then dried}]{\text{NaOH (0.1M)}}$	nickel hydroxide and ferric oxyhydroxide

Scheme 5. Synthesis of nickel hydroxide and ferric oxyhydroxide.

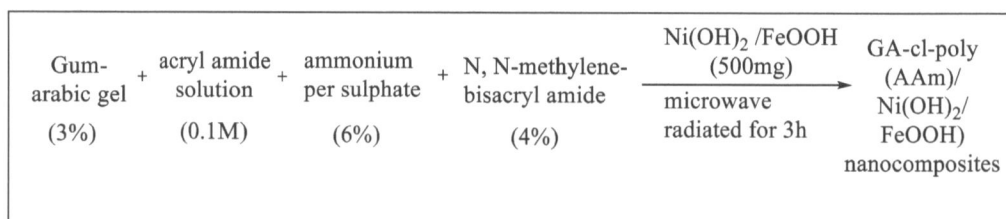

Gum-arabic gel (3%)	+	acryl amide solution (0.1M)	+	ammonium per sulphate (6%)	+	N, N-methylene-bisacryl amide (4%)	$\xrightarrow[\text{microwave radiated for 3h}]{\text{Ni(OH)}_2\text{ /FeOOH (500mg)}}$	GA-cl-poly (AAm)/ Ni(OH)$_2$/ FeOOH) nanocomposites

Scheme 6. Synthesis of GA-cl-poly(AAm)/Ni(OH)$_2$/FeOOH Nanocomposites

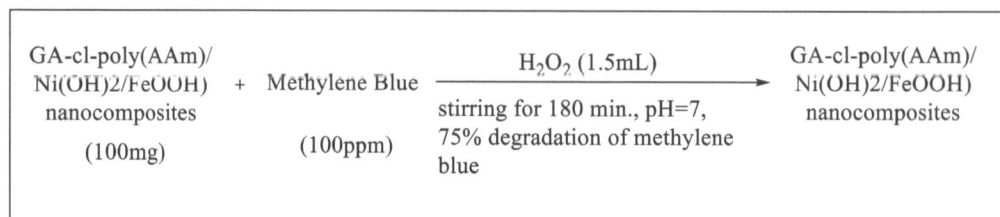

GA-cl-poly(AAm)/ Ni(OH)2/FeOOH) nanocomposites (100mg)	+	Methylene Blue (100ppm)	$\xrightarrow[\substack{\text{stirring for 180 min., pH=7,}\\\text{75\% degradation of methylene}\\\text{blue}}]{H_2O_2 \text{ (1.5mL)}}$	GA-cl-poly(AAm)/ Ni(OH)2/FeOOH) nanocomposites

Scheme 7. Degradation of dyes.

Aydoghmish, S. M. *et al.* synthesized nanocomposites for the degradation of dyes present in wastewater (Schemes **8-10**) [23].

A two-stage precipitation process was used to create the synthesized nanocomposites. The increase in light absorption caused by the inclusion of silver nanoparticles was ascribed to the metal's plasmon surface resonance. Degradation kinetics decreased as a result of the sites where electrons formed and electron holes were recombined and produced by silver nanoparticles. According to the degradation process, adding silver to nanocomposites up to 1 weight percent increased dye degradation to 94%, while adding more silver caused degradation to decrease.

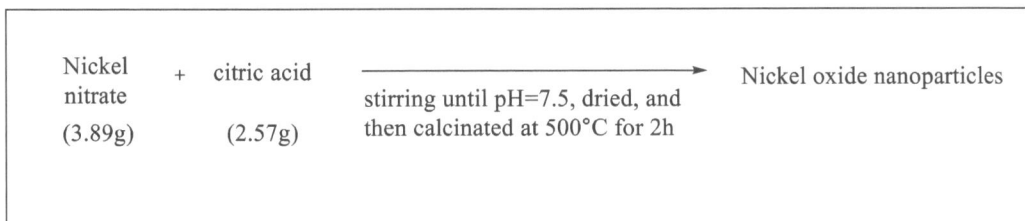

Nickel nitrate (3.89g) + citric acid (2.57g) → stirring until pH=7.5, dried, and then calcinated at 500°C for 2h → Nickel oxide nanoparticles

Scheme 8. Synthesis of Nickel Oxide Nanoparticles.

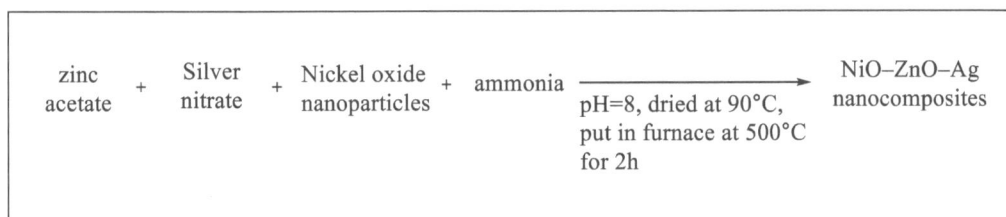

zinc acetate + Silver nitrate + Nickel oxide nanoparticles + ammonia → pH=8, dried at 90°C, put in furnace at 500°C for 2h → NiO–ZnO–Ag nanocomposites

Scheme 9. Synthesis of NiO–ZnO–Ag Nanocomposites.

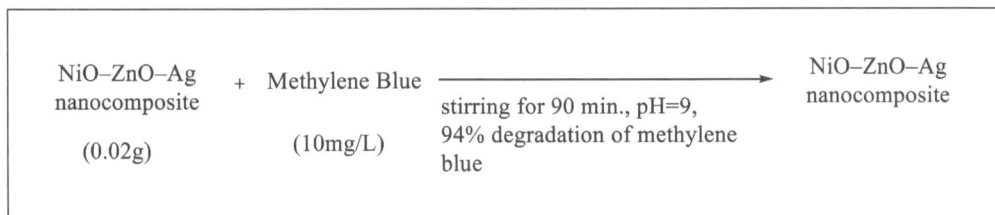

NiO–ZnO–Ag nanocomposite (0.02g) + Methylene Blue (10mg/L) → stirring for 90 min., pH=9, 94% degradation of methylene blue → NiO–ZnO–Ag nanocomposite

Scheme 10. Degradation of dyes.

Das, S. *et al.* synthesized nanocomposite films for the degradation of dyes present in wastewater (Schemes **11-14**) [24].

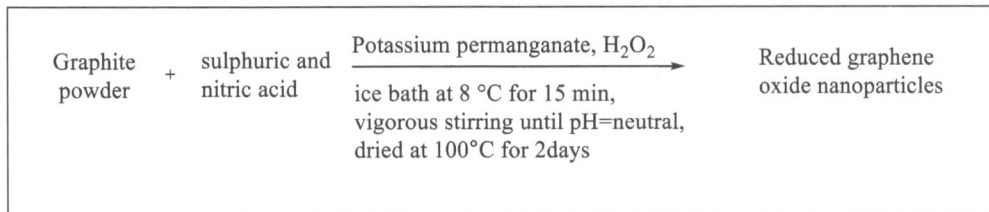

Graphite powder + sulphuric and nitric acid → Potassium permanganate, H_2O_2 ice bath at 8 °C for 15 min, vigorous stirring until pH=neutral, dried at 100°C for 2days → Reduced graphene oxide nanoparticles

Scheme 11. Synthesis of Reduced graphene oxide nanoparticles.

Reduced graphene oxide (2g/5g)	+	Titanium dioxide	+	powder melamine (7-42g)	put in furnace at 520°C for 4h ⟶	$TiO_2/rGO/g-C_3N_4$ nanocomposites

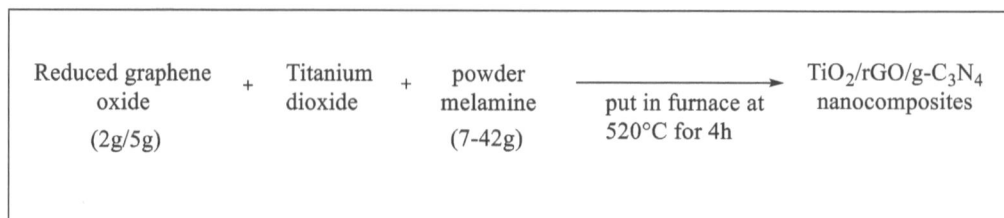

Scheme 12. Synthesis of $TiO_2/rGO/g-C_3N_4$ nanocomposites.

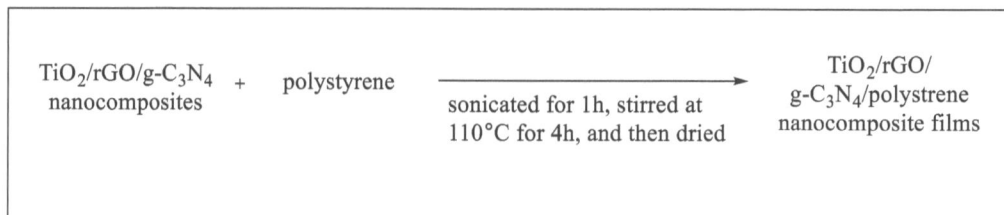

$TiO_2/rGO/g-C_3N_4$ nanocomposites	+	polystyrene	sonicated for 1h, stirred at 110°C for 4h, and then dried ⟶	$TiO_2/rGO/$ $g-C_3N_4/polystrene$ nanocomposite films

Scheme 13. Immobilization of photocatalyst into polystyrene substrate.

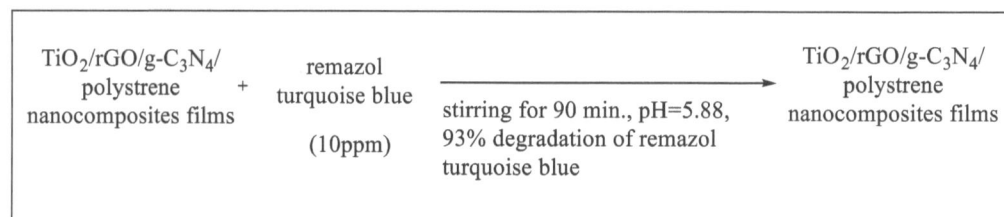

$TiO_2/rGO/g-C_3N_4/$ polystrene nanocomposites films	+	remazol turquoise blue (10ppm)	stirring for 90 min., pH=5.88, 93% degradation of remazol turquoise blue ⟶	$TiO_2/rGO/g-C_3N_4/$ polystrene nanocomposites films

Scheme 14. Degradation of dyes.

When the dye is photocatalyzed in ideal circumstances for 90 minutes, the degradation and decolorization detected by total organic carbon analysis are quite comparable. The dyes react with the highly oxidizing agent ($\cdot O_2^-$) to form simple molecules. In addition to attacking the dye molecules, the holes created in the HOMO react with the OH^- to make $\cdot OH$. Over the course of four reuses, the dye's decolorization somewhat decreased to around 50%, suggesting that the film is quite robust. Before being released into adjacent natural water sources, the immobilized nano-composite photocatalyst integrated into a multiphase airlift reactor is a highly promising approach to enhance water quality by reducing total organic carbon.

Anand, K. V. *et al.* fabricated nanocomposites for the degradation of organic dyes present in wastewater. (Scheme **15**) [25].

The green method was used to create the nanocomposites. The migration of light-induced charge carriers is faster in the synthesized nanocomposites. The produced nanocomposites were discovered to be an ideal photocatalyst, showing improved photocatalytic degradation performance compared to other nanoparticle samples. This might be because of a decreased recombination of charge carriers as well as a good synergistic impact.

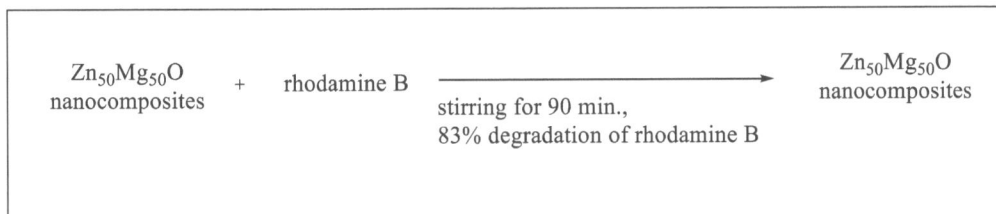

$Zn_{50}Mg_{50}O$ nanocomposites + rhodamine B $\xrightarrow{\text{stirring for 90 min.,} \atop \text{83\% degradation of rhodamine B}}$ $Zn_{50}Mg_{50}O$ nanocomposites

Scheme 15. Degradation of dyes.

Sakib, A. A. M. *et al.* synthesized nanocomposites for the photodegradation of harmful pollutants present in wastewater (Schemes **16**, **17**) [26].

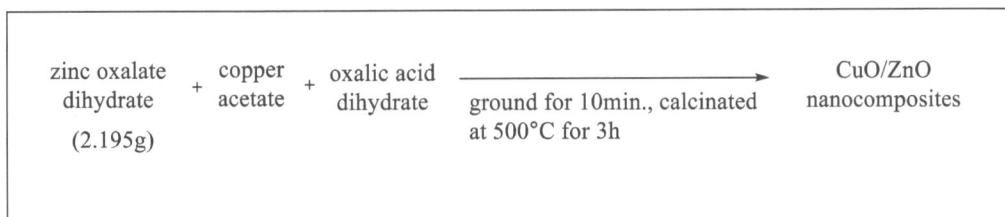

zinc oxalate dihydrate (2.195g) + copper acetate + oxalic acid dihydrate $\xrightarrow{\text{ground for 10min., calcinated} \atop \text{at 500°C for 3h}}$ CuO/ZnO nanocomposites

Scheme 16. Synthesis of CuO/ZnO nanocomposites.

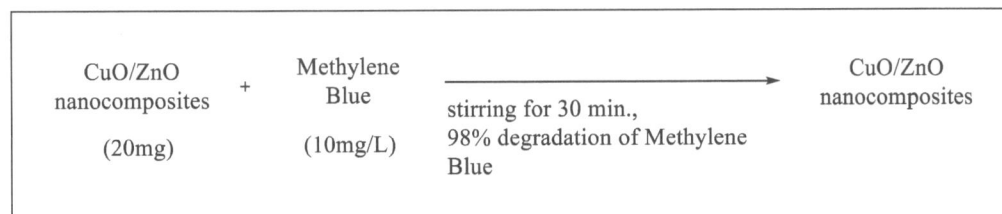

CuO/ZnO nanocomposites (20mg) + Methylene Blue (10mg/L) $\xrightarrow{\text{stirring for 30 min.,} \atop \text{98\% degradation of Methylene Blue}}$ CuO/ZnO nanocomposites

Scheme 17. Degradation of dyes.

The nanocomposites were created using a straightforward mechanochemical combustion process. Superoxide radical was produced by the reaction of oxygen molecules with electrons in the dye solution, while hydroxyl radical was produced by the reaction of holes with water. The photodegradation of dye is mostly mediated by the reactive radicals $\cdot O_2^-$ and $\cdot OH$, with the h^+ radical having a negligible effect. Under solar radiation, dye mineralizes to around 91% of its original state in 7 hours. Using sunlight to treat textile wastewater for photodegradation is a cheap, simple, and straightforward process. As a result, rather than using an artificial, costly mercury-xenon lamp for photodegradation, solar photodegradation technology might be a very effective way to treat wastewater.

Rani, A. *et al.* fabricated nanocomposites for the photocatalysis of organic effluents present in industrial wastewater (Schemes **18-20**) [27].

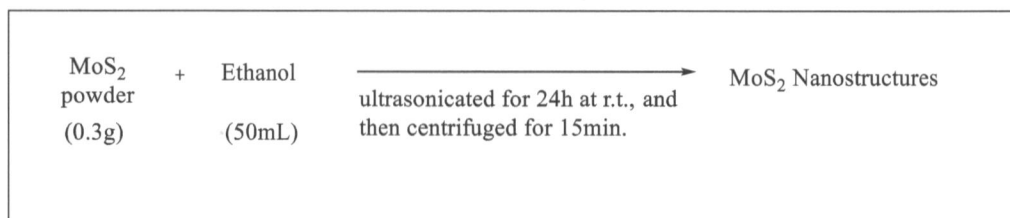

MoS$_2$ powder (0.3g) + Ethanol (50mL) → ultrasonicated for 24h at r.t., and then centrifuged for 15min. → MoS$_2$ Nanostructures

Scheme 18. Synthesis of MoS$_2$ Nanostructures

SnCl$_2$ .2H$_2$O (0.1g) + MoS$_2$ nanostructures (0.1g) → sonication for 24h, and refluxed for 2h at 80°C → SnO$_2$ decorated MoS$_2$ Nanostructures

Scheme 19. Synthesis of SnO$_2$ decorated MoS$_2$ Nanostructures.

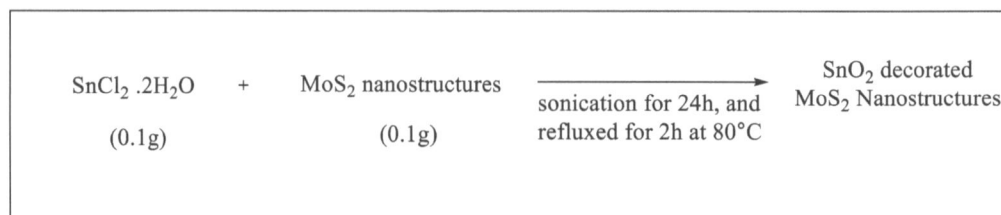

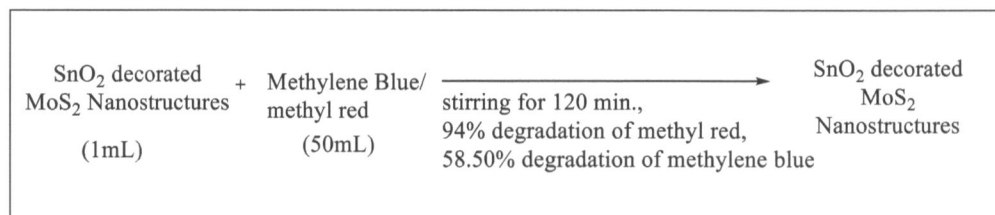

SnO$_2$ decorated MoS$_2$ Nanostructures (1mL) + Methylene Blue/ methyl red (50mL) → stirring for 120 min., 94% degradation of methyl red, 58.50% degradation of methylene blue → SnO$_2$ decorated MoS$_2$ Nanostructures

Scheme 20. Degradation of dyes.

The nanocomposites were created using an elementary two-step process. The composites' nanoparticles slow down the photocatalyst's electron hole's rate of recombination. Free radical production during the photocatalytic reaction and the number of chromophore sites in the dye were connected to the dye's degradation with nanocomposites. The chromophore groups in dyes are destabilized by the free radical ions O$_2^-$ and •OH, which operate as attacking species produced in nanocomposites under visible light. Furthermore, these holes immediately oxidize the dye molecules that have been deposited on the catalyst surface. When MR dye is used instead of conventional dyes, the dye degrades substantially more efficiently in two hours. As a result, the created nanocomposites have applications for environmental remediation problems and make excellent candidates for photocatalytic degradation of hazardous organic contaminants in wastewater utilizing visible light.

Khan, S. A. *et al.* fabricated nanocomposites for the effective degradation of dyes (Schemes **21-24**) [28].

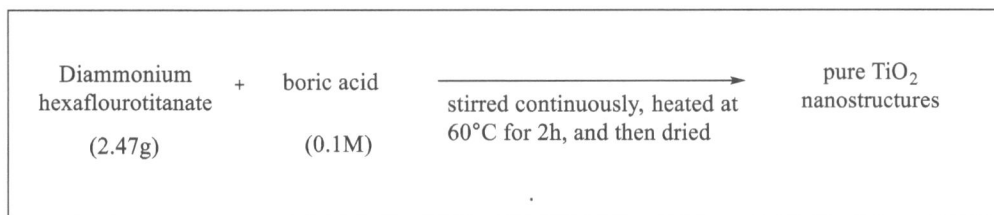

Diammonium hexaflourotitanate	+ boric acid	→ pure TiO_2 nanostructures
(2.47g)	(0.1M)	stirred continuously, heated at 60°C for 2h, and then dried

Scheme 21. Synthesis of pure TiO_2 nanostructures.

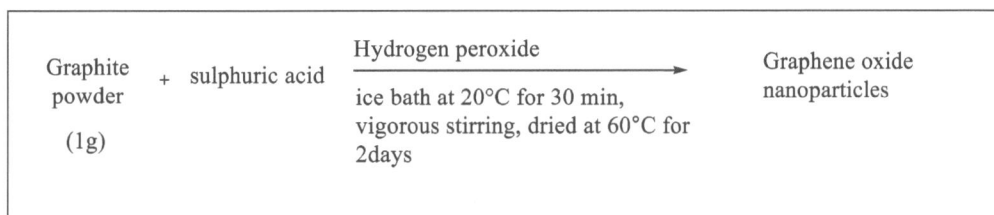

Graphite powder	+ sulphuric acid	Hydrogen peroxide → Graphene oxide nanoparticles
(1g)		ice bath at 20°C for 30 min, vigorous stirring, dried at 60°C for 2days

Scheme 22. Synthesis of graphene oxide nanoparticles.

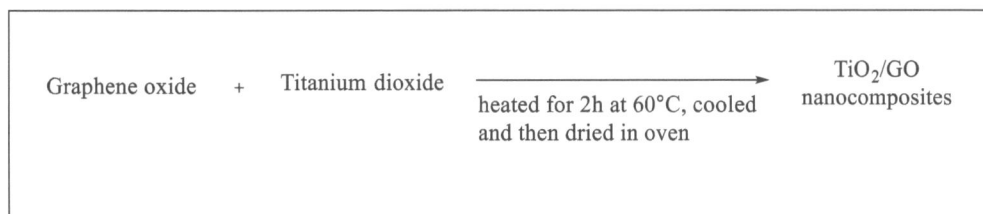

Graphene oxide	+ Titanium dioxide	→ TiO_2/GO nanocomposites
		heated for 2h at 60°C, cooled and then dried in oven

Scheme 23. Synthesis of TiO_2/GO nanocomposites

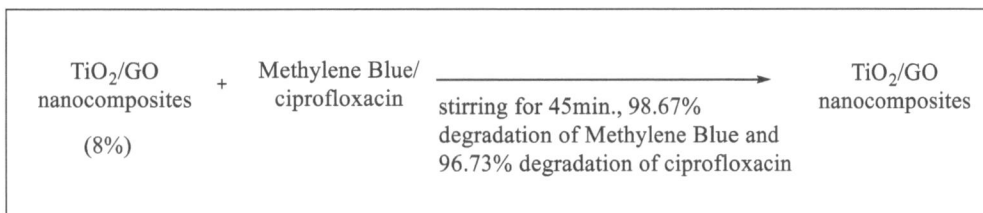

TiO_2/GO nanocomposites	+ Methylene Blue/ ciprofloxacin	→ TiO_2/GO nanocomposites
(8%)		stirring for 45min., 98.67% degradation of Methylene Blue and 96.73% degradation of ciprofloxacin

Scheme 24. Degradation of dyes.

The liquid phase deposition approach was utilized to synthesize the nanocomponents. Because of its π conjugation structure, graphene acts as an electron acceptor for excited electrons in nanostructures, rapidly transferring them from the conduction band of the nanostructure to graphene oxide nanoparticles.

This suppresses the recombination of charge carriers generated by photolysis, resulting in the production of reactive oxidative species that aid in dye degradation. It was discovered that the rate at which dyes degraded with the formed composites was greater than that of pure nanoparticles. The rate of deterioration by composites is increased by adding more graphene oxide nanoparticles. The photocatalyst's high stability and recyclability were proved by the fact that even at the sixth cycle, the photocatalytic disintegration of both dyes remained over 91%. The developed nanocomponents may prove to be useful photocatalytic systems that protect our environment from the harmful waste products that are generated.

Kamari, S. *et al.* synthesized nanocomposite membranes for the degradation of dyes present in wastewater. (Schemes **25-28**) [29]. Phase inversion was a technique employed to prepare the membranes. The modified membranes containing 0.5 weight percent of composites demonstrated the maximum removal effectiveness of dyes (97%) and metal ions (93%). During long-term filtering, there was a slight decrease in permeate flux with no change in dye retention, confirming the durability of the modified membrane. The chosen membrane's capacity to be reused for the removal of metal ions demonstrated an 86% efficiency for the fifth cycle. This little drop in removal efficiency confirms that the modified membrane may be reused successfully. As a result, this efficient membrane was discovered to be ideal for environmental applications.

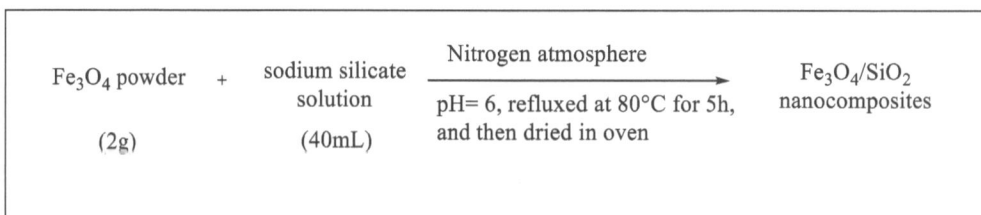

Fe_3O_4 powder $+$ sodium silicate solution $\xrightarrow[\text{pH= 6, refluxed at 80°C for 5h, and then dried in oven}]{\text{Nitrogen atmosphere}}$ Fe_3O_4/SiO_2 nanocomposites

(2g) (40mL)

Scheme 25. Synthesis of Fe_3O_4/SiO_2 nanocomposites.

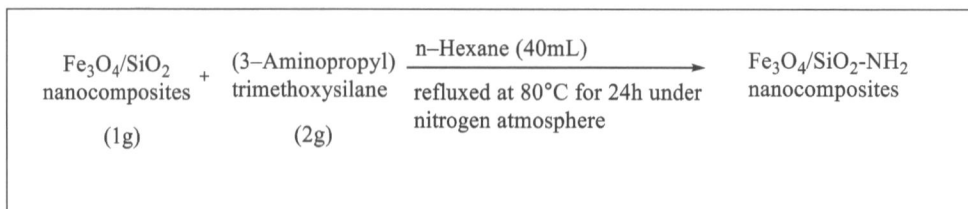

Fe_3O_4/SiO_2 nanocomposites $+$ (3–Aminopropyl) trimethoxysilane $\xrightarrow[\text{refluxed at 80°C for 24h under nitrogen atmosphere}]{\text{n–Hexane (40mL)}}$ Fe_3O_4/SiO_2-NH_2 nanocomposites

(1g) (2g)

Scheme 26. Synthesis of Fe_3O_4/SiO_2-NH_2 nanocomposites.

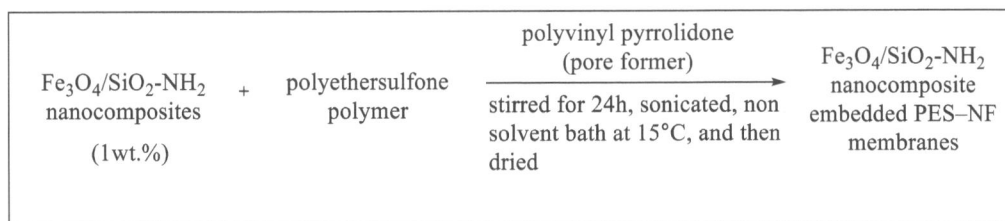

| Fe$_3$O$_4$/SiO$_2$-NH$_2$ nanocomposites (1wt.%) | + | polyethersulfone polymer | polyvinyl pyrrolidone (pore former) stirred for 24h, sonicated, non solvent bath at 15°C, and then dried | Fe$_3$O$_4$/SiO$_2$-NH$_2$ nanocomposite embedded PES–NF membranes |

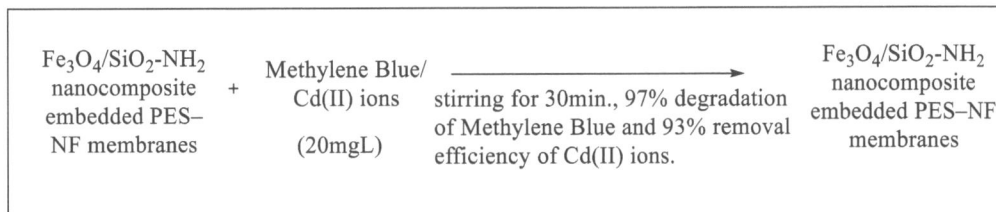

Scheme 27. Synthesis of Fe$_3$O$_4$/SiO$_2$-NH$_2$ nanocomposite embedded PES-NF membrane.

| Fe$_3$O$_4$/SiO$_2$-NH$_2$ nanocomposite embedded PES–NF membranes | + | Methylene Blue/ Cd(II) ions (20mgL) | stirring for 30min., 97% degradation of Methylene Blue and 93% removal efficiency of Cd(II) ions. | Fe$_3$O$_4$/SiO$_2$-NH$_2$ nanocomposite embedded PES–NF membranes |

Scheme 28. Degradation of dyes.

Sharma, M. *et al.* synthesized hybrid nanocomposites for the degradation of organic pollutants. (Schemes **29**, **30**) [30].

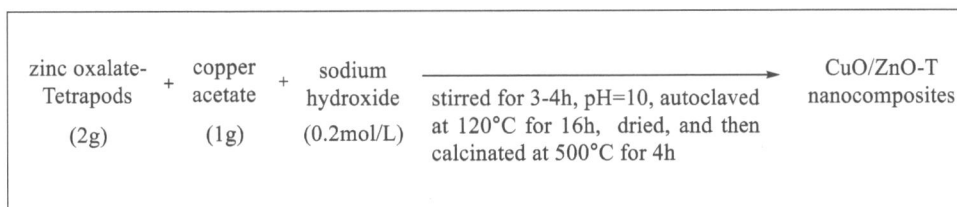

| zinc oxalate-Tetrapods (2g) | + | copper acetate (1g) | + | sodium hydroxide (0.2mol/L) | stirred for 3-4h, pH=10, autoclaved at 120°C for 16h, dried, and then calcinated at 500°C for 4h | CuO/ZnO-T nanocomposites |

Scheme 29. Synthesis of CuO/ZnO-T nanocomposites.

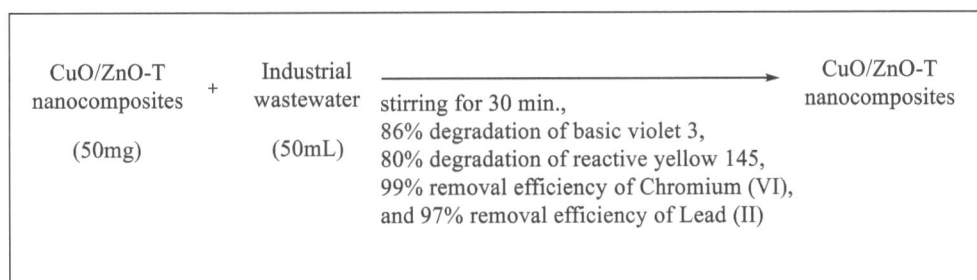

| CuO/ZnO-T nanocomposites (50mg) | + | Industrial wastewater (50mL) | stirring for 30 min., 86% degradation of basic violet 3, 80% degradation of reactive yellow 145, 99% removal efficiency of Chromium (VI), and 97% removal efficiency of Lead (II) | CuO/ZnO-T nanocomposites |

Scheme 30. Degradation of dyes.

The nanocomposites were created using a straightforward and reasonably priced hydrothermal method. Superoxide radical anion may be formed by e⁻s reacting with dissolved oxygens, whereas hydroxide radical can be produced by h⁺ reacting with water molecules. Eventually, the dye molecule may react with these reactive oxygen species, leading to degradation. The various kinetics and isotherm models were used to explain the mechanism for the reduction of dyes and heavy metal ions. After four cycles, the percentage of photodegradation did not significantly reduce. The nanocomposites show promise as an effective adsorbent and photocatalyst for the removal of different pollutants from wastewater.

Gapusan, R. B. *et al.* fabricated nanocomposites for the removal of organic pollutants present in . wastewater (Schemes **31-33**) [31].

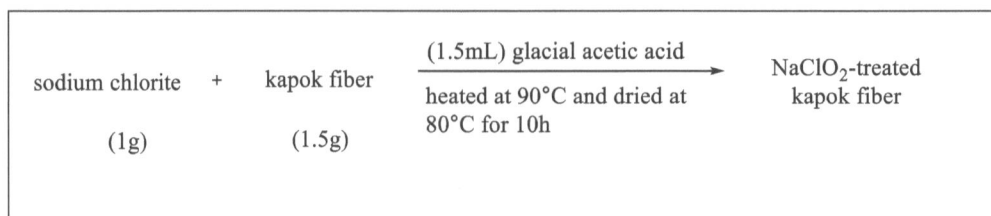

sodium chlorite	+	kapok fiber	(1.5mL) glacial acetic acid	NaClO$_2$-treated
			heated at 90°C and dried at 80°C for 10h	kapok fiber
(1g)		(1.5g)		

Scheme 31. Pre-treatment of kapok fibers.

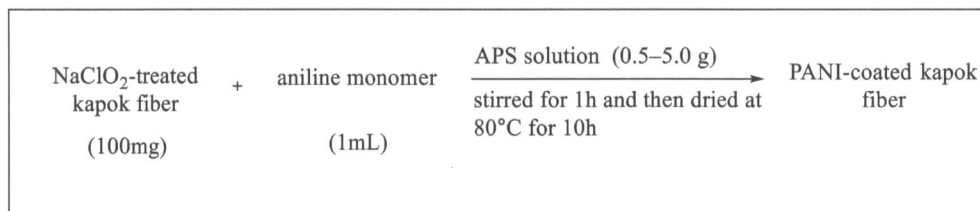

NaClO$_2$-treated kapok fiber	+	aniline monomer	APS solution (0.5–5.0 g)	PANI-coated kapok fiber
			stirred for 1h and then dried at 80°C for 10h	
(100mg)		(1mL)		

Scheme 32. Synthesis of PANI-coated kapok fiber.

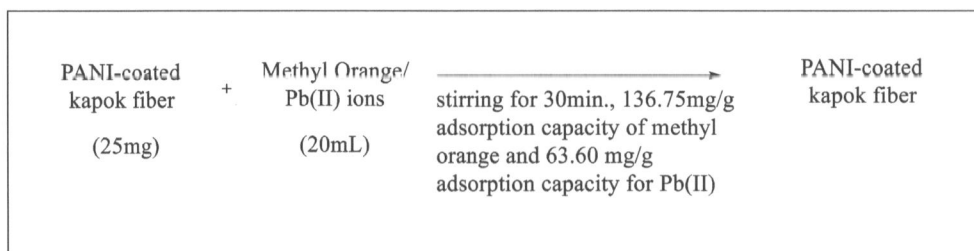

| PANI-coated kapok fiber | + | Methyl Orange/ Pb(II) ions | stirring for 30min., 136.75mg/g adsorption capacity of methyl orange and 63.60 mg/g adsorption capacity for Pb(II) | PANI-coated kapok fiber |
| (25mg) | | (20mL) | | |

Scheme 33. Degradation of dyes.

The monomer was polymerized *in situ* to create the nanocomposites. According to kinetic experiments, the pseudo-second-order kinetic model and the adsorption of contaminants onto the nanocomposites agree well. The Langmuir isotherm model is followed by the adsorption of both model pollutants, according to equilibrium isotherm research. Because of the electrostatic interaction between the protonated composite and the anionic sulfonate moiety of the dye, it was discovered that the adsorption of dyes onto the nanocomposites was more advantageous at pH values in the range of 4-6.

Preethi, S. *et al.* synthesized nanocomposites for the degradation of organic dyes present in wastewater (Schemes **34-36**) [32].

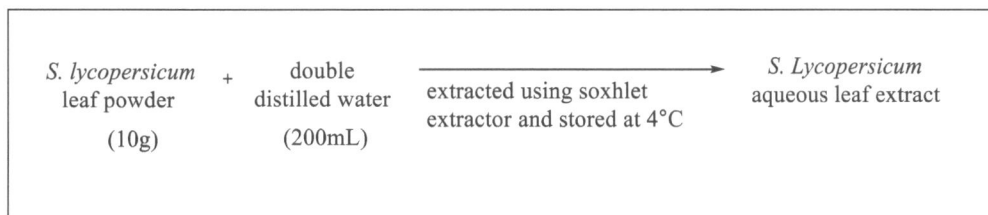

S. lycopersicum leaf powder	+	double distilled water	extracted using soxhlet extractor and stored at 4°C → *S. Lycopersicum* aqueous leaf extract
(10g)		(200mL)	

Scheme 34. Preparation of *S. Lycopersicum* aqueous leaf extract.

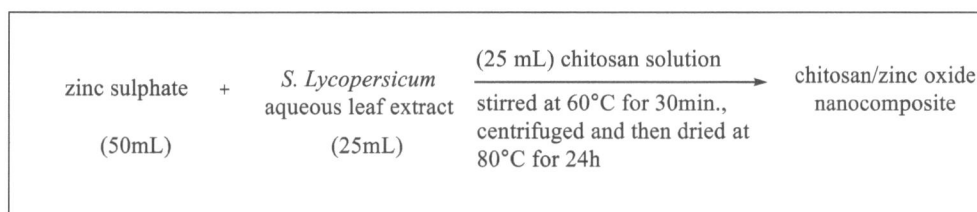

zinc sulphate	+	*S. Lycopersicum* aqueous leaf extract	(25 mL) chitosan solution stirred at 60°C for 30min., centrifuged and then dried at 80°C for 24h → chitosan/zinc oxide nanocomposite
(50mL)		(25mL)	

Scheme 35. Synthesis of chitosan/zinc oxide nanocomposite.

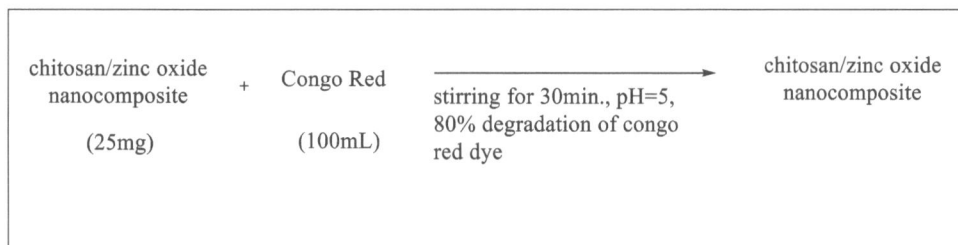

chitosan/zinc oxide nanocomposite	+	Congo Red	stirring for 30min., pH=5, 80% degradation of congo red dye → chitosan/zinc oxide nanocomposite
(25mg)		(100mL)	

Scheme 36. Degradation of dyes.

A bio-inspired technique was used to create the bio-nanocomposites. The nanocomposites are spherically formed and range in size from 21 to 47 nm on average. Organic dye molecules are broken down into CO_2 and H_2O by strong oxidants that are nonselective and result from the breakdown of photogenerated electron holes and dissolved oxygen in water. These radicals include superoxide and hydroxyl radicals. In order to remove dye pollutants, biosynthesized nanocomposites can be employed as a photocatalyst.

Al-Rawashdeh, N. A. *et al.* synthesized nanocomposites for the elimination of organic dyes from wastewater (Schemes **37-40**) [33].

Graphite powder + sulphuric acid $\xrightarrow[\text{ice bath at 0°C for 20 min, vigorous stirring at 40°C for 2h, centrifuged, and then dried at 60°C for 2days}]{\text{(2.5g) sodium nitrate, } H_2O_2}$ Graphene oxide nanoparticles

(4.5g)   (120mL)

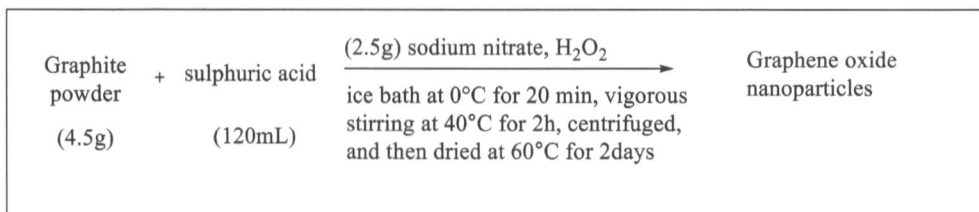

Scheme 37. Synthesis of graphene oxide nanoparticles.

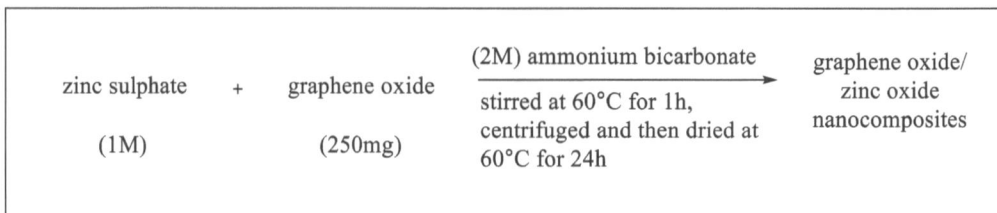

zinc sulphate + graphene oxide $\xrightarrow[\text{stirred at 60°C for 1h, centrifuged and then dried at 60°C for 24h}]{\text{(2M) ammonium bicarbonate}}$ graphene oxide/ zinc oxide nanocomposites

(1M)   (250mg)

Scheme 38. Synthesis of graphene oxide/zinc oxide nanocomposites.

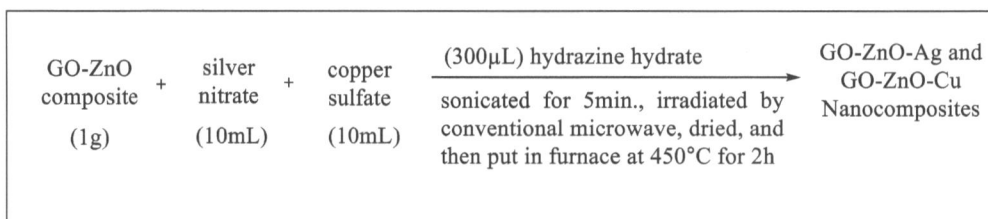

GO-ZnO composite + silver nitrate + copper sulfate $\xrightarrow[\text{sonicated for 5min., irradiated by conventional microwave, dried, and then put in furnace at 450°C for 2h}]{\text{(300µL) hydrazine hydrate}}$ GO-ZnO-Ag and GO-ZnO-Cu Nanocomposites

(1g)   (10mL)   (10mL)

Scheme 39. Synthesis of GO-ZnO-Ag and GO-ZnO-Cu Nanocomposites.

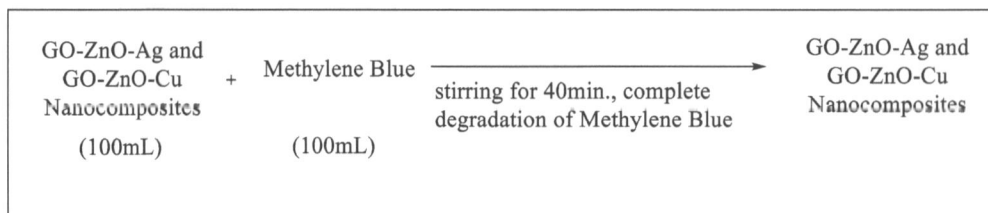

GO-ZnO-Ag and GO-ZnO-Cu Nanocomposites + Methylene Blue $\xrightarrow[\text{stirring for 40min., complete degradation of Methylene Blue}]{}$ GO-ZnO-Ag and GO-ZnO-Cu Nanocomposites

(100mL)   (100mL)

Scheme 40. Degradation of dyes.

Azar, B. E. *et al.* synthesized nanocomposites for the degradation of dyes. (Schemes **41-43**) [34].

The green sol-gel approach was utilized to synthesize the nanocomposites. The nanocomposite had a mean size of around 25–30 nm and was spherical in shape. Using the UV–vis–DRS spectrum, the band gap energy of the nanocomposite was found to be around 3.05 eV. The use of nanocomposites as effective heterogeneous photocatalysts for dye degradation has been studied.

zinc nitrate +   tragacanth gum + Aluminum nitrate

(1g)   (0.2g)   (2mmol)

→ stirred at 75°C for 60min., calcinated at 600°C for 4h

zinc aluminate nanoparticles

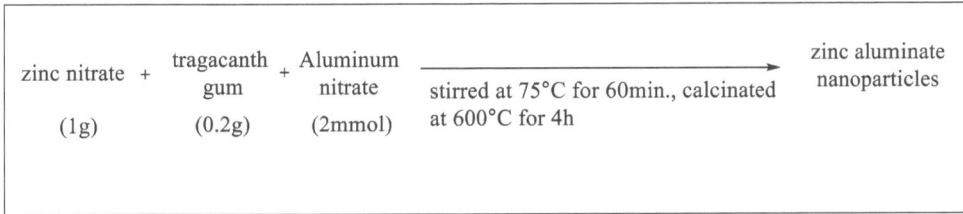

Scheme 41. Synthesis of zinc aluminate nanoparticles.

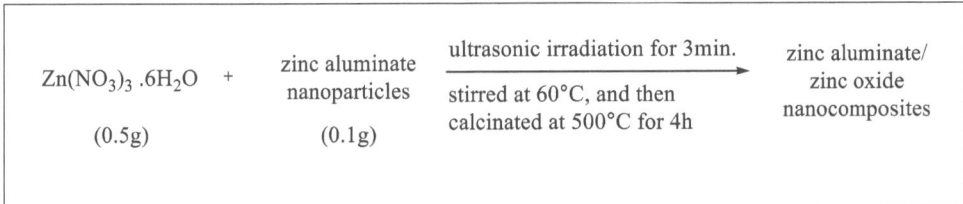

$Zn(NO_3)_3 \cdot 6H_2O$ + zinc aluminate nanoparticles

(0.5g)   (0.1g)

→ ultrasonic irradiation for 3min. stirred at 60°C, and then calcinated at 500°C for 4h

zinc aluminate/ zinc oxide nanocomposites

Scheme 42. Synthesis of zinc aluminate/zinc oxide nanocomposites.

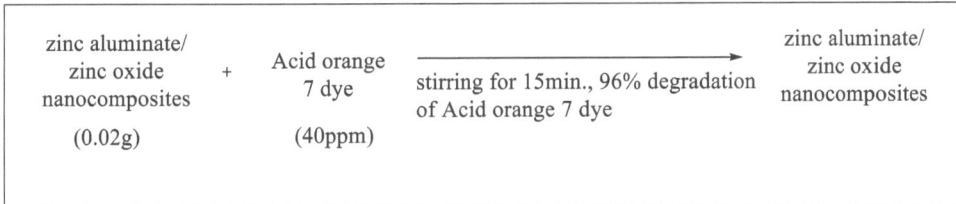

zinc aluminate/ zinc oxide nanocomposites + Acid orange 7 dye

(0.02g)   (40ppm)

→ stirring for 15min., 96% degradation of Acid orange 7 dye

zinc aluminate/ zinc oxide nanocomposites

Scheme 43. Degradation of dyes.

Potle, V. D. *et al.* fabricated nanocomposites for the degradation of organic dyes present in industrial wastewater (Schemes **44**, **45**) [35].

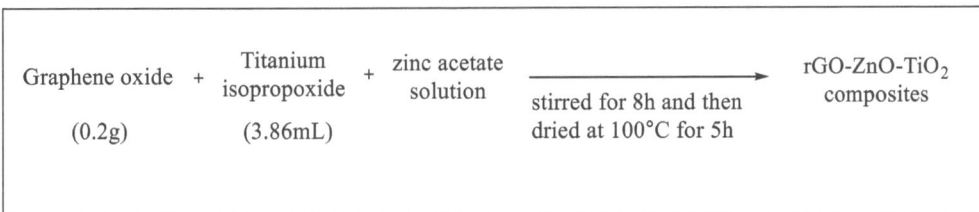

Graphene oxide + Titanium isopropoxide + zinc acetate solution

(0.2g)   (3.86mL)

→ stirred for 8h and then dried at 100°C for 5h

rGO-ZnO-TiO_2 composites

Scheme 44. Synthesis of rGO-ZnO-TiO_2 composites.

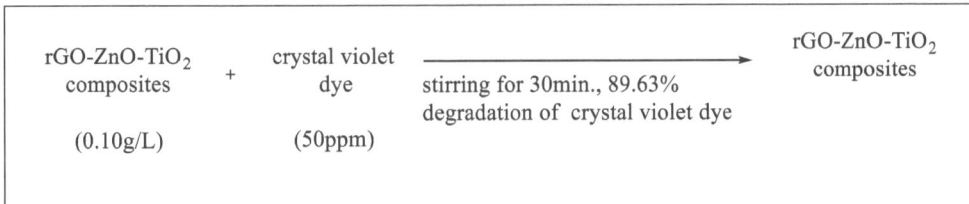

rGO-ZnO-TiO_2 composites + crystal violet dye

(0.10g/L)   (50ppm)

→ stirring for 30min., 89.63% degradation of crystal violet dye

rGO-ZnO-TiO_2 composites

Scheme 45. Degradation of dyes.

An ultrasound-assisted method is used to synthesize the nanocomposites. The produced electrons move to the conduction band, where they cause a hole formation in the valence band. The efficient separation of an electron-hole pair significantly lowers the recombination of electrons and holes produced by UV light irradiation. Superoxide and hydroxyl radicals were produced during the procedure in accordance with the reaction mechanism shown. It was discovered that the use of ultrasonic-aided synthesis was essential, and under comparable operating circumstances, the nanocomposite material generated using this approach degraded 15% more quickly than that created using a traditional technique.

Maruthupandy, M. *et al.* synthesized nanocomposites for the elimination of dyes present in industrial wastewater (Schemes **46-49**) [36].

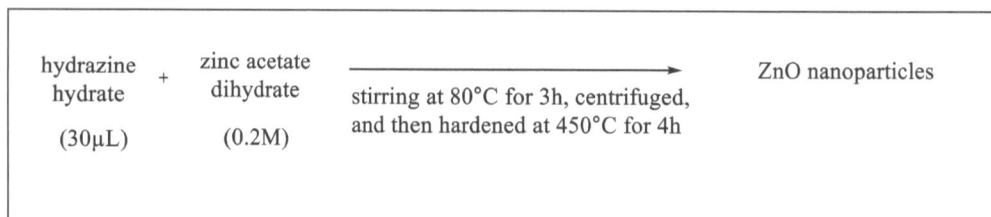

hydrazine   +   zinc acetate
hydrate   dihydrate $\xrightarrow{\text{stirring at 80°C for 3h, centrifuged, and then hardened at 450°C for 4h}}$ ZnO nanoparticles

(30μL)   (0.2M)

Scheme 46. Synthesis of ZnO nanoparticles.

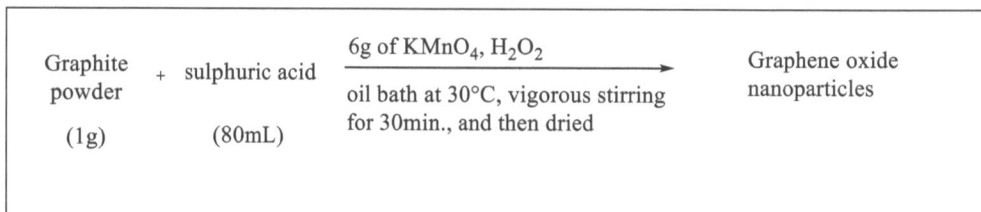

Graphite   +   sulphuric acid
powder   6g of KMnO$_4$, H$_2$O$_2$
                                      $\xrightarrow{\text{oil bath at 30°C, vigorous stirring for 30min., and then dried}}$ Graphene oxide
                                                                                                                             nanoparticles

(1g)   (80mL)

Scheme 47. Synthesis of graphene oxide nanoparticles.

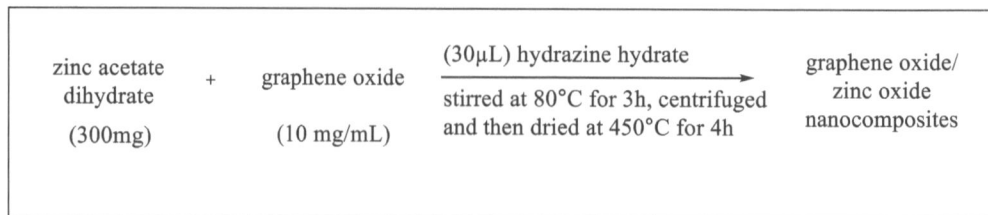

zinc acetate   +   graphene oxide
dihydrate   (30μL) hydrazine hydrate
                                      $\xrightarrow{\text{stirred at 80°C for 3h, centrifuged and then dried at 450°C for 4h}}$ graphene oxide/
                                                                                                                                zinc oxide
                                                                                                                                nanocomposites

(300mg)   (10 mg/mL)

Scheme 48. Synthesis of graphene oxide/zinc oxide nanocomposites.

graphene oxide/ zinc oxide nanocomposites (0.10g/L)	+	Methylene Blue/ Methyl orange/ rhodamine B (50ppm)	stirring for 90-120min., complete degradation of Methylene Blue, Methyl orange, and rhodamine B ⟶ graphene oxide/ zinc oxide nanocomposites

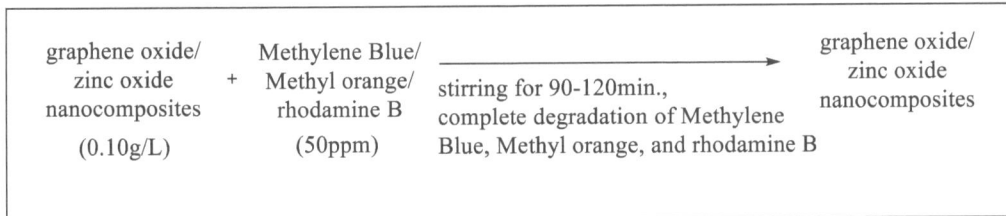

Scheme 49. Degradation of dyes.

The chemical precipitation approach was utilized to synthesize the nanocomposites. It is discovered that nanocomposites have a diameter of 8–12 nm. Thus, environmental dye remediation might take advantage of the synthesized nanocomposites. The hole has the ability to oxidize the adsorbed dye molecules directly, and the photoactive radicals produced during the photocatalytic process result in the production of CO_2, water, and other mineralized intermediates. When measured dye molecules were present, the photocatalytic efficacy of the nanocomposites was superior to that of pure nanoparticles. It is possible to conclude that nanocomposites are better absorbers and should be used in ecological remediation activities.

CONCLUSION

In conclusion, metal and metal oxide nanocomposites have been paid a lot of attention due to their adaptable and wonderful properties resulting from the nanometer end product. In this chapter, we have discussed a variety of nanocomposites extensively used in eradicating heavy metals and organic dye pollutants from polluted water. This chapter can corroborate to be extremely supportive for research scholars or chemists in the advancement of new methodologies for the enormous preparation of new nanocomposites for the application in the removal of heavy metals and organic dye pollutants from wastewater. It also summarizes the significance of nanocomposites in addressing wastewater pollution.

LIST OF ABBREVIATIONS

AMS   Mesoporous silica

APTES 3-Aminopropyltriethoxysilane

CMC   Carboxymethyl cellulose

CR   Congo Red

DMF   Dimethyl formamide

EDTA   Ethylene diamine tetra acetic acid

GA   Glutaraldehyde

GO	Graphite oxide
MB	Methylene Blue
NCs	Nanocomposites
PVA	Polyvinyl alcohol
RO	Reactive Orange
SA	Sodium alginate
TETA	Triethyllene tetramine
Zeo	Zeolite

ACKNOWLEDGEMENTS

The author, G. B. Dharma Rao, conveys his cordial gratefulness to the management of KPRIT (Autonomous), Hyderabad, India. Anirudh, Chinmay, and Vishaka state their heartfelt gratitude to the management of SGT University, Gurugram, Haryana, India, for providing the conveniences to write and submit this chapter for publication.

REFERENCES

[1] Verma, M.; Lee, I.; Oh, J.; Kumar, V.; Kim, H. Synthesis of EDTA-functionalized graphene oxide-chitosan nanocomposite for simultaneous removal of inorganic and organic pollutants from complex wastewater. *Chemosphere,* **2022**, *287*(Pt 4), 132385.
[http://dx.doi.org/10.1016/j.chemosphere.2021.132385] [PMID: 34597635]

[2] Shahi Khalaf Ansar, B.; Kavusi, E.; Dehghanian, Z.; Pandey, J.; Asgari Lajayer, B.; Price, G.W.; Astatkie, T. Removal of organic and inorganic contaminants from the air, soil, and water by algae. *Environ. Sci. Pollut. Res. Int.,* **2022**, *30*(55), 116538-116566.
[http://dx.doi.org/10.1007/s11356-022-21283-x] [PMID: 35680750]

[3] Chai, W.S.; Cheun, J.Y.; Kumar, P.S.; Mubashir, M.; Majeed, Z.; Banat, F.; Ho, S.H.; Show, P.L. A review on conventional and novel materials towards heavy metal adsorption in wastewater treatment application. *J. Clean. Prod.,* **2021**, *296*, 126589.
[http://dx.doi.org/10.1016/j.jclepro.2021.126589]

[4] Velusamy, S.; Roy, A.; Sundaram, S.; Kumar Mallick, T. A review on heavy metal ions and containing dyes removal through graphene oxide-based adsorption strategies for textile wastewater treatment. *Chem. Rec.,* **2021**, *21*(7), 1570-1610.
[http://dx.doi.org/10.1002/tcr.202000153] [PMID: 33539046]

[5] Solayman, H.M.; Hossen, M.A.; Abd Aziz, A.; Yahya, N.Y.; Leong, K.H.; Sim, L.C.; Monir, M.U.; Zoh, K.D. Performance evaluation of dye wastewater treatment technologies: A review. *J. Environ. Chem. Eng.,* **2023**, *11*(3), 109610.
[http://dx.doi.org/10.1016/j.jece.2023.109610]

[6] Kodispathi, T.; Mispa, K.J. Titanium-embedded nanocomposite material stimulates ion-exchange characteristics to deal with toxic heavy metal pollutants in aquatic environment. *Sep. Sci. Technol.,* **2023**, *58*(17-18), 2932-2949.
[http://dx.doi.org/10.1080/01496395.2023.2280507]

[7] Kumari, S.; Sharma, R.; Thakur, N.; Kumari, A. Removal of organic and inorganic effluents from wastewater by using degradation and adsorption properties of transition metal-doped nickel ferrite. *Environ. Sci. Pollut. Res. Int.,* **2023**, *31*(34), 46526-46545.

[http://dx.doi.org/10.1007/s11356-023-26567-4] [PMID: 36973621]

[8]   Babu Ponnusami, A.; Sinha, S.; Ashokan, H.; V Paul, M.; Hariharan, S.P.; Arun, J.; Gopinath, K.P.; Hoang Le, Q.; Pugazhendhi, A. Advanced oxidation process (AOP) combined biological process for wastewater treatment: A review on advancements, feasibility and practicability of combined techniques. *Environ. Res.,* **2023**, *237*(Pt 1), 116944.
[http://dx.doi.org/10.1016/j.envres.2023.116944] [PMID: 37611785]

[9]   Zhang, H.; Xue, K.; Wang, B.; Ren, W.; Sun, D.; Shao, C.; Sun, R. Advances in lignin-based biosorbents for sustainable wastewater treatment. *Bioresour. Technol.,* **2024**, *395*, 130347.
[http://dx.doi.org/10.1016/j.biortech.2024.130347] [PMID: 38242243]

[10]   Gao, R.; Mosquera-Romero, S.; Ntagia, E.; Wang, X.; Rabaey, K.; Bonin, L. Electrochemical separation of organic and inorganic contaminants in wastewater. *J. Electrochem. Soc.,* **2022**, *169*(3), 033505.
[http://dx.doi.org/10.1149/1945-7111/ac51f9]

[11]   Mishra, S.; Sundaram, B. A review of the photocatalysis process used for wastewater treatment. *Mater. Today Proc.,* **2023**.

[12]   Mahfooz, Y.; Yasar, A.; Islam, Q.U.; Rasheed, R.; Naeem, U.; Mukhtar, S. Field testing phytoremediation of organic and inorganic pollutants of sewage drain by bacteria assisted water hyacinth. *Int. J. Phytoremediation,* **2021**, *23*(2), 139-150.
[http://dx.doi.org/10.1080/15226514.2020.1802574] [PMID: 32757949]

[13]   Adeoye, J.B.; Tan, Y.H.; Lau, S.Y.; Tan, Y.Y.; Chiong, T.; Mubarak, N.M.; Khalid, M. Advanced oxidation and biological integrated processes for pharmaceutical wastewater treatment: A review. *J. Environ. Manage.,* **2024**, *353*, 120170.
[http://dx.doi.org/10.1016/j.jenvman.2024.120170] [PMID: 38308991]

[14]   Zadehnazari, A. Metal oxide/polymer nanocomposites: A review on recent advances in fabrication and applications. *Polymer-Plastics Technology and Materials,* **2023**, *62*(5), 655-700.
[http://dx.doi.org/10.1080/25740881.2022.2129387]

[15]   Umejuru, E.C.; Mashifana, T.; Kandjou, V.; Amani-Beni, M.; Sadeghifar, H.; Fayazi, M.; Karimi-Maleh, H.; Sithole, N.T. Application of zeolite based nanocomposites for wastewater remediation: Evaluating newer and environmentally benign approaches. *Environ. Res.,* **2023**, *231*(Pt 1), 116073.
[http://dx.doi.org/10.1016/j.envres.2023.116073] [PMID: 37164282]

[16]   AL-Salman, H.N.K.; Falih, M.; Deab, H.B.; Altimari, U.S.; Shakier, H.G.; Dawood, A.H.; Ramadan, M.F.; Mahmoud, Z.H.; Farhan, M.A.; Köten, H.; Kianfar, E. A study in analytical chemistry of adsorption of heavy metal ions using chitosan/graphene nanocomposites. *Case Studies in Chemical and Environmental Engineering,* **2023**, *8*, 100426.
[http://dx.doi.org/10.1016/j.cscee.2023.100426]

[17]   Anuma, S.; Mishra, P.; Bhat, B.R. Polypyrrole functionalized Cobalt oxide Graphene (COPYGO) nanocomposite for the efficient removal of dyes and heavy metal pollutants from aqueous effluents. *J. Hazard. Mater.,* **2021**, *416*, 125929.
[http://dx.doi.org/10.1016/j.jhazmat.2021.125929] [PMID: 34492859]

[18]   Ferri, M.; Campisi, S.; Polito, L.; Shen, J.; Gervasini, A. Tuning the sorption ability of hydroxyapatite/carbon composites for the simultaneous remediation of wastewaters containing organic-inorganic pollutants. *J. Hazard. Mater.,* **2021**, *420*, 126656.
[http://dx.doi.org/10.1016/j.jhazmat.2021.126656] [PMID: 34329080]

[19]   Zhang, W.; Ou, J.; Wang, B.; Wang, H.; He, Q.; Song, J.; Zhang, H.; Tang, M.; Zhou, L.; Gao, Y.; Sun, S. Efficient heavy metal removal from water by alginate-based porous nanocomposite hydrogels: The enhanced removal mechanism and influencing factor insight. *J. Hazard. Mater.,* **2021**, *418*, 126358.
[http://dx.doi.org/10.1016/j.jhazmat.2021.126358] [PMID: 34130162]

[20]   El-Nagar, D.A.; Massoud, S.A.; Ismail, S.H. Removal of some heavy metals and fungicides from

aqueous solutions using nano-hydroxyapatite, nano-bentonite and nanocomposite. *Arab. J. Chem.,* **2020**, *13*(11), 7695-7706.
[http://dx.doi.org/10.1016/j.arabjc.2020.09.005]

[21]   Isawi, H. Using Zeolite/Polyvinyl alcohol/sodium alginate nanocomposite beads for removal of some heavy metals from wastewater. *Arab. J. Chem.,* **2020**, *13*(6), 5691-5716.
[http://dx.doi.org/10.1016/j.arabjc.2020.04.009]

[22]   Naushad, M.; Sharma, G.; Alothman, Z.A. Photodegradation of toxic dye using Gum Arabic-crosslinked-poly(acrylamide)/Ni(OH)$_2$/FeOOH nanocomposites hydrogel. *J. Clean. Prod.,* **2019**, *241*, 118263.
[http://dx.doi.org/10.1016/j.jclepro.2019.118263]

[23]   Aydoghmish, S.M.; Hassanzadeh-Tabrizi, S.A.; Saffar-Teluri, A. Facile synthesis and investigation of NiO–ZnO–Ag nanocomposites as efficient photocatalysts for degradation of methylene blue dye. *Ceram. Int.,* **2019**, *45*(12), 14934-14942.
[http://dx.doi.org/10.1016/j.ceramint.2019.04.229]

[24]   Das, S.; Mahalingam, H. Dye degradation studies using immobilized pristine and waste polystyrene-TiO$_2$/rGO/g-C$_3$N$_4$ nanocomposite photocatalytic film in a novel airlift reactor under solar light. *J. Environ. Chem. Eng.,* **2019**, *7*(5), 103289.
[http://dx.doi.org/10.1016/j.jece.2019.103289]

[25]   Vijai Anand, K.; Aravind Kumar, J.; Keerthana, K.; Deb, P.; Tamilselvan, S.; Theerthagiri, J.; Rajeswari, V.; Sekaran, S.M.S.; Govindaraju, K. Photocatalytic Degradation of Rhodamine B Dye Using Biogenic Hybrid ZnO-MgO Nanocomposites under Visible Light. *ChemistrySelect,* **2019**, *4*(17), 5178-5184.
[http://dx.doi.org/10.1002/slct.201900213]

[26]   Sakib, A.; Masum, S.; Hoinkis, J.; Islam, R.; Molla, M. Synthesis of CuO/ZnO Nanocomposites and Their Application in Photodegradation of Toxic Textile Dye. *J. Compos. Sci.,* **2019**, *3*(3), 91.
[http://dx.doi.org/10.3390/jcs3030091]

[27]   Rani, A.; Singh, K.; Patel, A.S.; Chakraborti, A.; Kumar, S.; Ghosh, K.; Sharma, P. Visible light driven photocatalysis of organic dyes using SnO$_2$ decorated MoS2 nanocomposites. *Chem. Phys. Lett.,* **2020**, *738*, 136874.
[http://dx.doi.org/10.1016/j.cplett.2019.136874]

[28]   Khan, S.A.; Arshad, Z.; Shahid, S.; Arshad, I.; Rizwan, K.; Sher, M.; Fatima, U. *Synthesis of TiO2/Graphene Oxide Nanocomposites for Their Enhanced Photocatalytic Activity against Methylene Blue Dye and Ciprofloxacin*; Compos B Eng, **2019**, p. 175.

[29]   Kamari, S.; Shahbazi, A. Biocompatible Fe$_3$O$_4$@SiO$_2$-NH$_2$ nanocomposite as a green nanofiller embedded in PES–nanofiltration membrane matrix for salts, heavy metal ion and dye removal: Long–term operation and reusability tests. *Chemosphere,* **2020**, *243*, 125282.
[http://dx.doi.org/10.1016/j.chemosphere.2019.125282] [PMID: 31734593]

[30]   Sharma, M.; Poddar, M.; Gupta, Y.; Nigam, S.; Avasthi, D.K.; Adelung, R.; Abolhassani, R.; Fiutowski, J.; Joshi, M.; Mishra, Y.K. Solar light assisted degradation of dyes and adsorption of heavy metal ions from water by CuO–ZnO tetrapodal hybrid nanocomposite. *Mater. Today Chem.,* **2020**, *17*, 100336.
[http://dx.doi.org/10.1016/j.mtchem.2020.100336]

[31]   Gapusan, R.B.; Balela, M.D.L. Adsorption of anionic methyl orange dye and lead(II) heavy metal ion by polyaniline-kapok fiber nanocomposite. *Mater. Chem. Phys.,* **2020**, *243*, 122682.
[http://dx.doi.org/10.1016/j.matchemphys.2020.122682]

[32]   Preethi, S.; Abarna, K.; Nithyasri, M.; Kishore, P.; Deepika, K.; Ranjithkumar, R.; Bhuvaneshwari, V.; Bharathi, D. Synthesis and characterization of chitosan/zinc oxide nanocomposite for antibacterial activity onto cotton fabrics and dye degradation applications. *Int J Biol Macromol* **2020**, *164*, 2779–2787.

[33]   Al-Rawashdeh, N. A. F.; Allabadi, O.; Aljarrah, M. T. Photocatalytic activity of graphene oxide/zinc oxide nanocomposites with embedded metal nanoparticles for the degradation of organic dyes. *ACS Omega,* **2020**, *5*(43), 28046-28055.
[PMID: 33163787]

[34]   Eskandari Azar, B.; Ramazani, A.; Taghavi Fardood, S.; Morsali, A. Green synthesis and characterization of ZnAl2O4@ZnO nanocomposite and its environmental applications in rapid dye degradation. *Optik (Stuttg.),* **2020**, *208*, 164129.
[http://dx.doi.org/10.1016/j.ijleo.2019.164129]

[35]   Potle, V.D.; Shirsath, S.R.; Bhanvase, B.A.; Saharan, V.K. Sonochemical preparation of ternary rGO-ZnO-TiO$_2$ nanocomposite photocatalyst for efficient degradation of crystal violet dye. *Optik (Stuttg.),* **2020**, *208*, 164555.
[http://dx.doi.org/10.1016/j.ijleo.2020.164555]

[36]   Maruthupandy, M.; Qin, P.; Muneeswaran, T.; Rajivgandhi, G.; Quero, F.; Song, J.M. Graphene-zinc oxide nanocomposites (G-ZnO NCs): Synthesis, characterization and their photocatalytic degradation of dye molecules. *Mater. Sci. Eng. B,* **2020**, *254*, 114516.
[http://dx.doi.org/10.1016/j.mseb.2020.114516]

CHAPTER 8

Role of Nanosorbents in Wastewater Treatment

Roopa Rani[1,*], Jaya Tuteja[2], Akanshya Mishra[3], Jasaswini Tripathy[3] and **Arpit Sand[1]**

[1] *Department of Sciences, School of Sciences, Manav Rachna University, Faridabad, Haryana-121003, India*

[2] *School of Basic Sciences, Galgotias University, Greater Noida, India*

[3] *School of Applied Sciences (Chemistry), KIIT Deemed to be University, Bhubaneswar- 751024, India*

Abstract: Nanomaterials must be extensively considered in wastewater management to remove contaminants like heavy metals, pesticides, and dyes owing to their unique properties at the nanoscale. These nanomaterials have greater surface area and tunable surface chemistry, which make them appropriate for use as nanosorbents to gather the contaminations existing in water. The tunable surface properties allow nanosorbents to selectively absorb particular contaminants based on their functionalization. The application of nanomaterials to strengthen the adsorption phenomenon and eliminate numerous pollutants from wastewater has been considered. As the world is struggling for drinkable water and many cities are being declared as dry states, the urge to treat water is increasing continuously. Since the volume of wastewater is increasing day by day, the treatment of wastewater becomes a necessity. The present review includes the study of the properties of different nanosorbents and the method of sorption applied for the elimination of pollutants in wastewater.

Keywords: Adsorption isotherms, Nanomaterials, Nanosorbents, Treatment methods, Wastewater.

INTRODUCTION

Water, which is an indispensable component of the living system, represents the essential component of the existence of life on Earth. It constitutes an important segment that serves as the cornerstone of economics and ecosystems. It is important for nurturing agriculture as it has a pivotal role in the powering industry. Considering the abundance of its utility, the availability of clean and drinkable water is a major concern. About three-fourths of the earth is comprised of water; however, the availability of fresh water for the consumption of living

---

[*] **Corresponding author Roopa Rani:** Department of Sciences, School of Sciences, Manav Rachna University, Faridabad, Haryana-121003, India; E-mail: rooparani@mru.edu.in

Anjaneyulu Bendi (Ed.)

organisms is very limited. Mostly, the water content present in the ocean is saline and cannot be used for direct consumption. The treatment of such water systems is necessary to make them fit for domestic, industrial, commercial, and other uses. With continuous rise in populations, rapid urbanization, and industrialization, the demand for fresh and pure water is rising exponentially, resulting in exacerbating pressures on existing water sources. Thus, it generates the necessity to conserve, preserve, and treat used water to fulfill the demand.

Wastewater is generally obtained after utilizing freshwater or potable water for different purposes like washing, bathing, laundry, refinishing of furniture, beauty salons, *etc* [1]. The water obtained after its commercial, industrial, or domestic use (also known as wastewater) may also contain several contaminants and need to be further treated before reusing. Wastewater can also be referred to as used water from several fusions of agricultural, industrial, and commercial activities [2]. As per the surveys of the Central Board of Pollution Control, India, it has been estimated that more than 72 million liters of wastewater per day is generated across the country, out of which only 20% is treated. Treatment of wastewater not only helps solve the water crisis but also plays a significant role in safeguarding public health by reducing the risk of waterborne diseases due to the contaminants present in the wastewater. It supports the protection of the environment by effectively treating the contaminants before entering into rivers and lakes, thus safeguarding the aquatic ecosystems and biodiversity. The effective treatment of wastewater supports the good health of aquatic organisms, ensures safe water for sanitation and recreational purposes, facilitates the responsible management of water resources, and conserves freshwater supplies to reduce pressure on freshwater resources. The main significance of managing wastewater treatment can be focused on water security. Alternate sources of water and rainwater harvesting can be some methods that may satisfy the expected demands of industries and households [3]. The present chapter deals with the methods of treatment of wastewater and also the role of nanomaterials as biosorbents for such treatment.

Water and its Contaminants

Contaminants in water come in diverse forms, ranging from organic pollutants like pesticides and industrial chemicals to inorganic substances such as heavy metals and nitrates. These contaminants often infiltrate water sources through factory releases, agrarian overflow, improper methods of waste dumping, and natural processes like erosion. Once in the water, they provide substantial hazards to human health and the surroundings, potentially causing acute poisoning, chronic diseases, ecological imbalances, and even economic losses. Heavy metals like lead, mercury, and arsenic are predominantly vis-à-vis due to their perseverance and poisonousness, accumulating in aquatic ecosystems and posing long-term health risks to humans and wildlife. Similarly, nutrient pollutants such as phosphates and nitrates, primarily from fertilizers and sewage, can lead to harmful algal blooms, oxygen depletion, and ecosystem degradation in water bodies. Additionally, microbial contaminants like bacteria, viruses, and parasites pose threats to waterborne diseases, especially in areas with inadequate sanitation and water treatment infrastructure. Addressing water contamination requires a multifaceted approach involving monitoring, regulation, treatment, and public awareness. Water quality monitoring programs help identify sources and trends of contamination, guiding regulatory efforts to establish and enforce standards for pollutant levels in water bodies. Treatment machinery such as separation tools, chemical disinfection, and oxidation processes are employed to remove or neutralize contaminants from water supplies, ensuring safe drinking water for communities.

METHODOLOGIES USED FOR WASTEWATER TREATMENT

Water comprises a major part of the Earth's surface. But most of that water consists of a lot of contaminants. These impurities are mostly waste from many industries, dyes, agricultural wastes, and heavy metals [4]. Numerous methodologies can be employed to treat wastewater, each focused on targeting specific contaminants. The initiation step of any treatment methodology includes the physical process involving filtration, screening, and sedimentation of large visible debris to remove it from wastewater. These impurities often show many adverse effects. Hence, it is important to segregate the impurities. There are many techniques used for the separation of waste from water medium. Several factors are typically taken into consideration when choosing a method, including cost, practicality, environmental impact, efficiency, dependability, and operational challenges [5]. The secondary treatment includes the biological processes to break down the microorganisms into environmentally suitable substances. The tertiary and last step for treatment is to disinfect the water with chlorine and other chemicals. Additional upgraded methods like UV disinfection and reverse osmosis can also be used to remove pathogens. Various approaches are adopted to degrade wastewater discharges and address the problems associated with pollutants. The wastewater treatment includes various methods, as represented below in Fig. (**1**) [6].

Direct Membrane Filtration

Direct membrane filtration is the progressive approach for wastewater treatment to filter out the contaminants without involving expensive pre-treatment steps. This method is used to separate impurities by using a membrane. The process involves the pumping of water through a semi-permeable membrane while retaining solids, bacteria, and other solids or contaminants on the other side of the membrane. Factors affecting membrane filtration include the size of the membrane pore, substrate size, the pressure applied, and the concentration of the solution [7]. This method is mostly used for treating organic pollutants. The direct membrane process rejects the organic pollutants in every concentration, and they can further be reused to produce renewable energy. This method does not involve any additional activated sludge process [8]. aAdirect membrane process is also used to remove the heavy metals. The significant advantage of using this methodology over traditional treatment methods is the high efficiency and efficacy to retain pharmaceuticals, microplastics, and trace organic compounds. This method can be easily integrated with a water treatment plant for the primary treatment of wastewater. The different types of filtration processes used are ultra-filtration, nano-filtration, reverse osmosis, and electrodialysis [9].

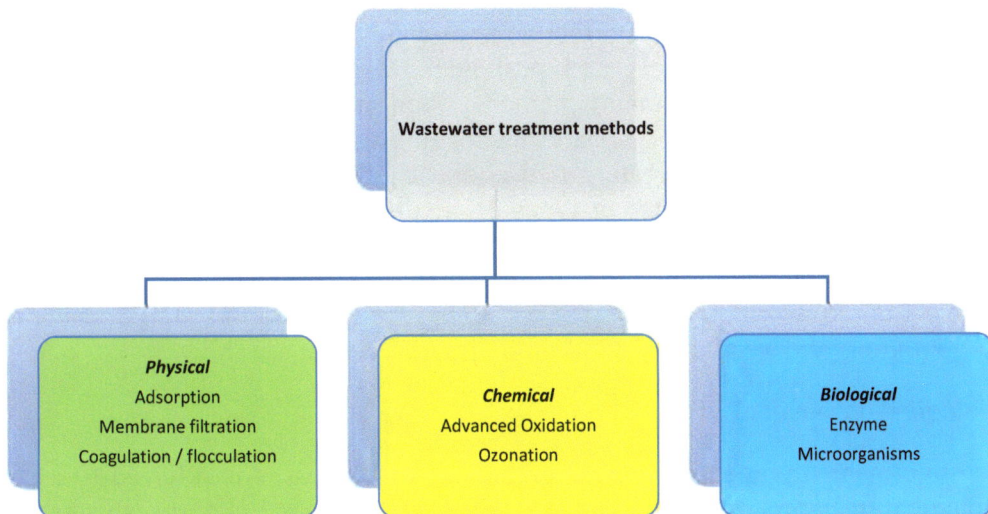

Fig. (1). Methods used in the treatment of wastewater.

Coagulation / Flocculation

Coagulation and flocculation are the processes that are used for the removal of suspended solids and other dissolved materials from water. The first step, coagulation, includes the addition of chemical coagulants like ferric chloride and aluminum sulfate to neutralize the surface charges of contaminants present in wastewater [10]. Following this, flocculants andvarious polymers are added to form larger aggregates that can be easily removed from water. This step involves the formation of flocs *i.e* ., larger and denser particles. Both together help remove impurities from wastewater. The various factors on which the efficacy of coagulants and flocculants depends are the dosage of chemicals used, the pH of wastewater, *etc*. This technique is effective for removing a wide range of contaminants. Studies have shown that for removing dyes and wastes from the paper and wood industry, the coagulation-flocculation method is the most useful [11]. This method consists of three mechanisms: (a) destabilizing the particle with a low coagulant quantity, (b) adding coagulants at very high concentrations, and (c) bridge formation [12]. The main advantages of the coagulation-flocculation method are that it is easy to operate, economical, and energy-efficient [13].

Advanced Oxidation

Advanced oxidation processes (AOPs) are known for the elimination of carbon-based pollutants and other persistent contaminants. The advanced oxidation methods are defined by oxidation and reactions associated with the production of reactive oxygen species [14]. The hydroxyl radicals ('OH) help in the degradation of organic compounds to environmentally less harmful products with water and carbon dioxide. The method is effective in processing the contaminants produced from personal care products and pesticides, which are hard to remove by traditional wastewater treatment methods. This method includes the attack of hydroxyl radicals on the organic effluents through different routes, including removal of hydrogen, integration and addition of radicals, and lastly, electron transfer [15]. There are different types of advanced oxidation methods, such as ozonization, electrochemical oxidation, photocatalytic oxidation, *etc* [16]. Despite being so advantageous, it is hard to implement the advanced oxidation method owing to a few challenges, such as high energy demand, large chemical consumption, and generation of harmful intermediates.

Biological Treatment

The treatment that uses biological organisms or processes is referred to as the biological treatment of wastewater [17]. Enzyme-producing microorganisms play a critical responsibility in wastewater treatment processes by facilitating the breakdown of organic matter and pollutants. These microorganisms produce a variety of enzymes that catalyze biochemical reactions involved in the degradation and transformation of the organic content of wastewater. It is one of the environmentally friendly approaches, efficient in removing organic pollutants and other harmful pollutants by converting or degrading them to simpler units. The significant treatment methods include aerobic and anaerobic processes with the objective of specific treatment to specific pollutants. The aerobic process involves the utilization of microorganisms to metabolize organic components into carbon dioxide or biogas. On the contrary, anaerobic biological treatment works on microbial activity without oxygen to degrade organic compounds through fermentation and anaerobic respiration. This approach utilizes a natural cell function and relies on microorganisms for the decomposition of biodegradable wastes [18]. Biological treatment has been divided into two groups: conventional and non-conventional methods. The traditional biological processes used are defined as conventional methods. These techniques are biofilm reactors, biofilters, microorganism-based, aerobic/anaerobic treatment, bioremediation, and biological nitrification and denitrification [19]. The application of non-traditional treatment methods *via* biological approaches is currently a growing focus of research. At present, prominent non-traditional bio-treatments include man-made wetlands, microbial electrochemical cells, biosorption, and membrane bioreactors [20].

Adsorption

Amongst all these methods, adsorption is preferred more as it has many advantages since it is more economical, flexible, and effective and also produces a more purified product. It offers an adaptable and efficient approach for the exclusion of variation among contaminants. In this process, impurities adhere to the surface of the adsorbent. The adsorbents can be selectively designed for the adsorption of particular impurities. The forces that drive the adhesion of impurities on the surface of the adsorbent are chemical interactions like hydrogen bonding or ion exchange and physical interactions like Van der Waals forces. The major advantage is that the adsorbents that are used in the process can be used multiple times [21]. Adsorption is the process characterized by a shift of the concentration of a molecule between the bulk solution and the surface layer. Using adsorption during the treatment process requires a good interaction between an adsorbate, adsorbent, and the wastewater. The adsorption technique is mostly controlled by the strong bonding between the adsorbate and adsorbent [22]. The adsorption processes involve different types of contacting systems. The batch reactors often use discontinuous treatment methods, and the fixed-bed type reactors use continuous treatment [23] (Fig. **2**).

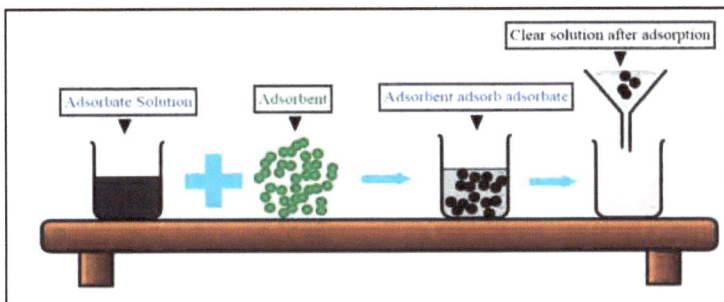

Fig. (2). Adsorption Process [24].

THE FUNCTION OF NANOPARTICLES IN THE ADSORPTION METHOD FOR WASTEWATER TREATMENT

Materials with dimensions ranging from 1 to 100 nm are called nanoparticles [25]. Studies have shown that nanoparticles have characteristics such as high reactivity, high surface plasmon resonance, high adsorption capacity, and catalytic activities, which result in different applications [26]. They represent a versatile tool in pollutant absorption from wastewater with enhanced efficiency and effectiveness. The small size of NPs with a high surface area makes them capable of capturing heavy metals, dyes, pesticides, and organic compounds from the wastewater. Furthermore, nanoparticles can be coordinated to possess specific surface characteristics and functional groups tailored to target particular contaminants. Surface modifications, such as coating nanoparticles with organic ligands or functionalizing them with reactive groups, can enhance their adsorption affinity towards specific pollutants, improving the efficiency and selectivity of the treatment process. There has been a lot of interest in the application of nanoparticles in wastewater treatment where several nanoparticles have played the role of sorbents. As nanoparticles have properties like good adsorption ability and combine small size and large surface area, they act as good nanosorbents and show efficient results [27]. Moreover, nanoparticles exhibit rapid adsorption kinetics due to their small size and high diffusivity, leading to shorter contact times and faster pollutant removal rates compared to conventional adsorbents. This rapid adsorption kinetics is particularly advantageous for treating large volumes of wastewater within shorter timeframes, making nanoparticles suitable for both batch and continuous-flow treatment systems. Emerging field nanotechnology offers a possible solution for water purification that is low-cost and highly effective at eliminating contaminants by the use of nanosorbents [28]. Additionally, nanoparticles offer the possibility of regeneration and reuse, extending their operational lifespan and reducing treatment costs. Through desorption techniques, adsorbed contaminants can be released from nanoparticles, allowing them to be regenerated and recycled for multiple treatment cycles, thereby minimizing waste generation and resource consumption (Fig. **3**).

TYPES OF NANOSORBENTS

Wastewater treatment is done using nanoparticles like CNT, graphene, activated carbon, ZnO, MgO, Fe_3O_4, Co_3O_4, *etc* [30]. Their different morphological forms, *e.g.* sheets, particles, rods, tubes, *etc*., can have variable extent of adsorption [31].

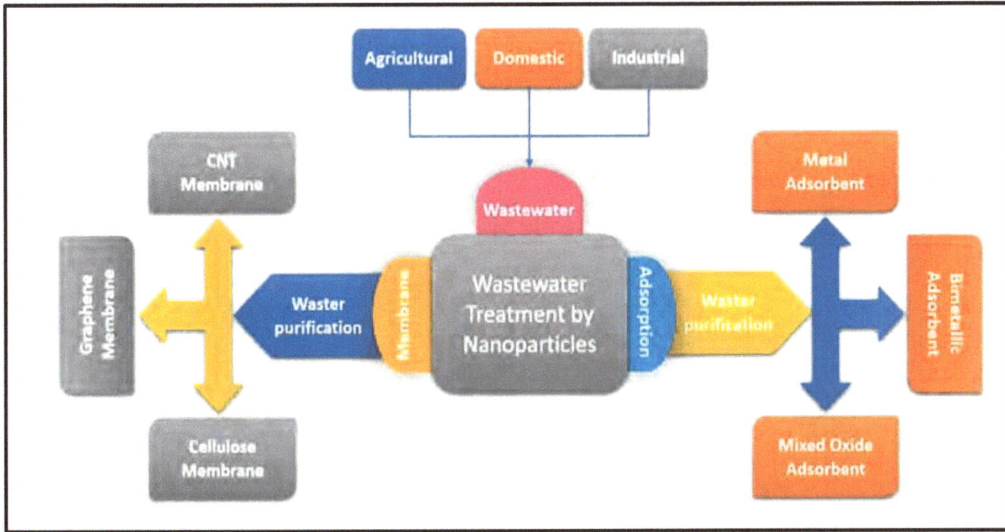

Fig. (3). Role of nanoparticles in treating wastewater [29].

Activated Carbon

Activated carbon was one the best methods to remove heavy metals and dyes, but their removal at the ppb level was a challenge; therefore, the use of CNTs, graphene, and fullerene came into existence. The synthesis of activated carbon through different precursors like wood, charcoal, coal, agricultural waste, and coconut shells shall result in some specific features such as high porosity and greater surface area [32]. Activated carbon (AC) is used frequently as an adsorbent due to its large surface dimension, good chemical reactivity, and microporous structure. The exterior part of the AC contains several functional groups [33]. They serve many advantages, such as they effectively adsorp contaminants and produce a good output. The property of weak acidic ion exchange character also helps to remove organic and inorganic impurities and heavy metal pollutants from wastewater effluent streams [34]. The different precursors used for the manufacture of AC are coconut shells, agricultural wastes, and chestnut shells. It removes the pollutants in the form of trace metals due to weak ion exchange characteristics. AC used in the adsorption process is also combined with other physical and chemical treatments for the production of a better yield [35]. The same was evidenced through the study of granular activated carbon acting as sorbent and used for the removal of As(V) [36]. A study was also carried out on the activated carbon prepared using the coconut tree sawdust, which was an effective adsorbent to remove Cr(VI) from wastewater [37]. Removing Hg(II) using activated carbon was also reported by Chatterjee *et al.* in 2010 [38]; however, Pb(II) and Cu(II) were extractedfrom the amorphous form of activated carbon produced using the bark of Eucalyptus camaldulensis [39].

Recent studies also show that to increase the sorption capacity on the surface of activated carbon, grafting with sodium polyacrylate was carried out through gamma irradiation, resulting in the development of an effective, simple, and cheap method for the removal of heavy metals and dyes [40]. The AC also has certain disadvantages. The higher quality of AC is often quite expensive and not that effectively selective. Adsorption by activated carbons (ACs) also comes with other limitations. For instance, there is a potential reduction in adsorption efficiency following the regeneration process. Additionally, regeneration is often necessary once the adsorption capacity is depleted. Another drawback is the possibility of secondary pollution since contaminants are separated from the activated carbons but not necessarily destroyed in the process.

Metal Oxide-based Nanosorbents

Nanosized metal oxides (NMO) are better adsorbents than activated carbons as they can attack the target contaminants at a faster rate [41]. FeO, TiO_2, MnO_2, and Al_2O_3 are eco-friendly and very effective for the degradation of heavy metals from wastewater. These NMOs can function as both Lewis acids and bases, demonstrating an ability to bind transition metals [42]. Metal nanomaterials are regarded as superior materials for eliminating harmful dyes, particularly azo dyes, owing to their extensive surface area and effective catalytic function [43]. Metal oxide nanoparticles are also a good source of removal of heavy metals and dyes from water systems like Fe_3O_4 [44], MnO_2 [45], TiO_2 [46], MgO [47], CdO [48], and ZnO [49], since they possess high surface area and greater affinity. Since metal oxides have the least environmental impact, they can be easily used to remove heavy metals from aqueous systems [50]. It can be demonstrated through the fact that the adsorption capacity of Fe_3O_4 metal oxide for the heavy metals of Ni(II), Cu(II), Cd(II), and Cr(VI) is dependent on the adsorbent amount, pH, incubation time, and temperature [51]. Modification in the features of metal oxide can lead to a change in adsorption capacity, like a change in functionality based on different functional groups, such as the introduction of COOH, NH_2, or SH group can also help in the removal of dyes or heavy metals *via* the process of chelation, *etc* [52].

The metal oxide nanoparticles are easy to reuse, have more adsorption sites, are compressible without effectively reducing surface area, and have a large specific surface area and a short internal diffusion distance [53]. The precursors used in these nanoparticles are purely made up of metal oxides [54]. These nanoparticles have a high selectivity for heavy metals and good removal capacity. The different types of metal oxides considered are titanium, zinc, copper, and iron oxides, among others. The size, aggregation, and form of these nanoparticles are among the many variables that affect them [55] (Fig. **4**).

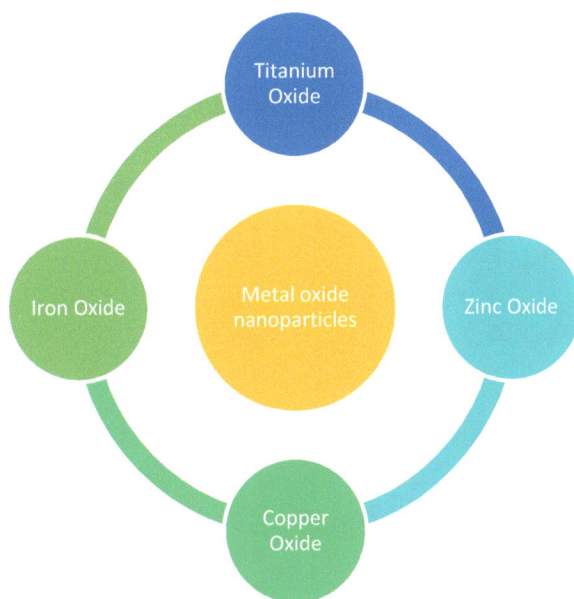

Fig. (4). Different types of metal oxide nanoparticles.

Titanium Oxide NP

Due to their high chemical stability, reasonable price, and low toxicity, TiO_2 NPs are used as photocatalysts in the treatment of wastewater [56]. TiO_2 nanoparticles exhibit minimal selectivity, rendering them well-suited for the process of degradation of various contaminants [57]. The TiO_2 NPs are very efficient in eliminating different types of viruses such as hepatitis B virus [76], Herpes simplex virus, MS_2 bacteriophages [58], *etc.*

Zinc Oxide NP

ZnO nanoparticles are considered very effective for wastewater treatment due to their unique attributes, such as potent oxidation capabilities, excellent photocatalytic properties, and a broadband gap within the UV spectral range [59]. Zinc oxides are used as a disinfectant to treat wastewater. In the disinfectant method, a good photocatalyst has a significant impact, as shown in studies [60].

Copper Oxide NP

Copper (II) oxide exhibits a monoclinic crystal structure and possesses numerous advantageous chemical and physical properties. These include high-temperature superconductivity, efficiency in solar energy utilization, stability, affordability, and antibacterial properties [61]. Studies have shown that CuO NPs have been prepared by using copper oxide as the precursor and sodium hydroxide as the reducing agent [62]. The utilization of copper and its compounds as bacterial disinfectants stands out as a significant application, which is crucial for minimizing toxicity levels due to human and biological activities in wastewater [63].

Iron Oxide NP

Extensive research has been dedicated to the utilization and creation of iron oxide nanoparticles driven by their considerable surface area, superparamagnetism, strong tensile strength, enhanced membrane characteristics, and nanoscale dimensions [64]. Iron oxide-containing nanoparticles have been engineered to modify their adsorption properties [65]. The synthesized iron oxide nanoparticles have demonstrated effective dye degradation results across a wide range of pH levels [66].

Zero-Valent Nanoparticles

Zero-valent nanoparticles are often used to treat toxic oxidized materials or reducing agents [67]. Zerovalent iron nanoparticles (nZVI) have garnered significant attention as promising nano adsorbents for wastewater treatment because of their high surface area and tunable properties. Zn nanoparticles used halogenation reactions to treat wastewater, as shown in many studies [68]. It also obstructs the growth process of the nitrifying bacteria using the sequencing batch reactor [69].

Iron Nanoparticles

Fe nanoparticles use the reductive dehalogenation processes for the degradation of wastewater [70]. These nanoparticles, composed of elemental iron in its zero oxidation state, exhibit exceptional adsorption capabilities and catalytic activities that make them effective for removing various contaminants from wastewater. Fe nanoparticles show a broad range of organic contaminants by transmitting $2e^-$ to an oxygen molecule, resulting in the formation of hydrogen peroxide, which subsequently undergoes reduction to yield a water molecule [71].

Silver Nanoparticles

AgNPs have emerged as promising nano adsorbents for wastewater treatment because of their exclusive physico-chemical properties, high surface area, and antimicrobial activity. These nanoparticles possess excellent adsorption capabilities and can effectively remove various contaminants from wastewater, including heavy metals, organic pollutants, and pathogens. Due to their good antimicrobial properties, AgNPs are used as a disinfectant [72]. The antimicrobial properties of AgNPs make them particularly suitable for disinfection and microbial control in wastewater treatment. Upon contact with microbial cells, AgNPs release silver ions (Ag^+), which disrupt cell membranes, inhibit enzyme functions, and interfere with microbial metabolism, leading to the inactivation and elimination of bacteria, viruses, and other pathogens. This antimicrobial action helps prevent the spread of waterborne diseases and ensures the safety of treated wastewater for reuse or discharge into the environment. Various studies have shown that there has been a significant inclination for the immobilization of Ag nanoparticles on membranes and ceramic materials due to their promising potential for mitigating biofouling and enhancing disinfection in wastewater treatment [73].

Biosorbents

The biosorbents lead to the bacteria becoming immobile on absorbent materials, enabling them to seize impurities from the wastewater and produce a pure product [74]. Biosorption offers several advantages, such as cost-effective, reusable, and efficient, as well as provides the potential for metal recovery when compared to other adsorption techniques [75]. There are several types of biosorbents, such as fungi, yeast, bacteria, algae, peat, chitosan, and polysaccharides [76].

C-based Nanoparticles

In recent decades, the removal of heavy metals and dyes has been widely achieved using distinct types of carbon-based nanomaterials on account of their nontoxic nature, easy method of synthesis, abundance, penetrability, greater sorption capacities, and more surface area [77, 78].

Carbon Nanotubes (CNTs)

The second category of nanomaterial used for wastewater treatment is carbon nanotubes (CNTs). The unique structural features, electronic and optical properties, and semi-conductor behavior of the material have led to its increasing usage in the field of wastewater treatment to remove dyes as well as heavy metal content [79]. CNTs have a high absorption capacity for dyes and thus are widely used for their removal from wastewater effluent of industrial discharge [80]. CNTs can be single-walled (SWCNTs) and multi-walled (MWCNTs), and it has been observed that MWCNTs have a higher capacity to remove contaminants from wastewater [81] than SWCNTs (Fig. **5**).

The removal of dyes from wastewater with the use of CNTs has been less widely studied [106 - 108]. The best feature of the use of CNTs in wastewater treatment is that they can be directly used for the process of removal without any further modification [82, 83]. However, mild oxidation, including the introduction of hydroxyl and carbonyl groups to the walls of CNTs, can result in greater potential for the adsorption of heavy metals to different sites of CNTs. They can lead to the effective removal of certain harmful dyes, such as methylene red and methylene blue, from wastewater [84].

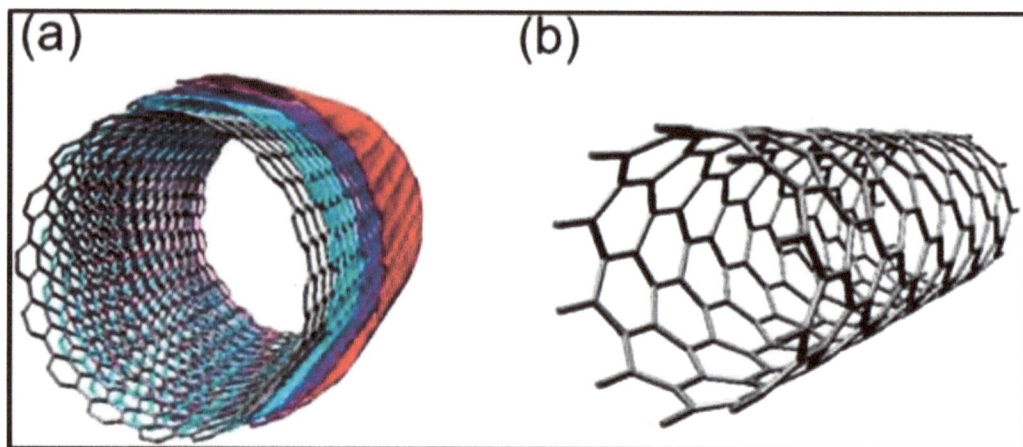

Fig. (5). a) Multi-walled CNTS b) Single-walled CNTs.

Graphene

Graphene (Fig. **6**), having a structure of single atomic layered carbon atoms, has the unique property of removing toxins from wastewater and has excellent electrical and optical properties [85]. Recent studies show that the modification of graphene sheets can also be a better source for altering chemical and physical properties, which can lead to further improvement of the adsorption capacity of graphene. The modification may be of reduced graphene oxide (rGONSs) or graphene oxide nanosheets (GONSs) [86]. Because of the opened-up sheet structure, rGONSs display quicker adsorption kinetics than CNTs [87]. Among all other carbon-based nanomaterials, rGONSs revealed enhanced adsorption abilities for two synthetic organic compounds, phenanthrene and biphenyl, in wastewater [88]. Other dyes that can be easily removed with the use of rGONSs are cationic red X-GRL [89], methylene blue [90], methyl orange [91], *etc*. The adsorption of three pesticides (chlorpyrifos (CP), endosulfan (ES), and malathion (ML)) have also been reported in recent studies with the use of graphenes from wastewater as great as 1200, 1100, and 800 mg g^{-1}, respectively [92].

TYPES OF ADSORPTIONS STUDIES

The adsorption phenomenon is mainly of two types - physical and chemical. In physical adsorption, the molecules are linked to each other by Van der Waals forces of interaction. This type is reversible but less specific. While in chemical adsorption, the molecules are bound by either ionic or covalent forces. This type is irreversible and more specific [93].

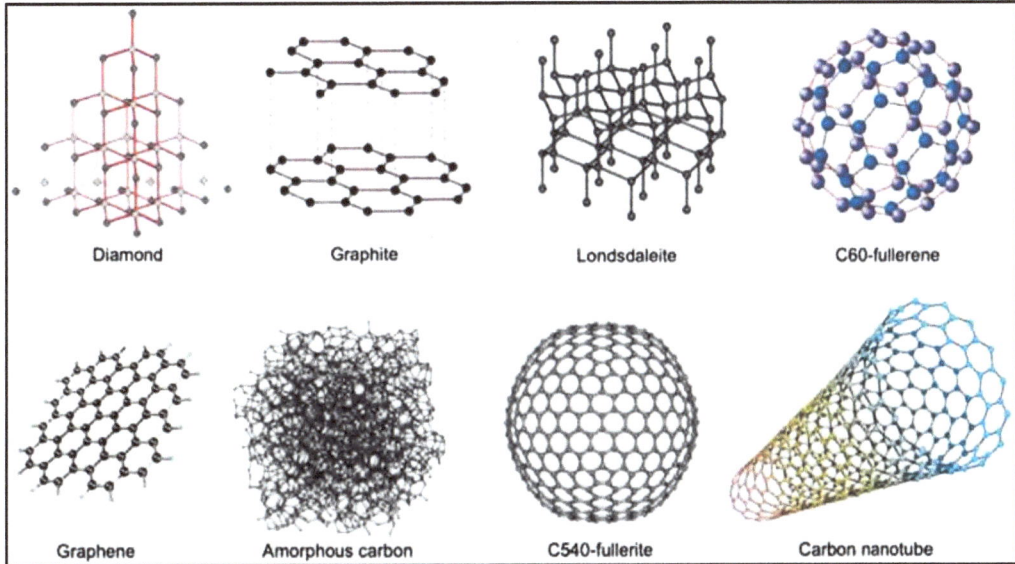

Fig. (6). Different forms of carbon used as adsorbents for the removal of heavy metals and dyes from wastewater.

Adsorption Isotherms

Adsorption isotherms can be explained as the equation that correlates the value of adsorbate present in the adsorbent's interface and the concentration of the adsorbent. It can also be associated with partial pressure under constant temperature [94].

Types

Langmuir Adsorption

This isotherm is used to explain gas-solid adsorption. It is monolayer in nature. The adsorbent surface is homogeneous. It can be obtained from the following equation:

$$\frac{C_e}{q_e} = \frac{1}{q_m KL} + \frac{C_e}{q_m}$$

Where Ce is the adsorbate concentration, qm is the number of particles adsorbed on the surface of the adsorbent at any time and related to its maximum adsorption capacity, andKLis the Langmuir constant.

Freundlich Isotherm

This isotherm is multilayer in nature. The surface is heterogeneous [95]. The Freundlich isotherm is calculated as:

$$\log q_m = \log K_F + \frac{1}{n} \log C_e$$

Where Kf is the Freundlich constant and n is the surface heterogeneity extent.

Temkin Isotherm

This isotherm describes that the interaction is indirect between the adsorbate and the heat of adsorption of the surfaces [96]. The Temkin isotherm is calculated by using the formula:

$$q_m = \frac{RT}{b} \ln(K_T C_{e.})$$

Where K_T is the isotherm constant, b is the Temkin constant, R is the gas constant, and T is temperature.

Brunauer, Emmett, and Teller (BET) Isotherm

BET also considers multilayer adsorption. It is used mostly for the porosity of the adsorbents and the gas-solid system [97].

Redliche -Peterson Isotherm

The mathematical formula used to obtain the Peterson isotherm is:

$$q_e = \frac{k_R C_e}{1 + \alpha_R C_e^{\beta}}$$

Where, KR and aR - Redlich Peterson isotherm constant and β - the exponent between 0-1 [98].

Koblee - Corrigan Isotherm

It is obtained through a mathematical formula:

$$q_e = \frac{aC_e^n}{1+bC_e^n}$$

Where a, b, and n are the Koble Corrigan parameters.

Sips Isotherm

It can be obtained through the formula:

$$q_e = \frac{Q_{max}K_SC_e^{1/n}}{1+K_SC_e^{1/n}}$$

Where KS - Sips constant and Qmax - Sips maximum adsorption capacity.

Toth Isotherm

It is obtained through the equation:

$$q_e = \frac{fC_e}{[g+(C_e)^d]^{1/4}}$$

Where, f, g, and d are Toth constants [99].

Dubinine - Radushkevich Isotherm

It can be expressed mathematically as: $q_e = q_m e^{-\beta \varepsilon^{2.}}$

Where, $\varepsilon = RT \ln(1 + \frac{1}{C_e})$, ε - the Polanyi potential.

Fritze - Schlunder Isotherm

It is calculated using the equation:

$$q_e = \frac{K_{FS}q_m C_e}{1+ q_m C_e^{mFS}}$$

Where q_{mFS} is the maximum Fritze-Schlunder adsorption capacity and K_{FS} and m_{FS} are the Fritze- Schlunder parameters [100].

Radkee - Prausnitz Isotherm

This isotherm can be expressed mathematically as:

$$q_e = \frac{K_{RP}q_m C_e}{(1+K_{RP}C_e)^{mRP.}}$$

Where, qmRP - maximum adsorption capacity and KRP - RadkeePrausnitz isotherm model constant.

Dubinin - Radushkevich Isotherm

It can be obtained with the expression:

$$\ln q_e = \ln q_S K_D \varepsilon^2$$

Where, qs -theoretical isotherm saturation capacity; KD and ε - Dubinin-Radushkevich constant [101].

POLLUTANTS AND THEIR REMOVAL USING NANO-ADSORBENTS

Nano-adsorbents hold promise for the removal of various pollutants from wastewater. Here is a breakdown of some common pollutants and how specific nano-adsorbents can target them:

Heavy Metals

Nano-adsorbents such as metal oxide nanoparticles (*e.g* ., iron oxide, titanium dioxide) and carbon-based nanomaterials (*e.g* ., carbon nanotubes, graphene oxide) are effective in adsorbing heavy metals like lead, cadmium, mercury, and arsenic from wastewater. The high surface area and reactive sites of these materials facilitate strong interactions with metal ions, leading to their removal from aqueous solutions [102].

Organic Compounds

Nano-adsorbents have shown promising results in removing organic pollutants such as dyes, pharmaceuticals, pesticides, and hydrocarbons from wastewater. Functionalized carbon nanomaterials, in particular, exhibit high affinity towards organic molecules due to π-π interactions and hydrogen bonding. Additionally, nanocomposites incorporating metal nanoparticles or organic-inorganic hybrids enhance the adsorption capacity and selectivity for specific organic contaminants.

Nutrients

Nutrient pollution, particularly from nitrogen and phosphorus compounds, poses significant environmental challenges in wastewater treatment. Nano-adsorbents functionalized with specific ligands or metal ions can selectively adsorb phosphate and ammonium ions from wastewater, thereby mitigating eutrophication and algal blooms in receiving water bodies. Materials like metal-organic frameworks (MOFs) and zeolites have demonstrated potential for nutrient removal due to their well-defined pore structures and adjustable surface chemistry [103].

Emerging Pollutants

Nano-adsorbents offer a promising approach to removing emerging contaminants. Advanced nanomaterials with tailored properties, including mesoporous silica nanoparticles, magnetic nanoparticles, and dendrimers, exhibit high adsorption capacities and efficient removal of emerging pollutants, thus addressing emerging environmental concerns [104].

Pathogens

Nano-adsorbents functionalized with antimicrobial agents or surface-modified with specific ligands can effectively capture and remove pathogens (*e.g* ., bacteria, viruses, protozoa) from wastewater. Silver nanoparticles, for instance, possess inherent antimicrobial properties and can be immobilized onto various substrates to inhibit microbial growth and enhance pathogen removal efficiency in water treatment applications [105].

The proposed mechanisms for the removal of pollutants by nano adsorbents are:

Electrostatic Interactions

Metal oxide nanoparticles have charged surfaces that can attract oppositely charged contaminants like heavy metals.

Complexation

This involves the formation of a weak bond between the metal oxide nanoparticle and the pollutant molecule.

Chemisorption

This is a stronger form of adsorption where a chemical bond forms between the pollutant and the nano-adsorbent surface.

π-π Interactions

Graphene, with its network of carbon atoms, can interact with organic pollutants through a specific type of attractive force (pi-pi stacking) due to their electron configuration.

Size Exclusion

Zeolites have a well-defined pore structure. Pollutants larger than the pore size cannot enter, achieving a sieving effect.

LONG-TERM EFFECTS OF NANOSORBENTS IN THE ENVIRONMENT

Nanosorbents, which are materials designed at the nanometer scale to remove contaminants from wastewater, have been increasingly recognized for their effectiveness in addressing water pollution. However, their environmental impact presents a complex situation. Although nanosorbents offer numerous advantages, such as a high surface area and customizable chemical properties, there are growing concerns regarding their disposal and long-term environmental consequences [106]. Carbon nanotubes, graphene oxide, and magnetic nanoparticles are among the nanosorbents that have demonstrated potential in improving the treatment process. Despite their advantages, it is important to carefully consider the potential risks and challenges associated with the use of nanosorbents.

One issue that arises is the future of nanosorbents once they have been used in treating wastewater. Since nanosorbents are composed of different materials, such as metals, carbon, and polymers, disposing them of can be challenging. If not properly contained or recovered, nanosorbents may end up being released into the environment, where their small size can cause them to persist and spread. In aquatic environments, nanoparticles can be consumed by microorganisms, fish, and other wildlife, potentially disturbing food chains or causing toxicity [107]. Research indicates that certain nanosorbents, especially those containing heavy metals or toxic substances, can release harmful components into the environment over time, posing a threat to both ecosystems and human health [108].

The potential long-term environmental effects of nanosorbents are a cause for concern. Because of their unique characteristics, nanoparticles may exhibit different behaviors compared to larger materials, resulting in unpredictable interactions in the environment. For instance, nanosorbents might accumulate in soil or sediment, with uncertain impacts on microorganisms, plants, and other organisms. Furthermore, the possibility of bioaccumulation and biomagnification, where nanoparticles collect in the tissues of organisms and become increasingly concentrated as they move up the food chain, poses a significant concern. This situation can lead to unexpected ecological repercussions, potentially harming species crucial to ecosystem well-being.

Moreover, the synthesis and development of nanosorbents frequently require the use of harmful chemicals and energy-intensive procedures, which raises concerns about the sustainability of these materials. If nanosorbents are to be widely used, it is important to consider the environmental impact of their production. Sustainable synthesis methods, like green chemistry approaches, are being researched but are not fully established yet.

CONCLUSION

The ever-increasing burden of pollutants in wastewater poses a significant threat to global water resources and human health. Conventional wastewater treatment methods, while effective for basic pollutant removal, often struggle to address emerging contaminants like pharmaceuticals and industrial chemicals. In this context, nanosorbents have emerged as a revolutionary technology, offering a powerful and targeted approach to wastewater purification. Nanosorbents offer significant improvements over traditional adsorbents, achieving faster reaction rates, greater capacity, and specificity. However, challenges such as cost-effectiveness, large-scale application, and environmental impacts of nanoparticles need addressing for widespread adoption. Overall, nanosorbents represent a promising advancement in sustainable wastewater treatment, with ongoing research focused on optimizing their efficiency, safety, and scalability.

Nanosorbents, with their unique properties and exceptionally high surface area, can effectively capture a wide range of pollutants at the microscopic level. Carbon-based giants like carbon nanotubes (CNTs) and graphene act as sponges for organic pollutants and heavy metals due to their extensive porous structure and strong adsorption capabilities. Metal oxide nanoparticles, like iron oxide and titanium oxide, leverage electrostatic interactions and complexation to remove heavy metals from wastewater. Zeolites, with their well-defined pores, act as sieves, selectively filtering out pollutants based on size. Polymeric nanoparticles, tailored with specific functional groups, offer targeted removal of pollutants, while bio-based nanosorbents provide a sustainable and eco-friendly alternative.

The efficacy of nanosorbents extends beyond their exceptional adsorption capacity. Their properties can be customized to target specific pollutants. For instance, zeolites can be tailored to have pores that preferentially trap certain-sized molecules, while polymeric nanoparticles can be functionalized to attract specific contaminants. This level of control allows for a more targeted and efficient wastewater treatment process. Furthermore, some nanosorbents can be regenerated, making them reusable and reducing operational costs. This reusability aspect is crucial for long-term sustainability and economic feasibility. However, regeneration methods need further development for broader application across different nano-adsorbent types.

Despite the remarkable potential of nanosorbents, challenges remain. Cost-effective production and large-scale implementation are essential for widespread adoption. Additionally, the environmental impact and potential risks associated with the use and disposal of nanoparticles require careful consideration. Research efforts are underway to address these concerns, focusing on developing greener synthesis methods and responsible disposal strategies.

In conclusion, nanosorbents represent a revolutionary leap forward in wastewater treatment technology. Their ability to target a wide range of pollutants, coupled with their tailorable properties and reusability potential, positions them as a game-changer in ensuring clean water for future generations. As research continues to refine production methods, address environmental concerns, and optimize regeneration techniques, nanosorbents are poised to play a pivotal role in creating a more sustainable and water-secure world.

LIST OF ABBREVIATIONS

AC   Activated Carbon

AgNPs Silver Nanoparticles

AOP   Advanced Oxidation Process

CNTs   Carbon Nanotubes

MOFs   Metal-Organic Frameworks

NMOs   Nanosized Metal Oxides

NPs   Nanoparticles

rGO   Reduced Graphene Oxide

ACKNOWLEDGEMENTS

The authors would like to express their sincere thanks to the management of Manav Rachna University, Faridabad, Haryana, India, Galgotias University, Greater Noida, India, and KIIT Deemed to be University, Bhubaneswar, India, for providing the facilities to write and submit the book chapter for publication.

REFERENCES

[1]   Tchobanoglous, G.; Burton, F.L. *Stensel, David H. Wastewater engineering: treatment and reuse,* 4th ed; McGraw-Hill: Boston, **2003**.

[2]   Tilley, E.; Ulrich, L.; Lüthi, C.; Reymond, Ph.; Zurbrügg, C. *Compendium of Sanitation Systems and Technologies*; , **2014**.

[3]   Report, CPCB *Status of water supply, wastewater generation and treatment in Class I cities and Class II towns of India.,* **2009**.

[4]   Rashid, R.; Shafiq, I.; Akhter, P.; Iqbal, M.J.; Hussain, M. A state-of-the-art review on wastewater treatment techniques: the effectiveness of adsorption method. *Environ. Sci. Pollut. Res. Int.,* **2021**, *28*(8), 9050-9066.
[http://dx.doi.org/10.1007/s11356-021-12395-x] [PMID: 33483933]

[5]   Crini, G.; Lichtfouse, E. Advantages and disadvantages of techniques used for wastewater treatment. *Environ. Chem. Lett.,* **2019**, *17*(1), 145-155.
[http://dx.doi.org/10.1007/s10311-018-0785-9]

[6]   Xu, Z.; Wei, C.; Jin, J.; Xu, W.; Wu, Q.; Gu, J.; Ou, M.; Xu, X. Development of a novel mixed titanium, silver oxide polyacrylonitrile nanofiber as a superior adsorbent and its application for MB removal in wastewater treatment. *J. Braz. Chem. Soc.,* **2018**, *29*, 560-571.

[7]   Abdullah, N.; Yusof, N.; Lau, W.J.; Jaafar, J.; Ismail, A.F. Recent trends of heavy metal removal from water/wastewater by membrane technologies. *J. Ind. Eng. Chem.,* **2019**, *76*, 17-38.
[http://dx.doi.org/10.1016/j.jiec.2019.03.029]

[8]   Jin, Z.; Meng, F.; Gong, H.; Wang, C.; Wang, K. Improved low-carbon-consuming fouling control in long-term membrane-based sewage pre-concentration: The role of enhanced coagulation process and air backflushing in sustainable sewage treatment. *J. Membr. Sci.,* **2017**, *529*, 252-262.
[http://dx.doi.org/10.1016/j.memsci.2017.02.009]

[9]   Nasir, A.M.; Adam, M.R.; Mohamad Kamal, S.N.E.A.; Jaafar, J.; Othman, M.H.D.; Ismail, A.F.; Aziz, F.; Yusof, N.; Bilad, M.R.; Mohamud, R.; A Rahman, M.; Wan Salleh, W.N. A review of the potential of conventional and advanced membrane technology in the removal of pathogens from wastewater. *Separ. Purif. Tech.,* **2022**, *286*, 120454.
[http://dx.doi.org/10.1016/j.seppur.2022.120454] [PMID: 35035270]

[10]   Sahu, O.P.; Chaudhari, P.K. Review on Chemical treatment of Industrial Waste Water. *J. Appl. Sci. Environ. Manag.,* **2013**, *17*(2), 241-257.
[http://dx.doi.org/10.4314/jasem.v17i2.8]

[11]   Ahmad, A.; Wong, S.; Teng, T.; Zuhairi, A. Improvement of alum and PACl coagulation by polyacrylamides (PAMs) for the treatment of pulp and paper mill wastewater. *Chem. Eng. J.,* **2008**, *137*(3), 510-517.
[http://dx.doi.org/10.1016/j.cej.2007.03.088]

[12]   Kumar, M.M.; Karthikeyan, R.; Anbalagan, K.; Bhanushali, M.N. Coagulation process for tannery industry effluent treatment using *Moringa oleifera* seeds protein: Kinetic study, pH effect on floc characteristics and design of a thickener unit. *Sep. Sci. Technol.,* **2016**, *51*(12), 2028-2037.
[http://dx.doi.org/10.1080/01496395.2016.1190378]

[13]   Amuda, O.; Amoo, I.; Ajayi, O. Performance optimization of coagulant/flocculant in the treatment of wastewater from a beverage industry. *J. Hazard. Mater.,* **2006**, *129*(1-3), 69-72.
[http://dx.doi.org/10.1016/j.jhazmat.2005.07.078] [PMID: 16310308]

[14]   Ghernaout, D. Advanced oxidation phenomena in electrocoagulation process: a myth or a reality? *Desalination Water Treat.,* **2013**, *51*(40-42), 7536-7554.
[http://dx.doi.org/10.1080/19443994.2013.792520]

[15]   Garrido-Cardenas, J.A.; Esteban-García, B.; Agüera, A.; Sánchez-Pérez, J.A.; Manzano-Agugliaro, F. Wastewater treatment by advanced oxidation process and their worldwide research trends. *Int. J.*

Environ. Res. Public Health, **2019**, *17*(1), 170.
[http://dx.doi.org/10.3390/ijerph17010170] [PMID: 31881722]

[16] Ma, D.; Yi, H.; Lai, C.; Liu, X.; Huo, X.; An, Z.; Li, L.; Fu, Y.; Li, B.; Zhang, M.; Qin, L.; Liu, S.; Yang, L. Critical review of advanced oxidation processes in organic wastewater treatment. *Chemosphere,* **2021**, *275*, 130104.
[http://dx.doi.org/10.1016/j.chemosphere.2021.130104] [PMID: 33984911]

[17] Grandclément, C.; Seyssiecq, I.; Piram, A.; Wong-Wah-Chung, P.; Vanot, G.; Tiliacos, N.; Roche, N.; Doumenq, P. From the conventional biological wastewater treatment to hybrid processes, the evaluation of organic micropollutant removal: A review. *Water Res.,* **2017**, *111*, 297-317.
[http://dx.doi.org/10.1016/j.watres.2017.01.005] [PMID: 28104517]

[18] Mani, S.; Chowdhary, P.; Zainith, S. Microbes mediated approaches for environmental waste management. In: *Microorganisms for Sustainable Environment and Health*; Chowdhary, P., Ed.; Elsevier, **2020**; pp. 17-36.
[http://dx.doi.org/10.1016/B978-0-12-819001-2.00002-4]

[19] Girijan, S.; Kumar, M. Immobilized biomass systems: an approach for trace organics removal from wastewater and environmental remediation. *Curr. Opin. Environ. Sci. Health,* **2019**, *12*, 18-29.
[http://dx.doi.org/10.1016/j.coesh.2019.08.005]

[20] Ahmed, SF; Mofijur, M; Nuzhat, S; Chowdhury, AT; Rafa, N; Uddin, MA; Inayat, A; Mahlia, TM; Ong, HC; Chia, WY; Show, PL *Recent developments in physical, biological, chemical, and hybrid treatment techniques for removing emerging contaminants from wastewater.,* **2021**.
[http://dx.doi.org/10.1016/j.jhazmat.2021.125912]

[21] Pan, B.; Pan, B.; Zhang, W.; Lv, L.; Zhang, Q.; Zheng, S. Development of polymeric and polymer-based hybrid adsorbents for pollutants removal from waters. *Chem. Eng. J.,* **2009**, *151*(1-3), 19-29.
[http://dx.doi.org/10.1016/j.cej.2009.02.036]

[22] Furuya, e.g .; Chang, H.T.; Miura, Y.; Noll, K.E. A fundamental analysis of the isotherm for the adsorption of phenolic compounds on activated carbon. *Separ. Purif. Tech.,* **1997**, *11*(2), 69-78.
[http://dx.doi.org/10.1016/S1383-5866(96)01001-5]

[23] Crini, G. Recent developments in polysaccharide-based materials used as adsorbents in wastewater treatment. *Prog. Polym. Sci.,* **2005**, *30*(1), 38-70.
[http://dx.doi.org/10.1016/j.progpolymsci.2004.11.002]

[24] Ali, I. Water treatment by adsorption columns: evaluation at ground level. *Separ. Purif. Rev.,* **2014**, *43*(3), 175-205.
[http://dx.doi.org/10.1080/15422119.2012.748671]

[25] Stark, W.J.; Stoessel, P.R.; Wohlleben, W.; Hafner, A. Industrial applications of nanoparticles. *Chem. Soc. Rev.,* **2015**, *44*(16), 5793-5805.
[http://dx.doi.org/10.1039/C4CS00362D] [PMID: 25669838]

[26] Khan, I.; Saeed, K.; Khan, I. Nanoparticles: properties, applications and toxicities. *Arab. J. Chem.,* **2017**.

[27] Mauter, M.S.; Zucker, I.; Perreault, F.; Werber, J.R.; Kim, J.H.; Elimelech, M. The role of nanotechnology in tackling global water challenges. *Nat. Sustain.,* **2018**, *1*(4), 166-175.
[http://dx.doi.org/10.1038/s41893-018-0046-8]

[28] Baruah, S.; Najam Khan, M.; Dutta, J. Perspectives and applications of nanotechnology in water treatment. *Environ. Chem. Lett.,* **2016**, *14*(1), 1-14.
[http://dx.doi.org/10.1007/s10311-015-0542-2]

[29] Miao, X.S.; Bishay, F.; Chen, M.; Metcalfe, C.D. Occurrence of antimicrobials in the final effluents of wastewater treatment plants in Canada. *Environ. Sci. Technol.,* **2004**, *38*(13), 3533-3541.
[http://dx.doi.org/10.1021/es030653q] [PMID: 15296302]

[30] Zare, K.; Gupta, V.K.; Moradi, O.; Makhlouf, A.S.H.; Sillanpää, M.; Nadagouda, M.N.; Sadegh, H.;

Shahryari-ghoshekandi, R.; Pal, A.; Wang, Z.; Tyagi, I.; Kazemi, M. A comparative study on the basis of adsorption capacity between CNTs and activated carbon as adsorbents for removal of noxious synthetic dyes: a review. *J. Nanostructure Chem.,* **2015**, *5*(2), 227-236.
[http://dx.doi.org/10.1007/s40097-015-0158-x]

[31]   Gupta, V.K.; Moradi, O.; Tyagi, I.; Agarwal, S.; Sadegh, H.; Shahryari-Ghoshekandi, R.; Makhlouf, A.S.H.; Goodarzi, M.; Garshasbi, A. Study on the removal of heavy metal ions from industry waste by carbon nanotubes: Effect of the surface modification: a review. *Crit. Rev. Environ. Sci. Technol.,* **2016**, *46*(2), 93-118.
[http://dx.doi.org/10.1080/10643389.2015.1061874]

[32]   Machida, M.; Mochimaru, T.; Tatsumoto, H. Lead(II) adsorption onto the graphene layer of carbonaceous materials in aqueous solution. *Carbon,* **2006**, *44*(13), 2681-2688.
[http://dx.doi.org/10.1016/j.carbon.2006.04.003]

[33]   Chatterjee, S.; Chatterjee, T.; Lim, S.R.; Woo, S.H. Effect of the addition mode of carbon nanotubes for the production of chitosan hydrogel core–shell beads on adsorption of Congo red from aqueous solution. *Bioresour. Technol.,* **2011**, *102*(6), 4402-4409.
[http://dx.doi.org/10.1016/j.biortech.2010.12.117] [PMID: 21277770]

[34]   Rodríguez, A.; Ovejero, G.; Sotelo, J.L.; Mestanza, M.; García, J. Adsorption of dyes on carbon nanomaterials from aqueous solutions. *J. Environ. Sci. Health Part A Tox. Hazard. Subst. Environ. Eng.,* **2010**, *45*(12), 1642-1653.
[http://dx.doi.org/10.1080/10934529.2010.506137] [PMID: 20730657]

[35]   Machado, F.M.; Bergmann, C.P.; Fernandes, T.H.M.; Lima, E.C.; Royer, B.; Calvete, T.; Fagan, S.B. Adsorption of Reactive Red M-2BE dye from water solutions by multi-walled carbon nanotubes and activated carbon. *J. Hazard. Mater.,* **2011**, *192*(3), 1122-1131.
[http://dx.doi.org/10.1016/j.jhazmat.2011.06.020] [PMID: 21724329]

[36]   Monser, L.; Adhoum, N. Modified activated carbon for the removal of copper, zinc, chromium and cyanide from wastewater. *Separ. Purif. Tech.,* **2002**, *26*(2-3), 137-146.
[http://dx.doi.org/10.1016/S1383-5866(01)00155-1]

[37]   Selvi, K.; Pattabhi, S.; Kadirvelu, K. Removal of Cr(VI) from aqueous solution by adsorption onto activated carbon. *Bioresour. Technol.,* **2001**, *80*(1), 87-89.
[http://dx.doi.org/10.1016/S0960-8524(01)00068-2] [PMID: 11554606]

[38]   Joss, A.; Zabczynski, S.; Göbel, A.; Hoffmann, B.; Löffler, D.; McArdell, C.S.; Ternes, T.A.; Thomsen, A.; Siegrist, H. Biological degradation of pharmaceuticals in municipal wastewater treatment: Proposing a classification scheme. *Water Res.,* **2006**, *40*(8), 1686-1696.
[http://dx.doi.org/10.1016/j.watres.2006.02.014] [PMID: 16620900]

[39]   Lissemore, L.; Hao, C.; Yang, P.; Sibley, P.K.; Mabury, S.; Solomon, K.R. An exposure assessment for selected pharmaceuticals within a watershed in Southern Ontario. *Chemosphere,* **2006**, *64*(5), 717-729.
[http://dx.doi.org/10.1016/j.chemosphere.2005.11.015] [PMID: 16403551]

[40]   Chai, W.S.; Cheun, J.Y.; Kumar, P.S.; Mubashir, M.; Majeed, Z.; Banat, F.; Ho, S.H.; Show, P.L. A review on conventional and novel materials towards heavy metal adsorption in wastewater treatment application. *J. Clean. Prod.,* **2021**, *296*, 126589.
[http://dx.doi.org/10.1016/j.jclepro.2021.126589]

[41]   Shahryari-ghoshekandi, R.; Sadegh, H. Kinetic study of the adsorption of synthetic dyes on graphene surfaces. *Jordan J. Chem.,* **2014**, *9*(4), 267-278.

[42]   Denis, P.A.; Iribarne, F. A first-principles study on the interaction between alkyl radicals and graphene. *Chemistry,* **2012**, *18*(24), 7568-7574.
[http://dx.doi.org/10.1002/chem.201103711] [PMID: 22532481]

[43]   Yu, J.G.; Zhao, X.H.; Yang, H.; Chen, X.H.; Yang, Q.; Yu, L.Y.; Jiang, J.H.; Chen, X.Q. Aqueous adsorption and removal of organic contaminants by carbon nanotubes. *Sci. Total Environ.,* **2014**, *482-*

483, 241-251.
[http://dx.doi.org/10.1016/j.scitotenv.2014.02.129] [PMID: 24657369]

[44]   Kotov, N.A. Carbon sheet solutions. *Nature,* **2006**, *442*(7100), 254-255.
[http://dx.doi.org/10.1038/442254a] [PMID: 16855576]

[45]   Li, Y.; Liu, T.; Du, Q.; Sun, J.; Xia, Y.; Wang, Z.; Zhang, W.; Wang, K.; Zhu, H.; Wu, D. *Adsorption of cationic red X-GRL from aqueous solutions by graphene: equilibrium, kinetics and thermodynamics study.,* **2012**.

[46]   Li, B.; Cao, H.; Yin, G. Mg(OH)$_2$@reduced graphene oxide composite for removal of dyes from water. *J. Mater. Chem.,* **2011**, *21*(36), 13765-13768.
[http://dx.doi.org/10.1039/c1jm13368c]

[47]   Li, N.; Zheng, M.; Chang, X.; Ji, G.; Lu, H.; Xue, L.; Pan, L.; Cao, J. Preparation of magnetic CoFe2O4-functionalized graphene sheets *via* a facile hydrothermal method and their adsorption properties. *J. Solid State Chem.,* **2011**, *184*(4), 953-958.
[http://dx.doi.org/10.1016/j.jssc.2011.01.014]

[48]   Maliyekkal, S.M.; Sreeprasad, T.S.; Krishnan, D.; Kouser, S.; Mishra, A.K.; Waghmare, U.V.; Pradeep, T. Graphene: a reusable substrate for unprecedented adsorption of pesticides. *Small,* **2013**, *9*(2), 273-283.
[http://dx.doi.org/10.1002/smll.201201125] [PMID: 23001848]

[49]   Zhao, G.; Li, J.; Ren, X.; Chen, C.; Wang, X. Few-layered graphene oxide nanosheets as superior sorbents for heavy metal ion pollution management. *Environ. Sci. Technol.,* **2011**, *45*(24), 10454-10462.
[http://dx.doi.org/10.1021/es203439v] [PMID: 22070750]

[50]   Chandra, V.; Park, J.; Chun, Y.; Lee, J.W.; Hwang, I.C.; Kim, K.S. Water-dispersible magnetite-reduced graphene oxide composites for arsenic removal. *ACS Nano,* **2010**, *4*(7), 3979-3986.
[http://dx.doi.org/10.1021/nn1008897] [PMID: 20552997]

[51]   Zhao, G.; Wen, T.; Yang, X.; Yang, S.; Liao, J.; Hu, J.; Shao, D.; Wang, X. Preconcentration of U(VI) ions on few-layered graphene oxide nanosheets from aqueous solutions. *Dalton Trans.,* **2012**, *41*(20), 6182-6188.
[http://dx.doi.org/10.1039/C2DT00054G] [PMID: 22473651]

[52]   Gao, C.; Zhang, W.; Li, H.; Lang, L.; Xu, Z. Controllable fabrication of mesoporous MgO with various morphologies and their absorption performance for toxic pollutants in water. *Cryst. Growth Des.,* **2008**, *8*(10), 3785-3790.
[http://dx.doi.org/10.1021/cg8004147]

[53]   Tadjarodi, A.; Imani, M.; Kerdari, H. Adsorption kinetics, thermodynamic studies, and high performance of CdO cauliflower-like nanostructure on the removal of Congo red from aqueous solution. *J. Nanostructure Chem.,* **2013**, *3*(1), 51.
[http://dx.doi.org/10.1186/2193-8865-3-51]

[54]   Zare, K.; Sadegh, H.; Shahryari-ghoshekandi, R.; Asif, M.; Tyagi, I.; Agarwal, S.; Gupta, V.K. Equilibrium and kinetic study of ammonium ion adsorption by Fe$_3$O$_4$ nanoparticles from aqueous solutions. *J. Mol. Liq.,* **2016**, *213*, 345-350.
[http://dx.doi.org/10.1016/j.molliq.2015.08.045]

[55]   Ge, F.; Li, M.M.; Ye, H.; Zhao, B.X. Effective removal of heavy metal ions Cd^{2+}, Zn^{2+}, Pb^{2+}, Cu^{2+} from aqueous solution by polymer-modified magnetic nanoparticles. *J. Hazard. Mater.,* **2012**, *211-212*, 366-372.
[http://dx.doi.org/10.1016/j.jhazmat.2011.12.013] [PMID: 22209322]

[56]   Wang, Z.; Wu, D.; Wu, G.; Yang, N.; Wu, A. Modifying Fe$_3$O$_4$ microspheres with rhodamine hydrazide for selective detection and removal of Hg^{2+} ion in water. *J. Hazard. Mater.,* **2013**, *244-245*, 621-627.
[http://dx.doi.org/10.1016/j.jhazmat.2012.10.050] [PMID: 23177242]

[57]   Warner, C.L.; Addleman, R.S.; Cinson, A.D.; Droubay, T.C.; Engelhard, M.H.; Nash, M.A.; Yantasee, W.; Warner, M.G. High-performance, superparamagnetic, nanoparticle-based heavy metal sorbents for removal of contaminants from natural waters. *ChemSusChem,* **2010**, *3*(6), 749-757.
        [http://dx.doi.org/10.1002/cssc.201000027] [PMID: 20468024]

[58]   Naseem, T.; Durrani, T. The role of some important metal oxide nanoparticles for wastewater and antibacterial applications: A review. *Environmental Chemistry and Ecotoxicology,* **2021**, *3*, 59-75.
        [http://dx.doi.org/10.1016/j.enceco.2020.12.001]

[59]   Corsi, I.; Winther-Nielsen, M.; Sethi, R.; Punta, C.; Della Torre, C.; Libralato, G.; Lofrano, G.; Sabatini, L.; Aiello, M.; Fiordi, L.; Cinuzzi, F.; Caneschi, A.; Pellegrini, D.; Buttino, I. Ecofriendly nanotechnologies and nanomaterials for environmental applications: Key issue and consensus recommendations for sustainable and ecosafe nanoremediation. *Ecotoxicol. Environ. Saf.,* **2018**, *154*, 237-244.
        [http://dx.doi.org/10.1016/j.ecoenv.2018.02.037] [PMID: 29476973]

[60]   Qu, X.; Alvarez, P.J.J.; Li, Q. Applications of nanotechnology in water and wastewater treatment. *Water Res.,* **2013**, *47*(12), 3931-3946.
        [http://dx.doi.org/10.1016/j.watres.2012.09.058] [PMID: 23571110]

[61]   Rana, N.; Ghosh, K.S.; Chand, S.; Gathania, A.K. Investigation of ZnO nanoparticles for their applications in wastewater treatment and antimicrobial activity. *Indian J. Pure Appl. Phy.,* **2018**, *56*(1), 19-25.

[62]   Baruah, S.; Pal, S.K.; Dutta, J. Nanostructured Zinc Oxide for Water Treatment. *Nanosci. Nanotechnol. Asia,* **2012**, *2*(2), 90-102.
        [http://dx.doi.org/10.2174/2210681211202020090]

[63]   Zhu, J.; Li, D.; Chen, H.; Yang, X.; Lu, L.; Wang, X. Highly dispersed CuO nanoparticles prepared by a novel quick-precipitation method. *Mater. Lett.,* **2004**, *58*(26), 3324-3327.
        [http://dx.doi.org/10.1016/j.matlet.2004.06.031]

[64]   Anjaneyulu, B.; Rana, R.; Versha, Afshari, M.; Carabineiro, S.A. The use of magnetic porous carbon nanocomposites for the elimination of organic pollutants from wastewater. *Surfaces,* **2024**, *7*(1), 120-142.

[65]   Es'haghzade, Z.; Pajootan, E.; Bahrami, H.; Arami, M. Facile synthesis of Fe$_3$O$_4$ nanoparticles *via* aqueous based electro chemical route for heterogeneous electro-Fenton removal of azo dyes. *J. Taiwan Inst. Chem. Eng.,* **2017**, *71*, 91-105.
        [http://dx.doi.org/10.1016/j.jtice.2016.11.015]

[66]   Tsarev, S.; Collins, R.N.; Fahy, A.; Waite, T.D. Reduced uranium phases produced from anaerobic reaction with nanoscale zerovalent iron. *Environ. Sci. Technol.,* **2016**, *50*(5), 2595-2601.
        [http://dx.doi.org/10.1021/acs.est.5b06160] [PMID: 26840619]

[67]   Mahgoub, S.; Samaras, P. Nanoparticles from biowastes and microbes: focus on role in water purification and food preservation. *Proceedings of the 2nd international conference on sustainable solid waste management,* Athens**2014**, pp. 1-39.

[68]   Ghosh, A.; Nayak, A.K.; Pal, A. Nano-particle-mediated wastewater treatment: a review. *Curr. Pollut. Rep.,* **2017**, *3*(1), 17-30.
        [http://dx.doi.org/10.1007/s40726-016-0045-1]

[69]   Hou, L.; Xia, J.; Li, K.; Chen, J.; Wu, X.; Li, X. Removal of ZnO nanoparticles in simulated wastewater treatment processes and its effects on COD and NH4+-N reduction. *Water Sci. Technol.,* **2013**, *67*(2), 254-260.
        [http://dx.doi.org/10.2166/wst.2012.530] [PMID: 23168621]

[70]   Anjum, M.; Miandad, R.; Waqas, M.; Gehany, F.; Barakat, M.A. Remediation of wastewater using various nano-materials. *Arab. J. Chem.,* **2019**, *12*(8), 4897-4919.
        [http://dx.doi.org/10.1016/j.arabjc.2016.10.004]

[71]   Lee, C. Oxidation of organic contaminants in water by iron-induced oxygen activation: A short review. *Environ. Eng. Res.,* **2015**, *20*(3), 205-211.
[http://dx.doi.org/10.4491/eer.2015.051]

[72]   Amin, M.T.; Alazba, A.A.; Manzoor, U. A review of removal of pollutants from water/wastewater using different types of nanomaterials. *Adv. Mater. Sci. Eng.,* **2014**, *2014*, 1-24.
[http://dx.doi.org/10.1155/2014/825910]

[73]   Daneshvar, E.; Zarrinmehr, M.J.; Hashtjin, A.M.; Farhadian, O.; Bhatnagar, A. Versatile applications of freshwater and marine water microalgae in dairy wastewater treatment, lipid extraction and tetracycline biosorption. *Bioresour. Technol.,* **2018**, *268*, 523-530.
[http://dx.doi.org/10.1016/j.biortech.2018.08.032] [PMID: 30118973]

[74]   Šoštarić, T.D.; Petrović, M.S.; Pastor, F.T.; Lončarević, D.R.; Petrović, J.T.; Milojković, J.V.; Stojanović, M.D. Study of heavy metals biosorption on native and alkali-treated apricot shells and its application in wastewater treatment. *J. Mol. Liq.,* **2018**, *259*, 340-349.
[http://dx.doi.org/10.1016/j.molliq.2018.03.055]

[75]   Ahmed, S.F.; Mofijur, M.; Nuzhat, S.; Chowdhury, A.T.; Rafa, N.; Uddin, M.A.; Inayat, A.; Mahlia, T.M.I.; Ong, H.C.; Chia, W.Y.; Show, P.L. Recent developments in physical, biological, chemical, and hybrid treatment techniques for removing emerging contaminants from wastewater. *J. Hazard. Mater.,* **2021**, *416*, 125912.
[http://dx.doi.org/10.1016/j.jhazmat.2021.125912] [PMID: 34492846]

[76]   Apul, O.G.; Wang, Q.; Zhou, Y.; Karanfil, T. Adsorption of aromatic organic contaminants by graphene nanosheets: Comparison with carbon nanotubes and activated carbon. *Water Res.,* **2013**, *47*(4), 1648-1654.
[http://dx.doi.org/10.1016/j.watres.2012.12.031] [PMID: 23313232]

[77]   Li, Y.H.; Ding, J.; Luan, Z.; Di, Z.; Zhu, Y.; Xu, C.; Wu, D.; Wei, B. Competitive adsorption of Pb^{2+}, Cu^{2+} and Cd^{2+} ions from aqueous solutions by multiwalled carbon nanotubes. *Carbon,* **2003**, *41*(14), 2787-2792.
[http://dx.doi.org/10.1016/S0008-6223(03)00392-0]

[78]   Chatterjee, S.; Lee, M.W.; Woo, S.H. Adsorption of congo red by chitosan hydrogel beads impregnated with carbon nanotubes. *Bioresour. Technol.,* **2010**, *101*(6), 1800-1806.
[http://dx.doi.org/10.1016/j.biortech.2009.10.051] [PMID: 19962883]

[79]   Molinari, R ; Poerio, T.; Cassano, R.; Picci, N.; Argurio, P. Copper (II) removal from wastewaters by a new synthesized selective extractant and SLM viability. *Ind. Eng. Chem. Res.,* **2004**, *43*(2), 623-628.
[http://dx.doi.org/10.1021/ie030392t]

[80]   Al-Degs, Y.S.; El-Barghouthi, M.I.; Issa, A.A.; Khraisheh, M.A.; Walker, G.M. Sorption of Zn(II), Pb(II), and Co(II) using natural sorbents: Equilibrium and kinetic studies. *Water Res.,* **2006**, *40*(14), 2645-2658.
[http://dx.doi.org/10.1016/j.watres.2006.05.018] [PMID: 16839582]

[81]   Sadegh, H.; Shahryari-ghoshekandi, R.; Agarwal, S.; Tyagi, I.; Asif, M.; Gupta, V.K. Microwave-assisted removal of malachite green by carboxylate functionalized multi-walled carbon nanotubes: Kinetics and equilibrium study. *J. Mol. Liq.,* **2015**, *206*, 151-158.
[http://dx.doi.org/10.1016/j.molliq.2015.02.007]

[82]   Li, Y.H.; Wang, S.; Luan, Z.; Ding, J.; Xu, C.; Wu, D. Adsorption of cadmium(II) from aqueous solution by surface oxidized carbon nanotubes. *Carbon,* **2003**, *41*(5), 1057-1062.
[http://dx.doi.org/10.1016/S0008-6223(02)00440-2]

[83]   Stafiej, A.; Pyrzynska, K. Adsorption of heavy metal ions with carbon nanotubes. *Separ. Purif. Tech.,* **2007**, *58*(1), 49-52.
[http://dx.doi.org/10.1016/j.seppur.2007.07.008]

[84]   Xu, D.; Tan, X.; Chen, C.; Wang, X. Removal of Pb(II) from aqueous solution by oxidized

multiwalled carbon nanotubes. *J. Hazard. Mater.,* **2008**, *154*(1-3), 407-416.
[http://dx.doi.org/10.1016/j.jhazmat.2007.10.059] [PMID: 18053642]

[85]   Bahgat, M.; Farghali, A.A.; El Rouby, W.M.A.; Khedr, M.H. Synthesis and modification of multi-walled carbon nano-tubes (MWCNTs) for water treatment applications. *J. Anal. Appl. Pyrolysis,* **2011**, *92*(2), 307-313.
[http://dx.doi.org/10.1016/j.jaap.2011.07.002]

[86]   Hu, S.W.; Li, W.J.; Chang, Z.D.; Wang, H.Y.; Guo, H.C.; Zhang, J.H.; Liu, Y. [Removal of methyl orange from aqueous solution by magnetic carbon nanotubes]. *Guangpuxue Yu Guangpu Fenxi,* **2011**, *31*(1), 205-209.
[PMID: 21428089]

[87]   Ghaedi, M.; Khajehsharifi, H.; Yadkuri, A.H.; Roosta, M.; Asghari, A. Oxidized multiwalled carbon nanotubes as efficient adsorbent for bromothymol blue. *Toxicol. Environ. Chem.,* **2012**, *94*(5), 873-883.
[http://dx.doi.org/10.1080/02772248.2012.678999]

[88]   Yao, Y.; Xu, F.; Chen, M.; Xu, Z.; Zhu, Z. Adsorption of cationic methyl violet and methylene blue dyes onto carbon nanotubes. *Nano/Micro Engineered and Molecular Systems (NEMS),* **2010**.

[89]   Shahryari, Z.; Goharrizi, A.S.; Azadi, M. Experimental study of methylene blue adsorption from aqueous solutions onto carbon nanotubes. *Int. J. Water Res. Environ. Eng.,* **2010**, *2*(2), 16-28.

[90]   Salama, A. New sustainable hybrid material as adsorbent for dye removal from aqueous solutions. *J. Colloid Interface Sci.,* **2017**, *487*, 348-353.
[http://dx.doi.org/10.1016/j.jcis.2016.10.034] [PMID: 27794235]

[91]   Nadafi, K.; Mesdaghinia, A.; Nabizadeh, R.; Younesian, M.; Rad, M.J. The combination and optimization study on RB29 dye removal from water by peroxy acid and single-wall carbon nanotubes. *Desalination Water Treat.,* **2011**, *27*(1-3), 237-242.
[http://dx.doi.org/10.5004/dwt.2011.1980]

[92]   Li, Y.H.; Wang, S.; Wei, J.; Zhang, X.; Xu, C.; Luan, Z.; Wu, D.; Wei, B. Lead adsorption on carbon nanotubes. *Chem. Phys. Lett.,* **2002**, *357*(3-4), 263-266.
[http://dx.doi.org/10.1016/S0009-2614(02)00502-X]

[93]   Robati, D. Pseudo-second-order kinetic equations for modeling adsorption systems for removal of lead ions using multi-walled carbon nanotube. *J. Nanostructure Chem.,* **2013**, *3*(1), 55.
[http://dx.doi.org/10.1186/2193-8865-3-55]

[94]   Gao, Z.; Bandosz, T.J.; Zhao, Z.; Han, M.; Qiu, J. Investigation of factors affecting adsorption of transition metals on oxidized carbon nanotubes. *J. Hazard. Mater.,* **2009**, *167*(1-3), 357-365.
[http://dx.doi.org/10.1016/j.jhazmat.2009.01.050] [PMID: 19264402]

[95]   Wang, X.; Chen, C.; Hu, W.; Ding, A.; Xu, D.; Zhou, X. Sorption of 243Am(III) to multiwall carbon nanotubes. *Environ. Sci. Technol.,* **2005**, *39*(8), 2856-2860.
[http://dx.doi.org/10.1021/es048287d] [PMID: 15884386]

[96]   Tan, X.; Xu, D.; Chen, C.; Wang, X.; Hu, W. Adsorption and kinetic desorption study of 152 + 154Eu (III) on multiwall carbon nanotubes from aqueous solution by using chelating resin and XPS methods. *Radiochim. Acta,* **2008**, *96*, 23-29.

[97]   Rao, G.; Lu, C.; Su, F. Sorption of divalent metal ions from aqueous solution by carbon nanotubes: A review. *Separ. Purif. Tech.,* **2007**, *58*(1), 224-231.
[http://dx.doi.org/10.1016/j.seppur.2006.12.006]

[98]   Moradi, O.; Gupta, V.K.; Agarwal, S.; Tyagi, I.; Asif, M.; Makhlouf, A.S.H.; Sadegh, H.; Shahryari-ghoshekandi, R. RETRACTED: Characteristics and electrical conductivity of graphene and graphene oxide for adsorption of cationic dyes from liquids: Kinetic and thermodynamic study. *J. Ind. Eng. Chem.,* **2015**, *28*, 294-301.
[http://dx.doi.org/10.1016/j.jiec.2015.03.005]

[99]   Volesky, B. Biosorption process simulation tools. *Hydrometallurgy,* **2003**, *71*(1-2), 179-190.
[http://dx.doi.org/10.1016/S0304-386X(03)00155-5]

[100]   Benmessaoud, A.; Nibou, D. Mekatel, El Hadj, Amokrane, S. A comparative study of the linear and non-linear methods for determination of the optimum equilibrium isotherm for adsorption of Pb^{2+} ions onto algerian treated clay. *Iran. J. Chem. Chem. Eng.,* **2020**, *39*(4), 153-171.

[101]   Kumar, S.; Srivastava, A.N. *Application of carbon nanomaterials decorated electrochemical sensor for analysis of environmental pollutants*; Analytical Chemistry-Advancement Perspectives and Applications, IntechOpen, **2021**.
[http://dx.doi.org/10.5772/intechopen.96538]

[102]   R, J.; Gurunathan, B.; K, S.; Varjani, S.; Ngo, H.H.; Gnansounou, E. Advancements in heavy metals removal from effluents employing nano-adsorbents: Way towards cleaner production. *Environ. Res.,* **2022**, *203*, 111815.
[http://dx.doi.org/10.1016/j.envres.2021.111815] [PMID: 34352231]

[103]   Hu, Q.; Zhang, Z. Application of Dubinin–Radushkevich isotherm model at the solid/solution interface: A theoretical analysis. *J. Mol. Liq.,* **2019**, *277*, 646-648.
[http://dx.doi.org/10.1016/j.molliq.2019.01.005]

[104]   Ahmad, A.; Azam, T. Water purification technologies. In: *Bottled and Packaged Water*; Grumezescu, A.M.; Holban, A.M., Eds.; Elsevier, **2019**; pp. 83-120.
[http://dx.doi.org/10.1016/B978-0-12-815272-0.00004-0]

[105]   Radovic, L.R.; Moreno-Castilla, C.; Rivera-Utrilla, J. Carbon materials as adsorbents in aqueous solutions. *Chem. Phys. Carbon,* **2000**, *27*, 247-426.
[http://dx.doi.org/10.1201/9781482270129-11]

[106]   El-sayed, M.E.A. Nanoadsorbents for water and wastewater remediation. *Sci. Total Environ.,* **2020**, *739*, 139903.
[http://dx.doi.org/10.1016/j.scitotenv.2020.139903] [PMID: 32544683]

[107]   Ray, P.C.; Yu, H.; Fu, P.P. Toxicity and environmental risks of nanomaterials: challenges and future needs. *J. Environ. Sci. Health Part C Environ. Carcinog. Ecotoxicol. Rev.,* **2009**, *27*(1), 1-35.
[http://dx.doi.org/10.1080/10590500802708267] [PMID: 19204862]

[108]   Stander, L.; Theodore, L. Environmental implications of nanotechnology--an update. *Int. J. Environ. Res. Public Health,* **2011**, *8*(2), 470-479.
[http://dx.doi.org/10.3390/ijerph8020470] [PMID: 21556197]

CHAPTER 9

Nanofiltration in Wastewater Treatment

Roopa Rani[1], Harsha Devnani[1], A. Jayamani[1] and Arpit Sand[1,*]

[1] *Department of Sciences, School of Sciences, Manav Rachna University, Faridabad, Haryana, India*

Abstract: As pure water is considered to be the best medicine in the world, there are serious concerns about water scarcity and water pollution all over the world to find a sustainable way to treat and reuse water. Although traditional wastewater treatment methods are effective, they face problems in terms of adaptability, making them unsuitable and preventing further development. The use of membrane technology in wastewater treatment has received great attention with proven results for removing pollutants, color, and COD/BOD, as well as the reclamation of cleaning solutions. The nanofiltration technique is ahead of all other membrane technologies of wastewater treatment and has more scope for advancements. This book chapter emphasizes the nanofiltration process with more insight into nanofiltration membranes, basic principles involved, mechanism, mathematical modeling of the membrane, use of polymers, and reasons for fouling of membranes used in nanofiltration. The chapter extends its discussion with inputs on the current scenario in the implementation of this technique for surface water, groundwater, fouling control, and effective water reuse. The summary of this book chapter provides information on future aspects of the nanofiltration process, which is expected to address current and future water treatment challenges, paving the way for cleaner and safer water resources worldwide.

Keywords: Fouling control, Membrane filtration, Nanofiltration, Wastewater treatment.

INTRODUCTION

Nanofiltration (NF) in wastewater treatment can be described as a process of membrane filtration where the membrane acts as a barrier to drive the passage of only water molecules and restrict the movement of impurities such as organic, inorganic, and other contaminants [1]. It is a process that falls between reverse osmosis (RO) and ultrafiltration (UF) in terms of pore size. Since the quality of water is continuously degrading day by day, the process of membrane filtration can result in an effective solution for making it reusable [2]. Several studies were reported on the application of nanofiltration for the treatment of wastewater,

---

[*] **Corresponding author Arpit Sand:** Department of Sciences, School of Sciences, Manav Rachna University, Faridabad, Haryana, India, Email: sand.arpit@gmail.com

Anjaneyulu Bendi (Ed.)

making it suitable for its reuse [3 - 6]. The process of nanofiltration utilizes the concept of semi-permeable membranes to separate ions, molecules, and particles in the molten state. This process is more effective and can be used to treat both groundwater and surface water since it depends on the nature of the membrane, which guides the passage of some selective ions while restricting others. The nanofiltration process is effectively utilized in contamination removal from wastewater, where the removal of organic pollutants, microorganisms, and heavy metals can be done. Another use of the nanofiltration technique is for desalination purposes, especially for treating brackish water sources where the salt content is lower than seawater but still too high for direct consumption or industrial use [7, 8]. NF membranes can selectively remove divalent ions like calcium and magnesium while allowing monovalent ions like sodium and chloride to pass through. This selective permeability helps in achieving partial desalination of water. Nanofiltration can also be used to diminish color and turbidity in liquids. This is particularly important for surface water treatment, where the water may contain natural organic matter and sediments. Many times, NF is combined with RO and has been investigated for wastewater treatment to make it potable. Also, NF can be a pre-treatment step for the process of reverse osmosis and multi-stage flash desalination (RO-MSF), resulting in significant enhancement in the process of desalination, reduction in the load on RO membrane, and improvement of overall efficiency [9]. Nanofiltration also has the advantage of selective filtration or removal of certain ions since they can be easily engineered in this manner. This can be advantageous in treating groundwater contaminated with specific ions or pollutants. NF can also be used for water softening through the selective removal of calcium and magnesium ions responsible for water hardness. This is particularly beneficial in areas where hard water causes scaling in pipes and appliances. This chapter briefly describes the use of nanofiltration in wastewater treatment, along with the details of the steps involved.

FUNDAMENTALS OF MEMBRANE FILTRATION PROCESS

The effectiveness and selectivity of membrane filtering make it a popular separation method across many sectors. Membrane filtration works on the basis of the membrane's semi-permeable nature, which permits some molecules or particles to flow through while obstructing others due to differences in size, shape, charge, or chemical composition. The liquid or solution is forced to pass through the membrane by a pressure differential across it, which is what drives this process and separates the desired components from the feed stream. The size of the particles that can pass through a membrane is determined by the size of its pores; with varying membrane types having different pore diameters and selectivity, multiple filtration can be achieved. Furthermore, the concentration of the feed solution, the pH of the solution, and the temperature can all have an impact on membrane filtration and affect the process's selectivity and efficiency [10].

Membrane filtration is a vital procedure that, by providing effective and selective separation of materials at the molecular and ionic levels, has transformed a number of industries. The type of membrane to choose is important and relies on the demands of the particular application. Microorganisms like bacteria, viruses, and suspended particles are frequently removed using microfiltration (MF) membranes. Their pore diameters, which span from 0.1 to 10 μm, enable them to efficiently eliminate particles falling within this range. The pore sizes of ultrafiltration (UF) membranes range from 0.001 to 0.1 μm, making them effective for the efficient separation of proteins, colloids, and macromolecules from liquids [11, 12]. Nanofiltration (NF) membranes, which have even smaller holes than UF membranes, can be used to remove divalent ions and small organic molecules. Last but not least, desalination and the elimination of small ions and molecules are accomplished using reverse osmosis (RO) membranes, which have the smallest pore sizes—less than 0.001 μm [13].

Depending on the kind of membrane, multi-membrane filtering mechanisms exist. Size exclusion, in which particles bigger than the pore size are trapped on the membrane surface while smaller particles flow through, is the main method of particle removal in MF and UF. By utilizing extra mechanisms like electrostatic attraction and adsorption, NF and RO membranes enable the selective separation of ions and molecules according to their size and charge. This selectivity is essential in applications like water treatment and pharmaceutical manufacturing, where exact separation of particular chemicals is needed [14].

The optimization of membrane filtering operations is largely dependent on operating parameters such as feed concentration, temperature, feed flow rate, and trans-membrane pressure (TMP), which is the pressure differential across the membrane and a vital component that propels filtration. The rate of filtration is affected by the temperature and feed flow rate, with higher temperatures and flow rates typically resulting in higher filtration rates. Feed concentration has an impact on the membrane's performance as well; larger concentrations may result in more fouling [15, 16].

One of the main problems with membrane filtration is membrane fouling, which can shorten the membranes' lifespan and efficiency. It happens as a result of particles, microbes, or dissolved materials building up on the membrane surface and decreasing its permeability. The use of anti-fouling membranes, chemical cleaning, and routine backwashing are methods to reduce membrane fouling. These techniques lower the overall operating expenses of membrane filtering systems by extending the membranes' lifespan and preserving their efficiency [17].

Membrane filtration is an advanced and adaptable method of separation that provides exact control over the division of substances according to their size, charge, and solubility. Comprehending the foundations of membrane filtration, encompassing membrane kinds, mechanisms, operational factors, and ways to mitigate fouling, is imperative for process optimization and guarantees the process's triumphant implementation in diverse sectors. Future membrane technology research and development will continue to increase membrane technology's effectiveness and expand its range of uses.

NANOFILTRATION

The capacity of nanofiltration (NF), a membrane-based separation technique, to selectively remove ions and small molecules from a liquid stream while permitting water and specific other components to flow through has drawn a lot of attention. The narrow pore structure of NF membranes, which typically has pore diameters between 1 and 10 nanometers, enables them to reject divalent ions and tiny organic molecules while permitting monovalent ions and water molecules to pass through [18].

A combination of size exclusion, electrostatic repulsion, and adsorption underpins the separation mechanism of NF membranes. NF membranes' narrow pore structure functions as a molecular sieve, removing certain molecules and particles according to their size. Furthermore, NF membranes are frequently modified to add a negative charge to their surface, which repels negatively charged ions and molecules and permits them to flow across the membrane. NF membranes are especially good at removing small organic compounds and divalent ions like sulfate, calcium, and magnesium because of this combination of processes [19].

NF is a method through which some portion of the feed is passed through a semi-permeable membrane, as shown in Fig. (1). The inlet stream of feed pre-treatment is divided into two portions – one is permeate, which comes after filtration of the stream, and another one is the retentate, which is not filtered or is the rejected portion of the feed. This type of filtration is effective in removing the organic pollutants from wastewater. Inorganic functionalized membranes can be more effective in removing microorganisms, metal particles, *etc* [14].

There is a huge industrial application of nanofiltration among pharmaceutical industries, food processing industries, wastewater treatment industries, *etc*. NF is used in water treatment to remove organic matter, ions that cause hardness, and other impurities, resulting in high-quality water that is fit for industrial or drinking usage [20]. NF is utilized in the food and beverage industry for operations related to concentration, purification, and fractionation, like removing lactose from milk or concentrating fruit juices [21]. NF is used in pharmaceutical manufacturing to remove contaminants and purify medicinal compounds [22].

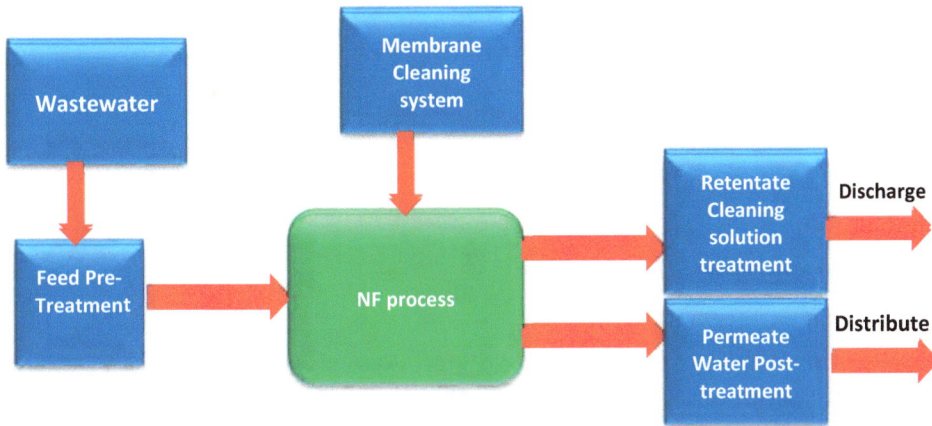

Fig. (1). Principle of Nanofiltration [14].

One of NF's main benefits is that, when compared to reverse osmosis (RO), it operates at a relatively low pressure, which saves energy and money in a variety of applications. On the other hand, fouling can cause NF membranes to lose some of their efficacy over time. Fouling can be reduced by using pretreatment techniques, including flocculation and coagulation, as well as routine membrane cleaning and maintenance [13].

Fundamentally, nanofiltration is a broad and efficient separation technique with numerous applications in various industries. Researchers and engineers may continue to develop and optimize NF techniques for effective and sustainable separation processes by knowing the fundamentals of NF and its applications.

Mechanism

Size exclusion, electrostatic interactions, and adsorption are the three guiding concepts of nanofiltration (NF), which is a membrane-based separation technique. Ions and small molecules are selectively separated from a liquid stream using this technique. The dense selective layer of NF membranes typically has pore diameters between 1 and 10 nanometers. This enables them to separate molecules and ions according to size preferentially. Smaller molecules go through the membrane, whereas larger molecules are trapped inside. Controlling the pore size distribution and structure of NF membranes by nanotechnology is essential for accurately adjusting the separation characteristics. A surface charge can be added to nanofiltration membranes to enhance the repulsion of similarly charged ions and molecules. This electrostatic repulsion can increase the transit of some selective ions and molecules as per their charge. NF membranes' surface charge is structured using nanotechnology to increase their efficacy, efficiency, accuracy, and selectivity. The addition of a functional group over the NF membrane surface can also have a major impact on further improving the property of selectivity, which can specifically adsorb particular ions or molecules and reject others [23 - 26].

The design and construction of membranes are two major steps in the NF mechanism that are impacted by nanotechnology. The performance and efficiency of nanofiltration (NF) membranes are largely dependent on how they are designed and manufactured. A number of design and fabrication-related variables can have a huge impact on the NF process. The selectivity and permeability of NF membranes are largely dependent on their pore size and shape. By precisely controlling the pore size and shape, nanotechnology makes it possible to create membranes with the best possible separation qualities [27].

The electrostatic interactions that NF membranes have with ions and molecules are influenced by their surface charge. Through the use of nanotechnology, the surface charge can be changed, improving the membrane's selectivity and separation efficiency. The permeability and selectivity of NF membranes are influenced by the thickness of their selective layer. Although selectivity may be compromised, thinner membranes typically have higher permeability. Selectivity and permeability can be balanced by creating thin, uniformly thick layers using nanotechnology [28, 29]. For NF performance, the material selection of the membrane is essential. With the use of nanotechnology, membrane performance and longevity can be increased by creating membranes with customized material qualities, such as increased chemical resistance, mechanical strength, and thermal stability.

The performance of NF membranes can be affected by their entire structure, which includes the selective and support layers. Asymmetric membranes with a thin selective layer supported by a porous substrate can be created with the help of nanotechnology, which improves separation efficiency. By applying nanotechnology to modify the surface of NF membranes, their overall performance, fouling resistance, and antibacterial qualities can be enhanced. The longevity and effectiveness of NF membranes can be increased by functionalizing membrane surfaces with certain groups to improve selectivity and decrease fouling. When salts and other impurities build up, NF membranes are vulnerable to scaling. Membrane longevity can be increased and maintained by designing membranes with enhanced scaling resistance [30 - 33].

NF membranes can be made to precisely match certain separation needs by optimizing their design and manufacturing, which makes them extremely useful for a variety of applications. The design and production of filtration membranes have been completely transformed by nanotechnology, which has resulted in the creation of several membranes based on nanomaterials that have upgraded properties.

Thin-Film Composite Membranes

These membranes allow for the selective passage of some ions and molecules while rejecting others according to their size, charge, and chemical makeup. They are composed of a thin selective layer supported by a porous substrate layer. The membrane's mechanical strength is derived from the porous substrate that supports the thin film layer. The selective layer's thickness and structure are controlled using nanotechnology, which enhances the selectivity and flow of the membrane. Typically, polyamide or polysulfone materials designed to have particular pore sizes and surface properties are used to create the selective layer of thin film NF membranes. Nanotechnology can be used to modify these characteristics in order to improve the membrane's permeability and selectivity [28].

The membranes made up of nanocomposites have the ability to remove impurities, including salts, heavy metals, and organic substances, from water and are frequently employed in desalination and wastewater treatment operations. Compared to other membrane types, thin film NF membranes provide a number of benefits, such as high permeability, resistance to fouling, and good selectivity. These membranes find extensive application in a multitude of industries, including electronics, pharmaceuticals, and food and beverage. They are essential in the manufacture of purified water and the extraction of valuable items from process streams [16].

A popular thin-film composite membrane uses polyamide as the selective layer. Usually, the process of creating this kind of membrane involves the interfacial polymerization of trimesoyl chloride and diamine on a porous substrate. Because of their excellent permeability and selectivity, these membranes have been extensively researched and used in a variety of nanofiltration applications [34]. Thin film composite (TFC) membranes with polyethersulfone (PES) as the substrate material are an additional example. PES is the perfect material for the substrate of TFC membranes since it has exceptional mechanical strength and chemical resistance. These membranes are frequently utilized in water treatment nanofiltration applications [35]. The goal of recent research has been to improve the performance of thin film composite membranes by adding nanomaterials like graphene oxide (GO). TFC membrane selectivity and permeability can be enhanced by GO's special qualities, which include strong mechanical strength and superior chemical stability. In a variety of nanofiltration applications, these membranes have demonstrated encouraging results [36].

Nanoporous Membranes

Advanced membrane architectures with nanopores in their selective layer are called nanoporous membranes for nanofiltration. These membranes are made to segregate molecules and ions according to their chemical makeup, size, and charge. Compared to conventional membranes, nanoporous membranes provide a number of benefits, such as increased selectivity, flux, and fouling resistance. Numerous methods, such as track-etching, anodization, and nanoparticle deposition, can be used to create nanoporous membranes. These methods provide customized separation performance by giving exact control over the membrane's pore size, distribution, and surface characteristics. The excellent selectivity of nanoporous membranes is one of their main benefits. Larger molecules are kept in the membrane's nanopores, which are engineered to only let smaller and selective molecules pass through it. Applications like water purification, pharmaceutical manufacturing, and target-oriented, selective separation benefit greatly from this discerning separation [37].

The high flux of nanoporous membranes is another benefit. Higher flow rates and more efficiency are achieved by the quick passage of molecules through the membrane, made possible by the nanopores' small size. In addition, nanoporous membranes have a higher fouling resistance than conventional membranes. Because the nanopores are small, more pollutant particles are restricted by the membrane, making it suitable for separation and also extending its lifespan with less maintenance requirement. In general, high selectivity, high flux, and fouling resistance are crucial for a variety of nanofiltration applications, such as chemical processing, pharmaceutical manufacturing, and water purification. For these applications, nanoporous membranes present a viable option [38].

Since polymeric nanoporous membranes have great permeability and selectivity, they are frequently used in a variety of applications. Track-etching and phase separation are two common fabrication methods for these membranes. For uses including gas separation and water purification, they have been thoroughly investigated [39]. A class of porous materials with a high surface area and adjustable pore size are MOF nanoporous membranes. These membranes have demonstrated potential for use in molecular sieving and gas separation. Usually,

MOF crystals are grown on a porous substrate to create them [40]. Aligned carbon nanotubes are used as the porous material to create CNT nanoporous membranes. These membranes are appropriate for use in gas separation and water desalination since they have demonstrated exceptional mechanical strength and chemical stability [41].

Nanocomposite Membranes

Nanoparticles are incorporated into the membrane matrix of nanocomposites, which resultsin sophisticated membrane structures used in nanofiltration. Usually, the purpose of these nanoparticles is to improve the stability, permeability, and selectivity of the membrane. Compared to conventional membranes, nanocomposite membranes provide a number of benefits, such as better chemical resistance, stronger mechanical strength, and better separation performance. The enhanced selectivity of nanocomposite membranes is one of their main benefits. The membrane's capacity to reject specific ions and molecules can be improved, and new separation pathways can be created by incorporating nanoparticles into the membrane matrix. When high purity is needed, like in the production of pharmaceuticals or water treatment, this enhanced selectivity is quite helpful [42].

In comparison to conventional membranes, nanocomposite membranes also offer greater mechanical strength. By serving as reinforcing agents, the nanoparticles improve the membrane structure and lessen the chance of a mechanical breakdown. Because of their enhanced strength, nanocomposite membranes can endure harsher environments and greater working pressures. The increased chemical resistance of nanocomposite membranes is another benefit. The membrane can be made more resistant to fouling and degradation by engineering the nanoparticles to be chemically inert or to interact with specific molecules in a targeted manner. The membrane lasts longer and requires less maintenance due to its increased chemical resistance [43, 44].

Membranes that are composite and comprise a thin selective layer embedded with nanoparticles are known as thin film nanocomposite (TFN) membranes. These membranes have a reputation for having improved permeability and selectivity. For instance, Lai *et al.* (2016) produced TFN membranes with enhanced salt rejection and water permeability by adding graphene oxide (GO) nanoparticles to the selective layer [45]. Improved membrane performance is also achieved by combining the characteristics of ceramics and polymers in polymer-ceramic nanocomposite membranes [46]. For example, by adding zirconium oxide (ZrO_2) nanoparticles to a polymeric matrix, Bossemghoun *et al.* (2020) created a polymer-ceramic nanocomposite membrane with improved mechanical strength and chemical resistance [47]. To improve separation performance, MOF nanocomposite membranes mix MOF components with a polymeric matrix. By adding MIL-101(Cr) MOF particles to a polymeric matrix, Ines *et al.* (2021) created a MOF nanocomposite membrane with improved adsorption and separation capabilities [48].

Nanotube Membranes

Carbon nanotubes (CNTs) or other forms of nanotubes can be used to create novel membrane architectures for nanofiltration. These membranes are promising options for a range of nanofiltration applications because of their special qualities, which include excellent mechanical strength, chemical stability, and adjustable pore size. The great mechanical strength of nanotube membranes is one of its main benefits. Carbon nanotubes are renowned for their extraordinary strength-t--weight ratio, which enables nanotube membranes to endure severe working conditions and high pressures without experiencing mechanical failure. They are perfect for situations where endurance is crucial because of this feature [49, 50].

Additionally, nanotube membranes have outstanding chemical stability, which means that exposure to a variety of substances will not cause them to break down. Because of their chemical stability, nanotube membranes can operate as good separators for longer stretches of time, which lowers the requirement for regular membrane replacement. The adjustable pore size of nanotube membranes is an additional benefit. The pore size of the membrane can be precisely tuned by adjusting the diameter of the nanotubes during manufacture. Because of their tunability, nanotube membranes can separate molecules according to their size, which makes them perfect for applications that need high selectivity. High surface area-to-volume ratios are another benefit of nanotube membranes, which can improve separation effectiveness. The overall performance of the separation is enhanced by the wide surface area, which gives the membrane and the molecules being separated plenty of pores to interact. Nanotube membranes hold a lot of potential for a variety of nanofiltration uses, such as organic molecule separation, desalination, and water purification. Their exceptional blend of chemical stability, mechanical strength, and adjustable pore size makes them an appealing option for applications requiring high-performance membranes [51, 52].

By aligning multi-walled carbon nanotubes (MWCNTs), Holt *et al.* (2006) created CNT membranes and showed how their high water permeability and salt rejection may be used in desalination applications [53]. Another kind of nanotube that has demonstrated promise for membrane applications is boron nitride nanotubes (BNNTs). Because of their excellent gas permeability and selectivity, BNNT membranes have been shown to have potential for use in water purification [54]. There have also been reports of composite membranes made of nanotubes mixed with other materials or with different kinds of nanotubes. In the study by Kim *et al.* (2016), they reported an MWCNT-silver (Ag) composite nanofilter, which is deemed to be highly permeable and has been demonstrated to effectively remove bacterial and viral pathogens from water at low pressure [55].

Nanofiber Membranes

Advanced membrane structures made of nanofibers, usually with diameters ranging from tens to hundreds of nanometers, are known as nanofiber membranes. These membranes are very useful for nanofiltration applications because of their special qualities, which include high porosity, a large surface area, and a customizable pore size distribution, which allow for greater filtration efficiency and fast fluid movement as the main advantages. There is plenty of surface area created by the interconnected network of nanofibers, which is ideal for the adsorption and separation of pollutants [56, 57].

Additionally, the pore size distribution of nanofiber membranes can be adjusted throughout the production process. This enables exact control over the selectivity of the membrane, enabling the selective separation of molecules according to their charge, size, and shape. Furthermore, nanofiber membranes have outstanding mechanical strength and resilience, which keeps them safe from tearing and harm. This characteristic guarantees the membrane's long-term stability and dependability under demanding working circumstances. Moreover, different surface treatments and coatings can be applied to nanofiber membranes to improve their ability to separate materials. These surface alterations can increase the membrane's chemical stability, fouling resistance, and selectivity, making it appropriate for a variety of uses. In general, nanofiber membranes exhibit considerable potential for use in nanofiltration processes such as chemical separation, wastewater treatment, and water treatment. Their distinctive blend of high porosity, adjustable pore size distribution, and mechanical robustness renders them immensely appealing for application in sectors where dependable and effective membrane filtration is vital [58 - 60].

Polymer nanofiber membranes are very porous and have a large surface area, making them useful in a variety of applications. Roche & Yalsinkaya (2019), for instance, created polyacrylonitrile (PAN) nanofiber membranes using electrospinning, showcasing their capacity for water purification because of their high flux and ability to reject impurities [61]. There have also been reports of composite nanofiber membranes made of many materials. For example, Chen *et al.* (2022) created a composite nanofiber membrane for oil-water separation that exhibited excellent durability and efficiency by combining graphene oxide (GO) with polyvinylidene fluoride (PVDF) [62]. Studies have been conducted on functionalized nanofiber membranes with improved characteristics. Senthil *et al.* (2022), for instance, reported the creation of functionalized nanofiber membranes for air filtration, showing enhanced antibacterial and particle capture efficiency [63].

Mathematical Modelling

Understanding the intricate transport phenomena involved in nanofiltration processes requires the use of mathematical models. These models support the prediction of membrane performance, the optimization of operational parameters, and the development of effective nanofiltration systems. A complete understanding of the underlying physical and chemical processes occurring at the membrane contact is necessary for the construction of accurate mathematical models. Numerous applications have made considerable use of mathematical models to optimize nanofiltration operations. For instance, researchers have forecast the rejection of particular solutes in water purification procedures using transport models. Operating methods and anti-fouling membranes have been designed using fouling models to reduce fouling. The flow rates and pressures of nanofiltration systems have been optimized for maximum efficiency using hydraulic models [62].

Models of transport mechanisms and irreversible thermodynamics are essential for comprehending nanofiltration procedures [63]. These models help in the design and improvement of nanofiltration systems by imposing light on the intricate transport phenomena that take place in nanofiltration membranes. A framework for examining nanofiltration processes is provided by irreversible thermodynamics, which takes the system's behavior away from equilibrium into account. It ignores the specific chemical mechanisms and considers the membrane as a "black box," concentrating only on the general transport phenomenon. Transport mechanism models are based on specific mechanisms of solute and solvent transport through the membrane. These models provide a more detailed understanding of nanofiltration processes compared to irreversible thermodynamics. Two main transport mechanisms are commonly considered in nanofiltration modeling: viscous pore flow and molecular diffusion. The pore-flow model, proposed by Lonsdale and colleagues (1964), suggests that water flow in the membrane is driven by a pressure gradient [63]. However, this model has been challenged for its applicability to nanofiltration membranes, particularly regarding the role of pore size and charge screening effects. In contrast, the solution-diffusion model, proposed by Lonsdale and colleagues (1970), assumes that solute transport occurs through molecular diffusion down a concentration gradient within a non-porous membrane. This model has been widely used to describe nanofiltration processes, although modifications have been made to incorporate pressure effects and other factors [64].

Irreversible thermodynamics and transport mechanism models are essential tools for understanding nanofiltration processes. While irreversible thermodynamics provides a macroscopic view of transport phenomena, transport mechanism models offer detailed insights into solute and solvent transport mechanisms. These models, when used in conjunction, can enhance our understanding of nanofiltration processes and aid in the development of more efficient nanofiltration systems.

The mass transfer of solutes across the membrane is described by transport models. These models take into account variables, including hydrodynamic circumstances, membrane characteristics, and solute concentration. The surface diffusion model, pore flow model, and solution-diffusion model are a few examples [65]. The deposition of foulants on the membrane surface, which over time may lower the membrane's performance, is described by fouling models. These models take into account variables, including operating conditions, membrane characteristics, and foulant concentration. Cake filtration model, standard blocking model, and intermediate blocking model are a few examples [66].

The solvent flow through the membrane and the pressure drop across the membrane are described by hydraulic models. These models take into account variables, including fluid characteristics, membrane geometry, and operational circumstances. The Navier-Stokes equation and the Darcy's law model are two examples. Biesheuvel, Elimelech, and associates presented a unique pore-flow transport model in an effort to provide an accurate model to explain transport processes in reverse osmosis (RO) membranes. This model, which is known as the solution-friction (SF) model, was created by carefully analyzing the force balances of the solution species as they moved through RO. This method makes it possible to investigate the interactions between the solutes (ions) and solvent (water) in the membrane structure in more detail. The SF model takes into account the frictional forces that the solutes encounter when they pass through the pores in the membrane. This is an important part of transport that is not sufficiently explained by Darcy's equation. The SF model's capacity to explain the linked movement of ions and water is one of its main features. This ability is crucial for comprehending the entire functionality of RO membranes. The efficacy of the SF model in explaining transport processes in RO membranes has been demonstrated by experimental and theoretical studies that have validated it [67 - 71].

By using cutting-edge technology to study water and salt transport mechanisms, advanced reverse osmosis (RO) membranes can be developed. Polyamide membranes have holes, as demonstrated by molecular simulations; nevertheless, little is known about how pore shape affects water transport. Pore dynamics and their implications for transport should be investigated in future simulations. Through training on quantum computations, machine learning can improve simulations. The mechanisms underlying the swelling and deswelling of RO membranes under different situations need to be examined. Exploring the behavior of water molecules and their interaction with the membrane matrix might enhance the creation and functioning of membranes. There is no complete model for compaction in RO membranes, particularly in the selective layer. Our comprehension of membrane performance can be improved by characterizing the elasticity of the membrane and modeling the impacts of compaction. Compaction and its function will be better understood by combining modeling, simulations, and tests [72].

Use of Polymers in Nanofiltration

The process of nanofiltration involves the membrane pore size in the range of 1-10nm. It operates on the principle of size exclusion, allowing smaller molecules to pass through while rejecting larger ones. Polymers produce porous membranes due to high mechanical strength and thus play a significant role in the fabrication and performance of nano-filtration membranes [73]. Ion selectivity, along with the transportation rate through the membrane pores, is governed by the viscosity flow and the size of the ions. Certain polymeric materials can be treated as a precursor for manufacturing nanofiltration membranes like polyamide, polyvenylidene fluoride, and polysulfone, since they possess good mechanical strength, high chemical resistance, high dielectric strength, resistance to organic solvents, and excellent compatibility with the membrane binders [74].

Polymers can also enhance the surface of the membrane and its performance by reducing fouling; for example application of polyethylene glycol grafted on the membrane surface can result in improved efficiency in terms of permeability and reduced protein fouling. The incorporation provides additional surface area to the foulant and prevents it from reaching the membrane surface.

Polymers have advantages over others; they generally provide diverse features that can cater to NF applications. Many times, they are easily fabricated over the membrane and are also cost-effective [75]. By manipulating polymer properties and fabrication techniques, it is possible to control the pore size distribution of nanofiltration membranes. This control allows for precise separation of molecules based on size and molecular weight. Polymers can be blended or crosslinked to tailor the membrane structure and optimize filtration performance for specific applications.

FILTRATION AND FOULING OF MEMBRANE

Filtration using membranes is a widely employed technique in various industries such as water treatment, pharmaceuticals, food and beverage, and biotechnology. However, membrane fouling is a significant challenge that can decrease filtration efficiency and increase operational costs [76]. The filtration process may involve:

Membrane Selection

Membranes are selected based on their pore size, material properties, and surface characteristics. Different membranes offer varying levels of selectivity and permeability depending on the application requirements.

Membrane Configuration

Membranes can be flat-sheet, tubular, spiral-wound, or hollow fiber, each with its advantages and limitations. The configuration choice depends on factors such as required surface area, operating conditions, and ease of cleaning.

Filtration Mechanisms

Membrane filtration primarily operates through several mechanisms, including size exclusion, adsorption, electrostatic interactions, and sieving. These mechanisms determine which particles or molecules are retained by the membrane and which pass through.

Operating Conditions

Parameters such as transmembrane pressure, crossflow velocity, temperature, and feed concentration influence filtration performance. Optimization of these parameters is essential for achieving the desired separation efficiency and throughput.

Membrane Fouling

Fouling refers to the accumulation of unwanted material on the membrane surface or within its pores, leading to decreased flux, increased pressure drop, and reduced separation efficiency [77]. There can be different types of fouling:

Particulate Fouling: Accumulation of suspended solids or colloids on the membrane surface;

Cake Formation: Build-up of a gel-like layer of retained particles on the membrane surface;

Scaling: Deposition of inorganic salts or minerals, leading to pore blockage; Biofouling: Growth of microorganisms such as bacteria, algae, or fungi on the membrane surface;

Organic Fouling: Adsorption or deposition of organic compounds, proteins, or oils on the membrane.

Fouling can have several origins, such as physical deposition, chemical interaction, and biological growth. Particles or aggregates are deposited on the membrane due to size exclusion or hydrodynamic effects [78]. In order to reduce fouling, identification of foulant is necessary, and once it is finalized, suitable adoption strategies can be planned to enhance the purification process. A brief description of foulant and necessary control strategy is described in Table **1** below:

Table 1. Foulants' nature along with their control strategies used in nanofiltration processes [65].

Foulant Nature	Control of Fouling
Organic	Ozone, activated C, ion exchange, increased coagulation process
Inorganic	Diminishes the pH of the system, reduces recovery rate and oxidation of metals to their oxides, and can even operate below solubility limits
Bio-molecular solids	Can be used for microfiltration, ultrafiltration, coagulation, and disinfection
Colloids	Microfiltration, ultrafiltration, coagulation
General	Chemical cleaning

MITIGATION STRATEGIES

Pre-treatment: Removal of particulates, colloids, and microorganisms from the feed stream through processes such as sedimentation, coagulation, or microfiltration [79 - 81].

Chemical Cleaning

Use of cleaning agents such as acids, bases, oxidants, or surfactants to dissolve or dislodge foulants from the membrane surface.

Backwashing

Reversal of flow or application of pressure pulses to remove accumulated material from the membrane surface.

Membrane Modification

Surface coatings, functionalization, or incorporation of anti-fouling additives to enhance fouling resistance.

Operational Optimization

Adjustment of operating conditions such as flux, crossflow velocity, and pH to minimize fouling propensity.

NANOFILTRATION IN SURFACE AND GROUNDWATER

Nanofiltration (NF) is a pressure-driven membrane process that removes impurities from raw water through a semi-permeable membrane with pore sizes ranging from 0.1 to 2 nanometers [82 - 84]. In surface water treatment, NF plays a crucial role by removing particles from the membrane that physically fit our colloidal particles, bacteria, viruses, *etc*. NF partially rejects organic molecules, reducing the content of dissolved organic carbon (DOC) and natural organic matter. The small pore size of NF membranes hinders the formation of biofilms on the membrane surface, reducing maintenance frequency. For groundwater NF membranes reject a significant portion of dissolved salts, making it a viable option for brackish groundwater desalination. NF is effective in removing nitrate ions, a potential health hazard, from groundwater sources. NF membranes can selectively remove heavy metals (*e.g* ., lead and arsenic) through complexation or electrostatic interactions. NF can remove dissolved iron and manganese, which can cause taste and odor problems.

Fouling Control

Fouling is a major challenge in membrane filtration processes, including NF. It occurs when contaminants accumulate on the membrane surface, reducing its enactment and efficiency [85]. Effective fouling control strategies are essential to maintain optimal membrane operation and extend its lifespan.

Fouling can be done in two ways:

Physical Fouling

It requires multiple steps like pre-filtration *i.e* ., removing suspended solids and particles from the feed water through pre-filtration steps like coagulation, sedimentation, and flocculation, which further reduces fouling on the NF membrane to a greater extent [86]. The next step required is membrane cleaning, where periodic cleaning of the NF membrane using chemical (like acids, bases, surfactants) or physical or biological (enzymes) methods help remove foulants and restore membrane permeability. Finally, backwashing the NF membrane with fresh or clean water or air helps dislodge loosely attached foulants and prevent their accumulation [87].

Chemical Fouling

The first method of chemical fouling is coagulation and flocculation, the addition of which to feed water promotes the formation of larger flocs that can be more easily removed by pre-filtration. Scaling due to the precipitation of inorganic salts on the membrane surface can be prevented by adding antiscalants to the feed water. These chemicals inhibit crystal formation and growth. Biofouling caused by microorganisms can be controlled by adding biocides to the feed water or by incorporating antimicrobial agents into the membrane material [88].

Membrane modification - Modifying the surface chemistry of the NF membrane can reduce its susceptibility to fouling. This can be achieved through grafting hydrophilic or antifouling polymers onto the membrane surface.

Nanoparticle incorporation - Incorporating nanoparticles into the NF membrane can enhance its fouling resistance. Nanoparticles can provide additional surface area for foulant attachment or create a barrier that prevents foulants from reaching the membrane surface.

CONCLUSION

The book chapter on wastewater treatment by nanofiltrationdiscusses the mechanism of the process, mathematical modeling, and nanomaterials used for membrane preparation. It also discusses factors influencing NF performance like membrane fouling, scaling, and operating conditions, and strategies to mitigate these challenges effectively, hence several key conclusions can be drawn:

Membrane filtration process: Membrane filtration is an advanced and adaptable method of separation that provides exact control over the division of substances according to their size, charge, and solubility. Depending on the types of membrane kinds, mechanisms, and methods of operation, the technique can be implemented in diverse sectors. Future membrane technology research and development will continue to increase effectiveness and explore a wide range.

Mechanism: Size exclusion, electrostatic interactions, and adsorption are basic processes in nanofiltration separation technique. Ions and tiny molecules are selectively separated by varying the pore diameters in such a way the smaller molecules pass the membrane and larger molecules are entombed. The charged ions and molecules are frequently altered with a surface charge that leads to selective transit of ions and molecules according to electrostatic repulsion. NF membranes are designed with the ability to adsorb specific ions and molecules on their surface with modification by functional groups, which increase the separation efficiency.

Mathematical Modeling: Mathematical modeling is used for understanding the intricate transport phenomena involved in nanofiltration processes and support the prediction of membrane performance, the optimization of operational parameters, and the development of effective nanofiltration systems. Numerous applications have made considerable use of mathematical models like transport models, surface diffusion models, pore flow models, solution-friction models, cake filtration models, standard blocking models, intermediate blocking models, *etc.*, for predicting membrane performance and optimizing nanofiltration operations.

Polymers in nanofiltration membranes: The usage of polymers offers a wide range of properties that can be tailored to specific NF applications. They are relatively inexpensive, easily processed into different membrane configurations and shapes, and resistant to a wide range of chemicals and solvents, ensuring long-term membrane performance.

Membrane fouling and its control: The fouling of nanofiltration membranes on regular usage occurs due to the physical deposition of particles or aggregates, chemical interaction by adsorbing of foulants onto the membrane, and biological growth on the membrane, which cause a significant challenge. Hence, careful membrane selection, process design, and fouling mitigation strategies are to be employed to ensure efficient and sustainable operation of the nanofiltration membrane.

LIST OF ABBREVIATIONS

BNNTs Boron Nitride Nanotubes

CNT Carbon Nanotubes

DOC Dissolved Organic Carbon

GO Graphene Oxide

MF Microfiltration

MOF Metal-organic Framework

NF Nanofiltration

PAN Polyacrylonitrile

PVDF Polyvinylidene Fluoride

RO Reverse Osmosis

TFC Thin Film Composite

TFN   Thin Film Nanocomposite

TMP   Trans-Membrane Pressure

UF   Ultrafiltration

ACKNOWLEDGEMENTS

The authors would like to express their sincere thanks to the management of Manav Rachna University, Faridabad, Haryana, India, for providing the facilities to write and submit the book chapter for publication.

REFERENCES

[1]   Chisti, Y. Principles of membrane separation processes. In: *Bioseparation and Bioprocessing: A handbook,* 2nd ed; Subramanian, G., Ed.; Wiley-VCH: New york, **2007**; 1, pp. 289-322.

[2]   Shannon, M.A.; Bohn, P.W.; Elimelech, M.; Georgiadis, J.G.; Mariñas, B.J.; Mayes, A.M. Science and technology for water purification in the coming decades. *Nature,* **2008**, *452*(7185), 301-310.
        [http://dx.doi.org/10.1038/nature06599] [PMID: 18354474]

[3]   Chao, W.; Yian, W.; Hui, Q.; Hua, L.; Kong, C. Application of microfiltration membrane technology in water treatment, iop conf. series: earth and environmental science 571 **2020**.

[4]   Song, Y.; Dong, B.; Gao, N.; Ma, X. Powder activated carbon pretreatment of a microfiltration membrane for the treatment of surface water. *Int. J. Environ. Res. Public Health,* **2015**, *12*(9), 11269-11277.
        [http://dx.doi.org/10.3390/ijerph120911269] [PMID: 26378552]

[5]   Matsui, Y.; Colas, F.; Yuasa, A. Removal of a synthetic organic chemical by PAC-UF systems. II: model application. *Water Res.,* **2001**, *35*(2), 464-470. a
        [http://dx.doi.org/10.1016/S0043-1354(00)00308-0] [PMID: 11229000]

[6]   Matsui, Y.; Yuasa, A.; Ariga, K. Removal of a synthetic organic chemical by PAC-UF systems—I: theory and modeling. *Water Res.,* **2001**, *35*(2), 455-463. b
        [http://dx.doi.org/10.1016/S0043-1354(00)00283-9] [PMID: 11228999]

[7]   Abdelkader, B.A.; Antar, M.A.; Khan, Z. Nanofiltration as a pretreatment step in desalination. A review. *Arab. J. Sci. Eng.,* **2018**, *43*(9), 4413-4432.
        [http://dx.doi.org/10.1007/s13369-018-3096-3]

[8]   Su, B.; Dou, M.; Gao, X.; Shang, Y. Study on seawater nanofiltration softening Technology for offshore oilfield water and polymer flooding. *Desalination Water Treat.,* **2013**, *297*(25-27), 30-37.

[9]   Berrabah, M.; Benyahia, K.; Zerfa, A. Hassiba b., feasibility of using nanofiltration membranes to delay reverse osmosis membrane fouling and to reduce energy consumption (BWC desalination plant). *J. Water Environ. Technol.,* **2023**, *8*(4), 384-395.

[10]   Obotey Ezugbe, E.; Rathilal, S. Membrane technologies in wastewater treatment: A review. *Membranes (Basel),* **2020**, *10*(5), 89.
        [http://dx.doi.org/10.3390/membranes10050089] [PMID: 32365810]

[11]   Charcosset, C. C. Charcosset, Microfiltration, in: C.B.T.-M.P. in B. and P. Charcosset (Ed.), Membr. Process. Biotechnol. Pharm., Elsevier, Amsterdam, 2012: pp. 101–141.
        [http://dx.doi.org/10.1016/B978-0-444-56334-7.00003-4]

[12]   Nasir, A.M.; Adam, M.R.; Mohamad Kamal, S.N.E.A.; Jaafar, J.; Othman, M.H.D.; Ismail, A.F.; Aziz, F.; Yusof, N.; Bilad, M.R.; Mohamud, R.; A Rahman, M.; Wan Salleh, W.N. A review of the potential of conventional and advanced membrane technology in the removal of pathogens from wastewater. *Separ. Purif. Tech.,* **2022**, *286*, 120454.

[http://dx.doi.org/10.1016/j.seppur.2022.120454] [PMID: 35035270]

[13] Baker, R.W. *Membrane Technology and Applications*; John Wiley & Sons, **2004**.
      [http://dx.doi.org/10.1002/0470020393]

[14] Mona, A. Abdel-fatah. Nitrofiltration systems and applications in wastewater treatment. *Review Article.,* **2018**, *9*, 3077-3092.

[15] Du, X.; Shi, Y.; Jegatheesan, V.; Haq, I.U. A review on the mechanism, impacts and control methods of membrane fouling in MBR system. *Membranes (Basel),* **2020**, *10*(2), 24.
      [http://dx.doi.org/10.3390/membranes10020024] [PMID: 32033001]

[16] Darvishmanesh, S.; Qian, X.; Wickramasinghe, S.R. Responsive membranes for advanced separations. *Curr. Opin. Chem. Eng.,* **2015**, *8*, 98-104.
      [http://dx.doi.org/10.1016/j.coche.2015.04.002]

[17] Van der Bruggen, B.; Vandecasteele, C. Distillation *vs.* membrane filtration: overview of process evolutions in seawater desalination. *Desalination,* **2002**, *143*(3), 207-218.
      [http://dx.doi.org/10.1016/S0011-9164(02)00259-X]

[18] Bowen, W.R.; Mohammad, A.W.; Hilal, N. Characterisation of nanofiltration membranes for predictive purposes–use of salts, uncharged solutes and atomic force microscopy. *J. Membr. Sci.,* **2002**, *209*(2), 375-390.

[19] Thanutamavong, K.; Bowman, R.S. Characterization of nanofiltration membranes by atomic force microscopy. *J. Membr. Sci.,* **2001**, *190*(2), 187-198.

[20] Mika, M.; Katja, V.; Nystrom, M. Nanofiltration of biologically treated effluents from the pulp and paper industry. *J. Membr. Sci.,* **2006**, *271*(1-2), 152-160.

[21] Gozalvez-Zafrilla, J.M.; Martin-Esteban, A.; Bowen, W.R. Nanofiltration of ferment demineralization in the dairy industry. *J. Membr. Sci.,* **2002**, *204*(1-2), 239-250.

[22] Drioli, E.; Giorno, L. *Comprehensive Membrane Science and Engineering*; Elsevier, **2010**.

[23] Nunes, S.P.; Peinemann, K.V., Eds. *Membrane Technology in the Chemical Industry*; Wiley-VCH, **2006**.
      [http://dx.doi.org/10.1002/3527608788]

[24] Baker, R.W.; Lokhandwala, K. Natural organic matter fouling of nanofiltration membranes: mechanisms, significance and control. *J. Membr. Sci.,* **2008**, *305*(1-2), 1-20.

[25] Koltuniewicz, A.B.; Wessling, M. Role of surface charge in nanofiltration: mechanisms and applications. *J. Membr. Sci.,* **2005**, *249*(1-2), 1-16.

[26] Goh, P.S.; Ismail, A.F. A review on inorganic membranes for desalination and wastewater treatment. *Desalination,* **2011**, *287*, 190-199.

[27] Li, D.; Livingston, A.G. Molecular simulation study of water and salt transport in thin-film composite membranes for desalination and pressure retarded osmosis. *J. Phys. Chem. C,* **2012**, *116*(8), 5100-5110.

[28] Mi, B.; Elimelech, M.; Sirkar, K.K. Membrane distillation using electrospun polyvinylidene fluoride nanofibers. *J. Membr. Sci.,* **2005**, *251*(1-2), 13-21.

[29] Wang, K.Y.; Chung, T.S.; Rajagopalan, R. Recent advances in polymeric membranes for solvent recovery in chemical and pharmaceutical industries. *Ind. Eng. Chem. Res.,* **2007**, *46*(10), 3121-3154.

[30] Yuan, H.; Liu, J.; Zhang, X.; Chen, L.; Zhang, Q.; Ma, L. Recent advances in membrane-based materials for desalination and gas separation. *J. Clean. Prod.,* **2023**, *387*, 135845.
      [http://dx.doi.org/10.1016/j.jclepro.2023.135845]

[31] Dalwani, M.; Benes, N.E.; Bargeman, G.; Stamatialis, D.; Wessling, M. Effect of pH on the performance of polyamide/polyacrylonitrile based thin film composite membranes. *J. Membr. Sci.,* **2011**, *372*(1-2), 228-238.

[http://dx.doi.org/10.1016/j.memsci.2011.02.012]

[32] Li, C.; Cai, X.; Xu, T.; Cheng, H.; Liu, G.; Zhu, H.; Guo, Y. Graphene oxide incorporated thin film nanocomposite nanofiltration membrane to enhance permeation and antifouling properties. *Desalination Water Treat.*, **2024**, *317*, 100090.
[http://dx.doi.org/10.1016/j.dwt.2024.100090]

[33] Priya, A.K.; Gnanasekaran, L.; Kumar, P.S.; Jalil, A.A.; Hoang, T.K.A.; Rajendran, S.; Soto-Moscoso, M.; Balakrishnan, D. Recent trends and advancements in nanoporous membranes for water purification. *Chemosphere*, **2022**, *303*(Pt 3), 135205.
[http://dx.doi.org/10.1016/j.chemosphere.2022.135205] [PMID: 35667502]

[34] Wang, Z.; Wu, A.; Colombi Ciacchi, L.; Wei, G. Recent advances in nanoporous membranes for water purification. *Nanomaterials (Basel)*, **2018**, *8*(2), 65.
[http://dx.doi.org/10.3390/nano8020065] [PMID: 29370128]

[35] Ensinger, W.; Sudowe, R.; Brandt, R.; Neumann, R. Gas separation in nanoporous membranes formed by etching ion irradiated polymer foils. *Radiat. Phys. Chem.*, **2010**, *79*(3), 204-207.
[http://dx.doi.org/10.1016/j.radphyschem.2009.08.045]

[36] Wang, D.; Su, H.; Han, S.; Tian, M.; Han, L. The role of microporous metal–organic frameworks in thin-film nanocomposite membranes for nanofiltration. *Separ. Purif. Tech.*, **2024**, *333*, 125859.
[http://dx.doi.org/10.1016/j.seppur.2023.125859]

[37] Nitodas, S.F.; Das, M.; Shah, R. Applications of Polymeric membranes with carbon nanotubes: A review. *Membranes (Basel)*, **2022**, *12*(5), 454.
[http://dx.doi.org/10.3390/membranes12050454] [PMID: 35629780]

[38] Ince, Muharrem; Ince, Olcay Kaplan Ince and olcay kaplan ince, preparation and applications of nanocomposite membranes for water/wastewater treatment **2021**.
[http://dx.doi.org/10.5772/intechopen.101905]

[39] Al Harby, N.F.; El-Batouti, M.; Elewa, M.M. Prospects of polymeric nanocomposite membranes for water purification and scalability and their health and environmental impacts: A review. *Nanomaterials (Basel)*, **2022**, *12*(20), 3637.
[http://dx.doi.org/10.3390/nano12203637] [PMID: 36296828]

[40] Nain, Amit; Sangili, Arumugam; Hu, Shun-Ruei; Chen, Chun-Hsien; Chen, Yen-Ling; Chang, Huan-Tsung recent Progress in nanomaterial-functionalized membranes for removal of pollutants **2022**, *25*(7), 104616.
[http://dx.doi.org/10.1016/j.isci.2022.104616]

[41] Lai, G.S.; Lau, W.J.; Goh, P.S.; Ismail, A.F.; Yusof, N.; Tan, Y.H. Graphene oxide incorporated thin film nanocomposite nanofiltration membrane for enhanced salt removal performance. *Desalination*, **2016**, *387*, 14-24.
[http://dx.doi.org/10.1016/j.desal.2016.03.007]

[42] Kotobuki, M.; Gu, Q.; Zhang, L.; Wang, J. Ceramic-polymer composite membranes for water and wastewater treatment: Bridging the big gap between ceramics and polymers. *Molecules*, **2021**, *26*(11), 3331.
[http://dx.doi.org/10.3390/molecules26113331] [PMID: 34206052]

[43] Boussemghoune, M.; Chikhi, M.; Balaska, F.; Ozay, Y.; Dizge, N.; Kebabi, B. Preparation of a zirconia-based ceramic membrane and its application for drinking water treatment. *Symmetry (Basel)*, **2020**, *12*(6), 933.
[http://dx.doi.org/10.3390/sym12060933]

[44] Ferreira, I.C.; Ferreira, T.J.; Barbosa, A.D.S.; de Castro, B.; Ribeiro, R.P.P.L.; Mota, J.P.B.; Alves, V.D.; Cunha-Silva, L.; Esteves, I.A.A.C.; Neves, L.A. Cr-based MOF/IL composites as fillers in mixed matrix membranes for CO2 separation. *Separ. Purif. Tech.*, **2021**, *276*, 119303.
[http://dx.doi.org/10.1016/j.seppur.2021.119303]

[45]   Li, Y.; Gao, H.; Song, Y.; Li, X. Carbon nanotube membranes: Synthesis, properties, and applications. *Ind. Eng. Chem. Res.,* **2019**, *58*(4), 1539-1563.

[46]   Majumder, M.; Chopra, N.; Andrews, R.; Hinds, B.J. Enhanced flow in carbon nanotubes. *Nature,* **2005**, *438*(7064), 44-44.
[http://dx.doi.org/10.1038/438044a] [PMID: 16267546]

[47]   Holt, J.K.; Park, H.G.; Wang, Y.; Stadermann, M.; Artyukhin, A.B.; Grigoropoulos, C.P.; Noy, A.; Bakajin, O. Fast mass transport through sub-2-nanometer carbon nanotubes. *Science,* **2006**, *312*(5776), 1034-1037.
[http://dx.doi.org/10.1126/science.1126298] [PMID: 16709781]

[48]   Fornasiero, F.; Park, H.G.; Holt, J.K.; Stadermann, M.; Grigoropoulos, C.P.; Noy, A.; Bakajin, O. Ion exclusion by sub-2-nm carbon nanotube pores. *Proc. Natl. Acad. Sci. USA,* **2008**, *105*(45), 17250-17255.
[http://dx.doi.org/10.1073/pnas.0710437105] [PMID: 18539773]

[49]   Liu, J.; Song, L.F.; Liu, J.; Liu, C. Highly selective nanofiltration through interlayer-expanded carbon nanotube membranes. *RSC Advances,* **2013**, *3*(44), 21894-21898.

[50]   García Doménech, N.; Purcell-Milton, F.; Sanz Arjona, A.; Casasín García, M.L.; Ward, M.; Cabré, M.B.; Rafferty, A.; McKelvey, K.; Dunne, P.; Gun'ko, Y.K. High-performance boron nitride-based membranes for water purification. *Nanomaterials (Basel),* **2022**, *12*(3), 473.
[http://dx.doi.org/10.3390/nano12030473] [PMID: 35159818]

[51]   Kim, J.P.; Kim, J.H.; Kim, J.; Lee, S.N.; Park, H.O. A nanofilter composed of carbon nanotube-silver composites for virus removal and antibacterial activity improvement. *J. Environ. Sci. (China),* **2016**, *42*, 275-283.
[http://dx.doi.org/10.1016/j.jes.2014.11.017] [PMID: 27090720]

[52]   Huang, Z.M.; Zhang, Y.Z.; Kotaki, M.; Ramakrishna, S. A review on polymer nanofibers by electrospinning and their applications in nanocomposites. *Compos. Sci. Technol.,* **2003**, *63*(15), 2223-2253.
[http://dx.doi.org/10.1016/S0266-3538(03)00178-7]

[53]   Li, D.; Xia, Y. Electrospinning of nanofibers: Reinventing the wheel? *Adv. Mater.,* **2004**, *16*(14), 1151-1170.
[http://dx.doi.org/10.1002/adma.200400719]

[54]   Subbiah, T.; Bhat, G.S.; Tock, R.W.; Parameswaran, S.; Ramkumar, S.S. Electrospinning of nanofibers. *J. Appl. Polym. Sci.,* **2005**, *96*(2), 557-569.
[http://dx.doi.org/10.1002/app.21481]

[55]   Audrey, F.; Ioannis, S.C. Polymer nanofibres assembled by electrospinning, current opinion in colloidal and interface science **2003**, *8*, 64-75.

[56]   Greiner, A.; Wendorff, J.H. Electrospinning: a fascinating method for the preparation of ultrathin fibers. *Angew. Chem. Int. Ed.,* **2007**, *46*(30), 5670-5703.
[http://dx.doi.org/10.1002/anie.200604646] [PMID: 17585397]

[57]   Roche, R.; Yalcinkaya, F. Electrospun polyacrylonitrile nanofibrous membranes for point-of-use water and air cleaning. *ChemistryOpen,* **2019**, *8*(1), 97-103.
[http://dx.doi.org/10.1002/open.201800267] [PMID: 30693173]

[58]   Yang, M.; Chen, K.; Wang, M.; Chen, H.; Ling, H.; Zhao, W.; Liu, H.; Xiao, C. Simple fabrication of polyvinylidene fluoride/graphene composite membrane with good lipophilicity for oil treatment. *ACS Omega,* **2022**, *7*(25), 21454-21464.
[http://dx.doi.org/10.1021/acsomega.2c00764] [PMID: 35785275]

[59]   Huang, X., Jiao, T. & Liu, Q. Hierarchical electrospun nanofibers treated by solvent vapor annealing as air filtration mat for high-efficiency PM2.5 capture. *Sci. China. Mater.* **2019**, *62*, 423–436.
[http://dx.doi.org/10.1038/s41598-022-12505-w] [PMID: 35589800]

[60] Zheng, W.; Chen, Y.; Xu, X.; Peng, X.; Niu, Y.; Xu, P.; Li, T. Research on the factors influencing nanofiltration membrane fouling and the prediction of membrane fouling. *J. Water Process Eng.*, **2024**, *59*, 104876.
[http://dx.doi.org/10.1016/j.jwpe.2024.104876]

[61] Roche, R.; Yalcinkaya, F. Electrospun polyacrylonitrile, nanofibrous membranes for point of use water and air cleaning. *ChemistryOpen*, **2019**, *8*(1), 97-103.
[http://dx.doi.org/10.1002/open.201800267] [PMID: 30693173]

[62] Chap, Tijing L., Yao, M., Ren, J., Kim C.S., Shon, H.K.,. nanofibres in water and wastewater Treatment: Recent Advances and Developments, Water and wastewater treatment technologies **2019**, 431-468.
[http://dx.doi.org/10.1007/978-981-13-3259-3_20]

[63] Senthil, R.; Sumathi, V.; Tamilselvi, A.; Kavukcu, S.B.; Aruni, A.W. Functionalized electrospun nanofibers for high efficiency removal of particulate matter. *Sci. Rep.*, **2022**, *12*(1), 8411.
[http://dx.doi.org/10.1038/s41598-022-12505-w] [PMID: 35589800]

[64] Heiranian, M.; Fan, H.; Wang, L.; Lu, X.; Elimelech, M. Mechanisms and models for water transport in reverse osmosis membranes: history, critical assessment, and recent developments. *Chem. Soc. Rev.*, **2023**, *52*(24), 8455-8480.
[http://dx.doi.org/10.1039/D3CS00395G] [PMID: 37889082]

[65] Silva, F.C. *Fouling of Nanofiltration Membranes*; InTech, **2011**.
[http://dx.doi.org/10.5772/intechopen.75353]

[66] Wang, L.; Cao, T.; Dykstra, J.e.; Porada, S.; Biesheuvel, P.M.; Elimelech, M. A concise tutorial review of Reverse Osmosis and Electrodialysis. *Environ. Sci. Technol.*, **2021**, *55*, 16665-16675.
[http://dx.doi.org/10.1021/acs.est.1c05649] [PMID: 34879196]

[67] Biesheuvel, P.M.; Porada, S.; Elimelech, M.; Dykstra, J.E. Tutorial review of reverse osmosis and electrodialysis. *J. Membr. Sci.*, **2022**, *647*, 120221.
[http://dx.doi.org/10.1016/j.memsci.2021.120221]

[68] Oren, Y.S.; Biesheuvel, P.M. Theory of ion and water transport in reverse osmosis membranes *Phys. Rev. Appl.*, **2018**, *9*(2), 024-034.
[http://dx.doi.org/10.1103/PhysRevApplied.9.024034]

[69] Bowen, W.R.; Welfoot, J.S. Modelling the performance of membrane nanofiltration—critical assessment and model development. *Chem. Eng. Sci.*, **2002**, *57*(7), 1121-1137.
[http://dx.doi.org/10.1016/S0009-2509(01)00413-4]

[70] Riscanu, D.; Favis, B.D.; Feng, C.; Matsuura, T. Thin-film membranes derived from co-continuous polymer blends: preparation and performance. *Polymer (Guildf.)*, **2004**, *45*(16), 5597-5609.
[http://dx.doi.org/10.1016/j.polymer.2004.05.067]

[71] Bryant, R.G. Polyimides in ullmann's encyclopedia of industrial chemistry, wiley-vch, verlag GmbH & co KGaA,, weinheim , **2014**.

[72] Jansen, J.C.; Darvishmanesh, S.; Tasselli, F.; Bazzarelli, F.; Bernardo, P.; Tocci, E.; Friess, K.; Randova, A.; Drioli, E.; Van der Bruggen, B. Influence of the blend composition on the properties and separation performance of novel solvent resistant polyphenylsulfone/polyimide nanofiltration membranes. *J. Membr. Sci.*, **2013**, *447*, 107-118.
[http://dx.doi.org/10.1016/j.memsci.2013.07.009]

[73] Ogawa, N.; Kimura, K.; Watanabe, Y. Membrane fouling in nanofiltration/reverse osmosis membranes coupled with a membrane bioreactor used for municipal wastewater treatment. *Desalination Water Treat.*, **2010**, *18*(1-3), 292-296.
[http://dx.doi.org/10.5004/dwt.2010.1795]

[74] Broeckmann, A.; Wintgens, T.; Schäfer, A.I. Removal and fouling mechanisms in nanofiltration of polysaccharide solutions. *Desalination*, **2005**, *178*(1-3), 149-159.

[http://dx.doi.org/10.1016/j.desal.2004.12.017]

[75]　Susanto, H.; Ulbricht, M. Influence of ultrafiltration membrane characteristics on adsorptive fouling with dextrans. *J. Membr. Sci.,* **2005**, *266*(1-2), 132-142.
[http://dx.doi.org/10.1016/j.memsci.2005.05.018]

[76]　Fane, A.G.; Beatson, P.; Li, H. Membrane fouling and its control in environmental applications. *Water Sci. Technol.,* **2000**, *41*(10-11), 303-308.
[http://dx.doi.org/10.2166/wst.2000.0667]

[77]　Wang, X.L.; Zhang, C.; Ouyang, P. The possibility of separating saccharides from a NaCl solution by using nanofiltration in diafiltration mode. *J. Membr. Sci.,* **2002**, *204*(1-2), 271-281.
[http://dx.doi.org/10.1016/S0376-7388(02)00050-9]

[78]　Wang, X.L.; Ying, A.L.; Wang, W.N. Nanofiltration of l-phenylalanine and l-aspartic acid aqueous solutions. *J. Membr. Sci.,* **2002**, *196*(1), 59-67.
[http://dx.doi.org/10.1016/S0376-7388(01)00570-1]

[79]　Li, S.; Li, C.; Liu, Y.; Wang, X.; Cao, Z. Separation of l-glutamine from fermentation broth by nanofiltration. *J. Membr. Sci.,* **2003**, *222*(1-2), 191-201.
[http://dx.doi.org/10.1016/S0376-7388(03)00290-4]

[80]　Wang, D.X.; Wang, X.L.; Tomi, Y.; Ando, M.; Shintani, T. Modeling the separation performance of nanofiltration membranes for the mixed salts solution. *J. Membr. Sci.,* **2006**, *280*(1-2), 734-743.
[http://dx.doi.org/10.1016/j.memsci.2006.02.032]

[81]　Wang, D.X.; Wu, L.; Liao, Z.D.; Wang, X.L.; Tomi, Y.; Ando, M.; Shintani, T. Modeling the separation performance of nanofiltration membranes for the mixed salts solution with Mg^{2+} and Ca^{2+}. *J. Membr. Sci.,* **2006**, *284*(1-2), 384-392.
[http://dx.doi.org/10.1016/j.memsci.2006.08.004]

[82]　Jiraratananon, R.; Sungpet, A.; Luangsowan, P. Performance evaluation of nanofiltration membranes of nanofiltration membranes for treatment of effluents containing reactive dye and salt. *Desalination,* **2000**, *130*, 177-183.
[http://dx.doi.org/10.1016/S0011-9164(00)00085-0]

[83]　Ducom, G.; Cabassud, C. Interests and limitations of nanofiltration for the removal of volatile organic compounds in drinking water production. *Desalination,* **1999**, *124*(1-3), 115-123.
[http://dx.doi.org/10.1016/S0011-9164(99)00095-8]

[84]　Galanakis, C.M.; Tornberg, E.; Gekas, V. Clarification of high-added value products from olive mill wastewater. *J. Food Eng.,* **2010**, *99*(2), 190-197.
[http://dx.doi.org/10.1016/j.jfoodeng.2010.02.018]

[85]　Galanakis, C.M.; Chasiotis, S.; Botsaris, G.; Gekas, V. Separation and recovery of proteins and sugars from Halloumi cheese whey. *Food Res. Int.,* **2014**, *65*, 477-483.
[http://dx.doi.org/10.1016/j.foodres.2014.03.060]

[86]　Muro, C.; Riera, F.; Del Carmen Diaz, M. Membrane separation process in wastewater treatment of food industry. In: *Food Industrial Processes—Methods and Equipment*; Valdez, B., Ed.; InTech: Rijeka, Croatia, **2012**; pp. 253-280.
[http://dx.doi.org/10.5772/31116]

[87]　Suárez, E.; Lobo, A.; Alvarez-Blanco, S.; Riera, F.A.; Álvarez, R. Utilization of nanofiltration membranes for whey and milk ultrafiltration permeate demineralization. *Desalination,* **2006**, *199*(1-3), 345-347.
[http://dx.doi.org/10.1016/j.desal.2006.03.081]

[88]　Tylkowski, B.; Tsibranska, I.; Kochanov, R.; Peev, G.; Giamberini, M. Concentration of biologically active compounds extracted from Sideritis ssp. L. by nanofiltration. *Food Bioprod. Process.,* **2011**, *89*(4), 307-314.
[http://dx.doi.org/10.1016/j.fbp.2010.11.003]

CHAPTER 10

Zeolites in Wastewater Treatment

Vishaka Chauhan[1], **Chinmay Mittal**[1], **Chanchal Vashisth**[2] and **Neera Raghav**[2,*]

[1] *Department of Chemistry, SGT University, Gurugram-122001, Haryana, India*

[2] *Department of Chemistry, Kurukshetra University, Kurukshetra-136119, Haryana, India*

Abstract: The increasing concern over water pollution and the scarcity of clean water resources, amplified by increased urbanization and industrialization, is accompanied by the accumulation of heavy metal contaminants within wastewater systems. This alarming trend poses significant threats to human and environmental health, intensifying the search for efficient and sustainable wastewater treatment technologies. Zeolites, a class of microporous aluminosilicate minerals, have gained attention as a favorable material for the management of wastewater because of their versatility, chemical stability, large surface area, considerable ion-exchange capacity, selective adsorption property, and environment-friendliness that enables them to effectively address several different kinds of contaminants present in wastewater streams. The objective of this book chapter is to bring light to the thorough investigation of the synthesis and application of synthetic and natural zeolites in efficiently removing various contaminants, such as organic dyes and heavy metals, from wastewater. The mechanisms of adsorption and ion exchange are discussed, highlighting the factors affecting the efficiency of zeolites, particularly surface charge, chemical composition, and pore dimension.

Keywords: Adsorption, Heavy metals, Organic dyes, Pollutants, Water treatment, Zeolites.

INTRODUCTION

The urgent worldwide problem of water contamination and shortage demands the creation of novel wastewater treatment technologies. Zeolites are a substance that has shown promise in many ways because of their special qualities and wide range of uses in water cleanup procedures. Zeolites, which are crystalline aluminosilicates with a porous structure and the ability to exchange ions, provide a variety of ways to remove and break down pollutants from aqueous solutions effectively [1 - 5].

* **Corresponding author Neera Raghav:** Department of Chemistry, Kurukshetra University, Kurukshetra-136119, Haryana, India; E-mail: nraghav.chem@gmail.com

Anjaneyulu Bendi (Ed.)

Environmental and public health risks arise from water pollution caused by urban activities, agricultural runoff, and industrial discharges. Alternative solutions are being investigated because traditional wastewater treatment procedures frequently fail to handle the wide variety of contaminants contained in wastewater streams. Zeolites' large surface area, adjustable pore size, and selective adsorption capabilities have drawn interest in their ability to successfully solve this issue [6 - 8].

Since zeolites can adsorb numerous kinds of contaminants, including nutrients, organic pollutants, and heavy metals, this is how they are mostly used in wastewater treatment [9 - 14]. Electrostatic interactions, ion exchange, and molecular sieving processes enable the selective adsorption of target molecules into the microporous structure of zeolites. Additionally, zeolite surfaces can be modified by functionalization or ion exchange procedures, which can improve their selectivity and adsorption capacity and increase their effectiveness in removing pollutants.

Zeolites are adsorbents, but they also have catalytic qualities that may be used to break down stubborn organic materials in wastewater [15 - 23]. Zeolites facilitate the effective conversion of contaminants into lesser hazardous or inert substances by oxidation, hydrolysis, or other chemical transformations by acting as catalyst supports or active sites for heterogeneous catalysis. The combination of adsorption and catalysis on zeolite surfaces provides a holistic method of treating wastewater, efficiently removing both soluble and insoluble pollutants.

Zeolites are also essential for sophisticated procedures for the treatment of wastewater, such as membrane filtration, nutrient recovery, and ion exchange. Because of their molecular sieving capabilities, zeolite-based membranes provide better separation performance by allowing contaminants to be selectively removed while keeping nutrients and necessary ions in place [24 - 33]. Furthermore, by using zeolites' ion exchangeability to remove certain ions or recover valuable materials from wastewater streams, resource conservation and the concepts of the circular economy may be supported (Fig. **1**).

To sum up, zeolites are a flexible and exciting family of materials that may be used to tackle the intricate problems involved in wastewater treatment. Zeolites provide efficient solutions for pollutant removal, catalytic degradation, and resource recovery across a range of wastewater treatment situations owing to their distinct physicochemical features and wide range of uses. The sustainability and effectiveness of wastewater treatment procedures might be greatly increased by carrying out ongoing research to explore new applications and optimize zeolite-based technology.

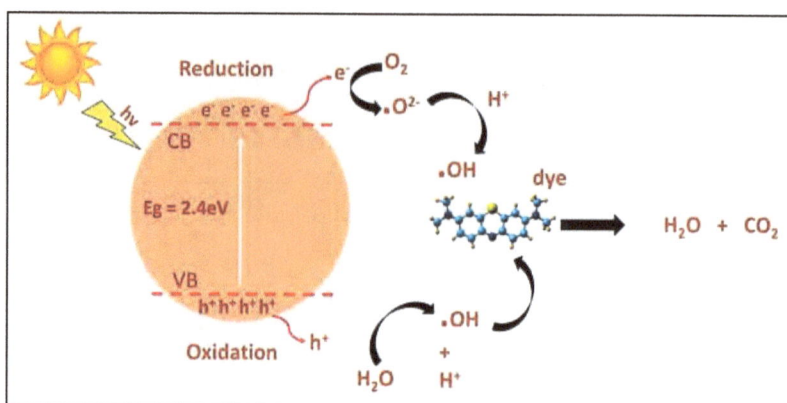

Fig. (1). Degradation of dyes present in wastewater. [30] .

This book chapter provides an in-depth analysis of the uses, underlying processes, and future possibilities of zeolites in wastewater treatment. The goal of this study is to promote sustainable and effective wastewater treatment solutions by clarifying the many functions that zeolites play in water remediation.

CHEMISTRY OF ZEOLITE COMPOUNDS FOR THE REMOVAL OF HEAVY METALS

Sivalingam, S. et al. created zeolite nanocrystalline to remove heavy metals from industrial effluents. (Schemes **1**, **2**) [34].

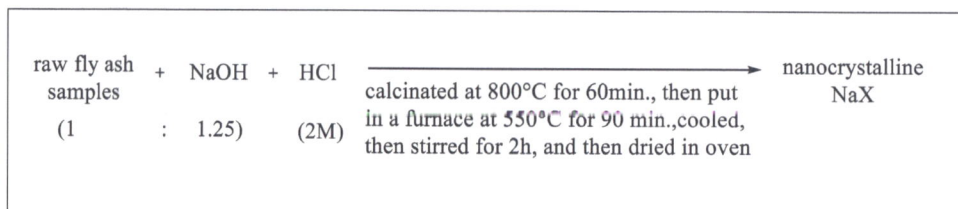

Scheme 1. Synthesis of nanocrystalline NaX.

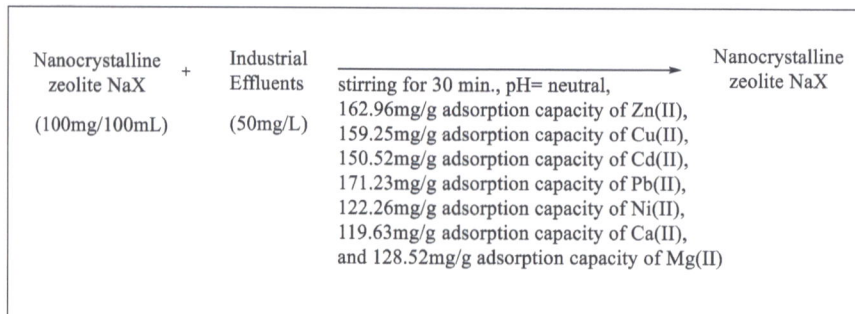

Scheme 2. Removal efficiency of heavy metal ions.

The sono-hydrothermal process is used to create the zeolites. The synthesized composites typically have crystal sizes between 19 and 27 nm. Of all the metal ions, lead was determined to have the highest adsorption capabilities. In order to recover and recycle clean water from industrial effluents, the final synthesized zeolite composites with high purity and a very high sorption capacity towards various divalent metal cations and organic dyes may be employed.

Ahmad, R. et al. fabricated zeolite bio-nanocomposite to eliminate heavy metal ions present in industrial wastewater. (Schemes **3, 4**) [35].

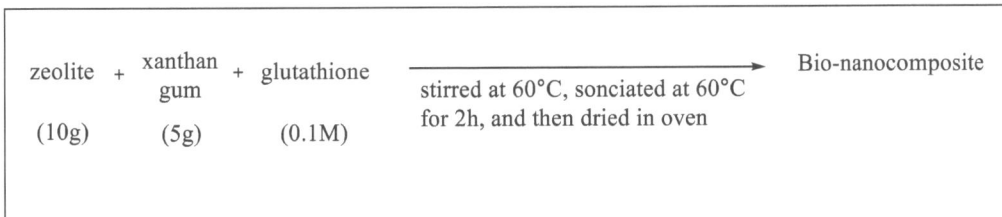

zeolite +	xanthan gum +	glutathione	⟶ stirred at 60°C, sonciated at 60°C for 2h, and then dried in oven	Bio-nanocomposite
(10g)	(5g)	(0.1M)		

Scheme 3. Synthesis of Bio-nanocomposite.

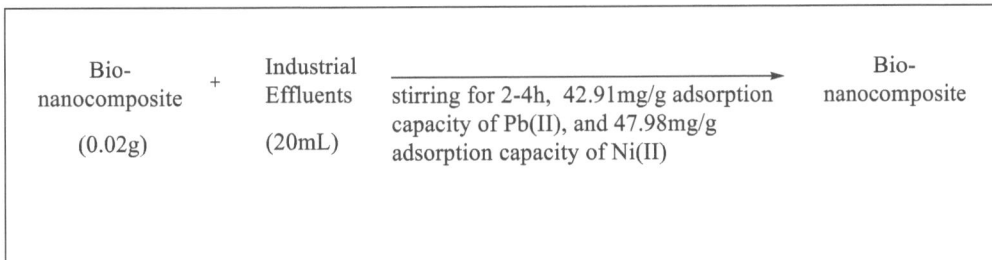

Bio-nanocomposite +	Industrial Effluents	⟶ stirring for 2-4h, 42.91mg/g adsorption capacity of Pb(II), and 47.98mg/g adsorption capacity of Ni(II)		Bio-nanocomposite
(0.02g)	(20mL)			

Scheme 4. Removal efficiency of heavy metal ions.

The bio-nanocomposite's point of zero charge, which is 8.2, reveals the fundamental characteristics of an adsorbent. Both the Freundlich isotherm models and pseudo-second-order provided the best description of the adsorption process. The uptake was observed as endothermic and spontaneous by thermodynamic studies. The most effective method for achieving desorption with regeneration in the fifth cycle for contaminants was using water. The findings demonstrate that the adsorption process includes ion exchange along with chemical and physical factors. In light of this, it has been demonstrated that the current bio-nanocomposite is a potential adsorbent that should be investigated in removing both inorganic and organic contaminants from industrial wastewater.

Weiwei, B. et al. synthesized zeolites to remove copper metal ions from wastewater. (Schemes **5, 6**) [36].

Oil shale + aluminium + sodium — muffle furnace at 600°C for 2h, → Na-A zeolite
ash   oxide   hydroxide   cooled, then ultrasonicated for 3h,
                                     and then dried in oven

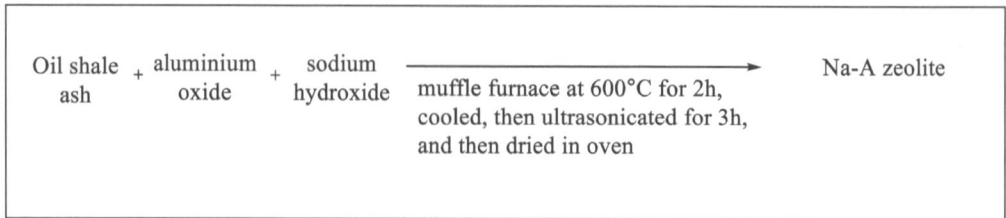

Scheme 5. Synthesis of Na-A zeolite.

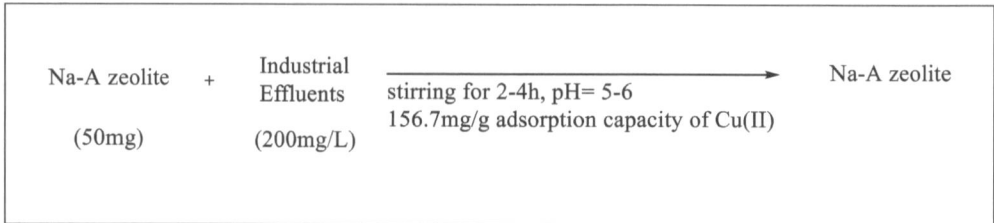

Na-A zeolite + Industrial — stirring for 2-4h, pH= 5-6 → Na-A zeolite
                  Effluents
(50mg)   (200mg/L)   156.7mg/g adsorption capacity of Cu(II)

Scheme 6. Removal efficiency of heavy metal ions.

The equilibrium adsorption data were matched using Freundlich and Langmuir models. The synthesized zeolite composite, derived from the Langmuir adsorption isotherm, has the greatest adsorption capacity of 156.7 mg/g of copper metal ions. The rise in pH level in the adsorption process indicates that an ion exchange mechanism underlies the heavy metals' absorption on the zeolite. The batch kinetic data and pseudo-second-order equation suit each other nicely. The spontaneous behavior of the adsorption processes is confirmed by negative ΔG values at various temperatures.

Chen, M. et al. created zeolites to absorb heavy metal ions found in industrial wastewater. (Schemes **7, 8**) [37].

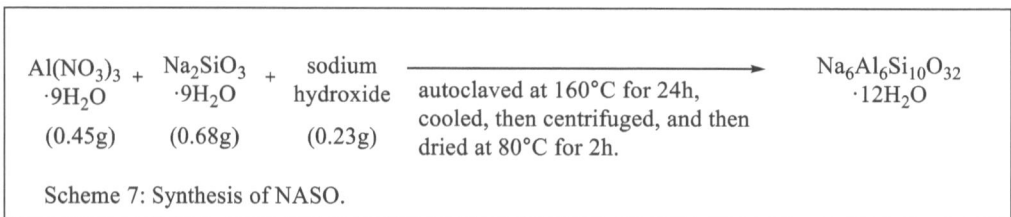

$Al(NO_3)_3$ + Na_2SiO_3 + sodium — autoclaved at 160°C for 24h, → $Na_6Al_6Si_{10}O_{32}$
$\cdot 9H_2O$ $\cdot 9H_2O$ hydroxide   cooled, then centrifuged, and then $\cdot 12H_2O$
(0.45g)   (0.68g)   (0.23g)   dried at 80°C for 2h.

Scheme 7: Synthesis of NASO.

Scheme 7. Synthesis of NASO.

NASO zeolite   +   Industrial
                   Effluents   stirring for 1h, pH= 5   ⟶   NASO
                                      649mg/g adsorption capacity of Pb(II),   zeolite
  (0.01g)   (130ppm)   210mg/g adsorption capacity of Cd(II),
                                      90mg/g adsorption capacity of Cu(II), and
                                      88mg/g adsorption capacity of Zn(II)

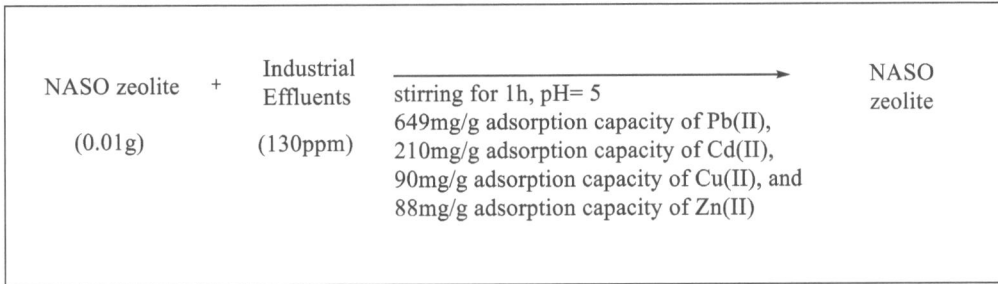

Scheme 8. Removal efficiency of heavy metal ions.

High adsorption capacity was made possible by the removal process, which was attributed to hydroxyl groups and ion exchange. To uncover adsorption behaviors, pseudo-first- and pseudo-second-order models were employed to calculate their adsorption rates. For industrialization and practical water treatment, zeolite was, therefore, promising. Lead metal ion concentration was seen to adsorb rapidly, dropping to 0.6 ppb in under two minutes. Even after five regeneration recycles, 97% of the lead metal ion was still removed. The synthetic composite is, therefore, a viable option for eliminating organic contaminants.

Chiang, Y. W. et al. synthesized zeolite composites to eliminate heavy metal ions. (Schemes **9, 10**) [38].

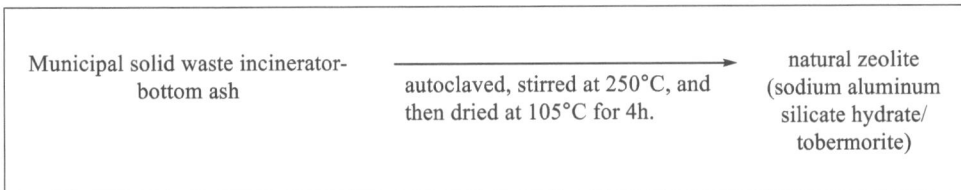

Municipal solid waste incinerator-   autoclaved, stirred at 250°C, and   natural zeolite
          bottom ash   then dried at 105°C for 4h.   ⟶   (sodium aluminum
                                                                                     silicate hydrate/
                                                                                        tobermorite)

Scheme 9. Synthesis of natural zeolite.

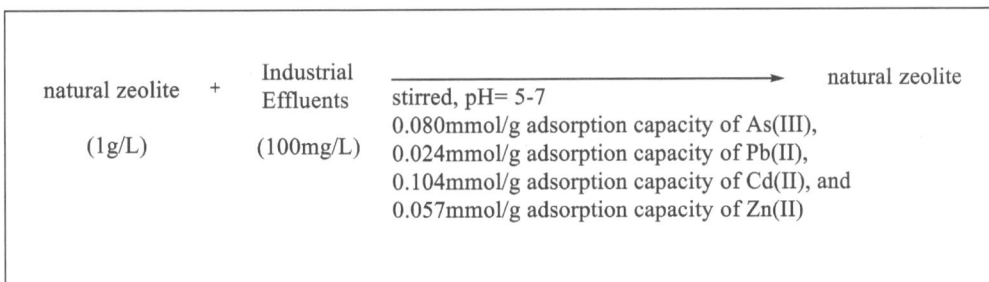

natural zeolite   +   Industrial
                      Effluents   stirred, pH= 5-7   ⟶   natural zeolite
                                         0.080mmol/g adsorption capacity of As(III),
  (1g/L)   (100mg/L)   0.024mmol/g adsorption capacity of Pb(II),
                                         0.104mmol/g adsorption capacity of Cd(II), and
                                         0.057mmol/g adsorption capacity of Zn(II)

Scheme 10. Removal efficiency of heavy metal ions.

The hydrothermal conversion process was used to create the final zeolite composites. Because of the creation of micropores and mesopores, the BET-specific surface area was found to be 22.1 m^2/g, which was a considerable increase over the bottom ash. Its efficiency was discovered to be better than that of primary clinoptilolite natural zeolite, resulting in larger sorption extents and improved sediment stabilization capacity. Converted ash reduced the concentration of cationic heavy metal in sediment porewater by more than 80% at a dose rate lower than that of natural zeolite. These findings point to a viable path for recycling municipal solid waste, which can significantly improve trash incineration technology's sustainability.

Figueroa-Torres, G. M. et al. fabricated zeolites immobilized with biomass for effectively removing heavy metal ions from water moieties. (Schemes **11, 12**) [39].

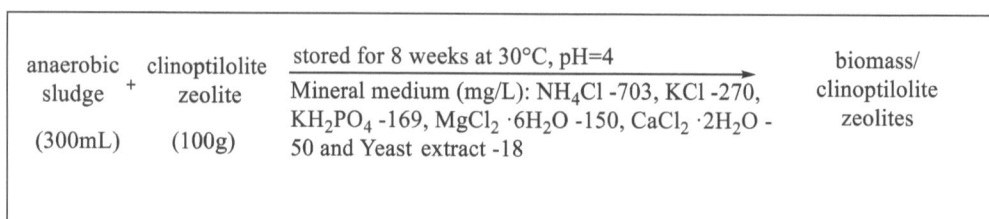

| anaerobic sludge (300mL) | + | clinoptilolite zeolite (100g) | stored for 8 weeks at 30°C, pH=4 $\longrightarrow$ Mineral medium (mg/L): NH_4Cl -703, KCl -270, KH_2PO_4 -169, $MgCl_2 \cdot 6H_2O$ -150, $CaCl_2 \cdot 2H_2O$ - 50 and Yeast extract -18 | biomass/ clinoptilolite zeolites |

Scheme 11. Synthesis of acidogenic biomass immobilized in clinoptilolite.

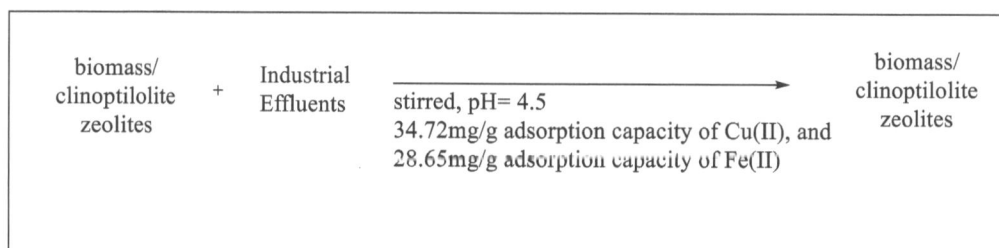

| biomass/ clinoptilolite zeolites | + | Industrial Effluents | stirred, pH= 4.5 $\longrightarrow$ 34.72mg/g adsorption capacity of Cu(II), and 28.65mg/g adsorption capacity of Fe(II) | biomass/ clinoptilolite zeolites |

Scheme 12. Removal efficiency of heavy metal ions.

The pseudo-second-order type process was employed to show how the zeolites absorb metal ions. The Langmuir adsorption model provided an appropriate description of the equilibrium findings. Despite this, the single system's expected maximal metal biosorption capacity was higher than the binary system's. It was discovered that the real biosorption capability of the biomass was reduced by the co-existence of metal ions. The outcomes demonstrated that, although binary metal mixes need to be carefully chosen to prevent suboptimal removal efficiencies in a biosorption-based wastewater treatment process, synthesized zeolites are a potential low-cost biosorbent for the elimination of heavy metals.

Pratti, L. M. et al. designed zeolites that effectively remove metal ions from industrial wastewater. (Schemes **13, 14**) [40].

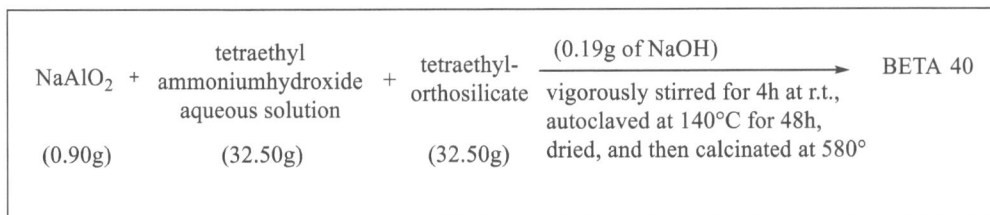

$NaAlO_2$ + tetraethyl ammoniumhydroxide aqueous solution + tetraethyl-orthosilicate $\xrightarrow[\text{dried, and then calcinated at 580°}]{\substack{(0.19g\ of\ NaOH)\\ \text{vigorously stirred for 4h at r.t.,}\\ \text{autoclaved at 140°C for 48h,}}}$ BETA 40

(0.90g)   (32.50g)   (32.50g)

Scheme 13. Synthesis of BETA 40.

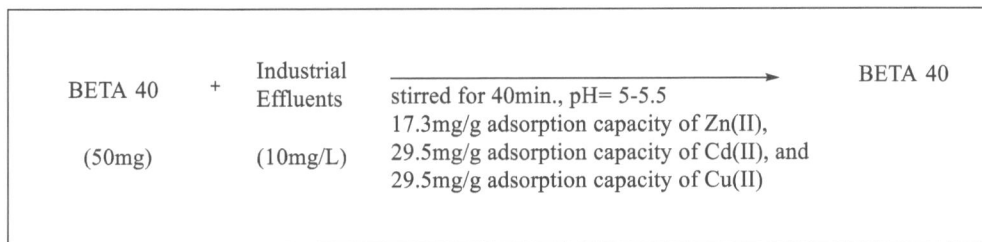

BETA 40 + Industrial Effluents $\xrightarrow[\substack{\text{17.3mg/g adsorption capacity of Zn(II),}\\ \text{29.5mg/g adsorption capacity of Cd(II), and}\\ \text{29.5mg/g adsorption capacity of Cu(II)}}]{\text{stirred for 40min., pH= 5-5.5}}$ BETA 40

(50mg)   (10mg/L)

Scheme 14. Removal efficiency of heavy metal ions.

The solid exhibiting a greater surface-to-volume ratio was also found to have a higher percentage of framework aluminum. An increased starting aluminum concentration has a direct impact on the β-structure's nucleation kinetics, guaranteeing a solid with a high density of ion exchange sites available for metal absorption. Because the synthesized composite sample had a greater framework Al content—that generates ion-exchange sites—along with a pore volume and larger surface area, it performed better in terms of removing heavy metal ions. The electronegativity, hydrated ion diameter, and hydration energy of the metals determine the order of uptake selectivity. The pseudo-second-order model explained the kinetic studies, while the adsorption isotherms were well described by the Langmuir model, indicating that ion exchange occurred on the zeolite monolayer surface throughout the adsorption process. As a result, a solid with certain physicochemical characteristics that maximize the quantity of adsorbed metal and its order of selectivity may be created.

CHEMISTRY OF ZEOLITE COMPOUNDS FOR THE REMOVAL OF ORGANIC DYES

Liu, M. et al. created zeolite composites to effectively degrade harmful organic dyes from wastewater. (Schemes **15-19**) [41].

sodium meta- + sodium + silica sol → (32.48mL) deionized water → NaY zeolite
aluminate   hydroxide   (4.5 mL) directing agent
                                       stirred for 30min., autoclaved at
(4.1g)   (3g)   (25g)   95°C for 24h, and then dried

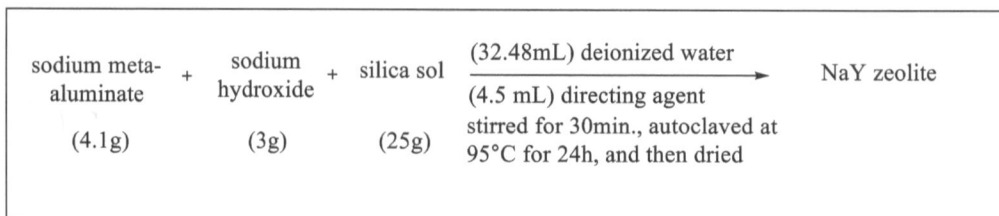

Scheme 15. Synthesis of NaY zeolite.

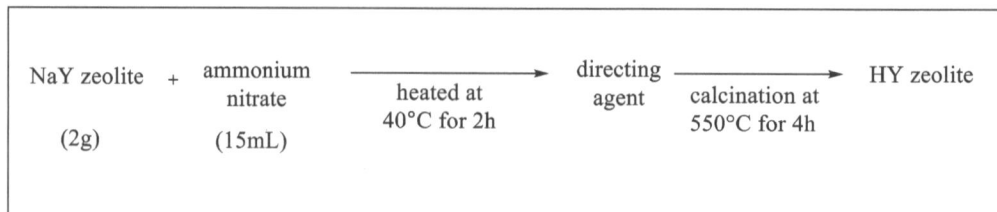

NaY zeolite + ammonium → heated at → directing → calcination at → HY zeolite
              nitrate   40°C for 2h   agent   550°C for 4h
(2g)   (15mL)

Scheme 16. Synthesis of HY zeolite.

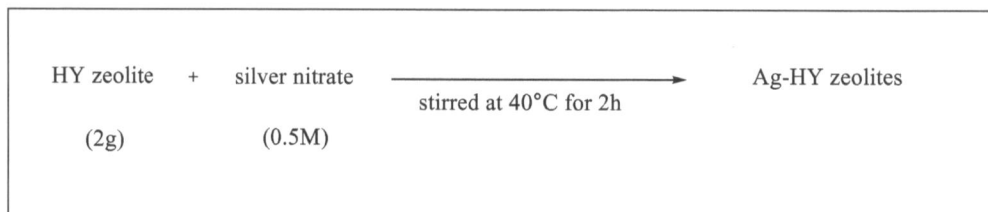

HY zeolite + silver nitrate → stirred at 40°C for 2h → Ag-HY zeolites
(2g)   (0.5M)

Scheme 17. Synthesis of Ag-HY zeolites.

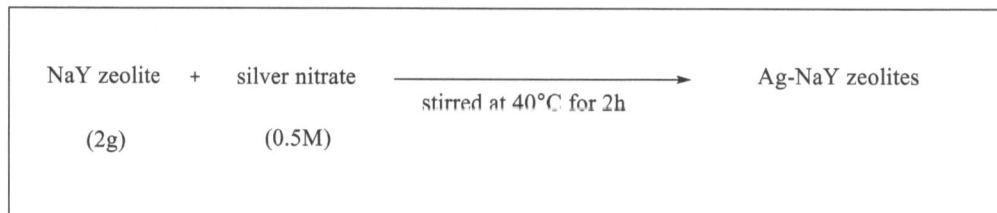

NaY zeolite + silver nitrate → stirred at 40°C for 2h → Ag-NaY zeolites
(2g)   (0.5M)

Scheme 18. Synthesis of Ag-NaY zeolites.

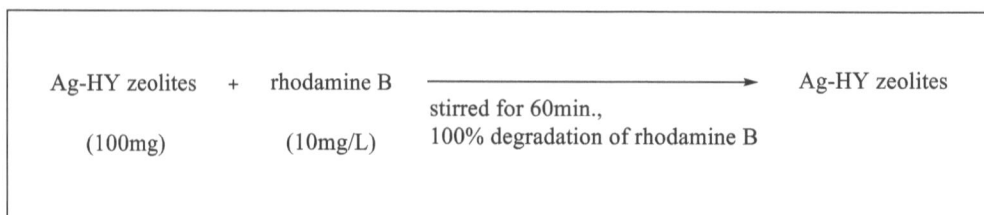

Ag-HY zeolites + rhodamine B → stirred for 60min., → Ag-HY zeolites
                                100% degradation of rhodamine B
(100mg)   (10mg/L)

Scheme 19. Degradation of dyes.

Ion exchange techniques were used to create the zeolite composites in one step and two steps. According to the findings, NaY zeolites were more capable of exchanging ions than HY zeolites. Ag-HY zeolites, on the other hand, have nearly twice as many exterior specific surface areas and mesopores as Ag-NaY zeolites. Additionally, compared to HY, both silver-doped zeolites showed reduced electron-hole recombination and improved absorbance. The photocatalytic activity of zeolites doped with silver was evaluated using a photodegrading dye. It has been suggested that the good zeolite surface adsorption, increased absorbance, and effective separation of photogenerated electron-hole pairs represent a potentially improved photocatalytic mechanism of silver-doped zeolite.

Znad, H. et al. created zeolite composites to degrade organic dyes from wastewater. (Schemes **20, 21**) [42].

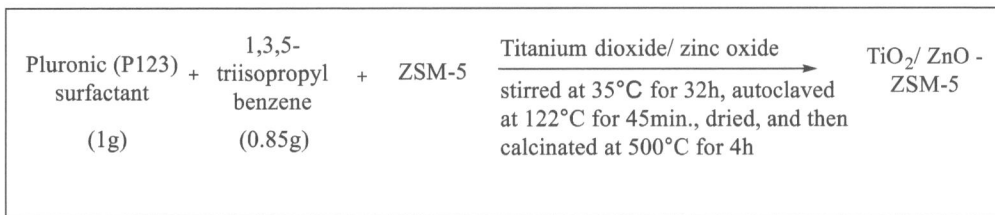

Pluronic (P123) surfactant (1g) + 1,3,5-triisopropyl benzene (0.85g) + ZSM-5 → Titanium dioxide/ zinc oxide, stirred at 35°C for 32h, autoclaved at 122°C for 45min., dried, and then calcinated at 500°C for 4h → TiO_2/ ZnO - ZSM-5

Scheme 20. Synthesis of TiO_2/ZnO-ZSM-5.

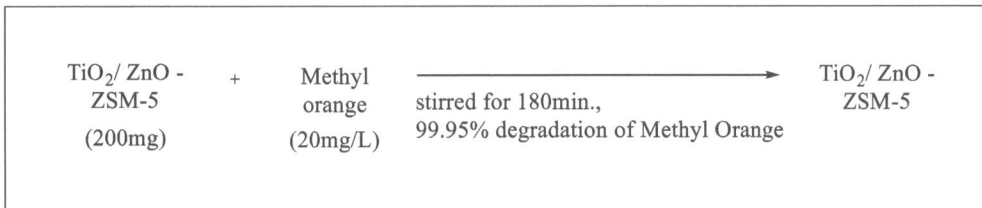

TiO_2/ ZnO - ZSM-5 (200mg) + Methyl orange (20mg/L) → stirred for 180min., 99.95% degradation of Methyl Orange → TiO_2/ ZnO - ZSM-5

Scheme 21. Degradation of dyes.

The direct templating method was used to create the zeolite composites. After the surfactant templates are removed, the mesoporous structure of the zeolite structure is preserved. The unique architecture of multilamellar vesicles can improve light utilization through various internal spaces in an efficient manner. Because of its recyclability, the photocatalysis kinetic fits the pseudo-second-order model quite well and offers a considerable deal of energy savings potential. There was just a little 4.39% decrease in degrading efficiency after the six cycles. As a result, the created composites offer promise as a photocatalysis solution for the quick removal and treatment of colored wastewater.

Aghajari, N. et al. developed zeolite nanocomposites for effective degradation of dye from industrial wastewater. (Schemes **22-25**) [43].

| sodium silicate (3.03g) | + | tetrapropyl ammonium bromide (1.48g) | + | iron (III) nitrate nonahydrate (0.22g) | autoclaved at 170°C for 72h, dried, and then calcinated at 550°C for 4h → | Fe -ZSM -5 zeolites |

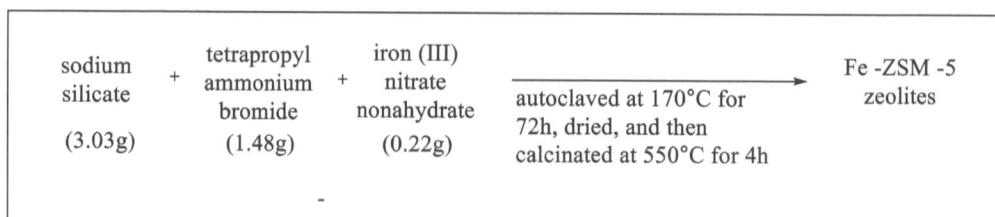

Scheme 22. Synthesis of Fe-ZSM-5 zeolites.

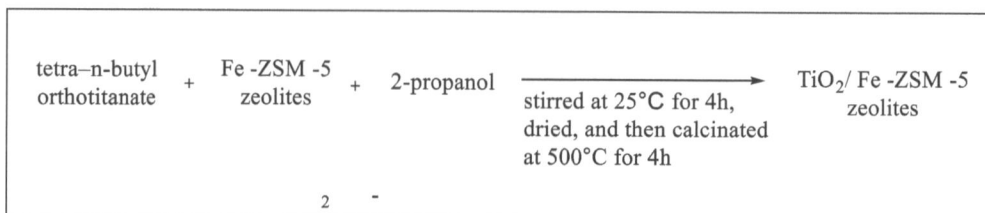

| tetra–n-butyl orthotitanate | + | Fe -ZSM -5 zeolites | + | 2-propanol | stirred at 25°C for 4h, dried, and then calcinated at 500°C for 4h → | TiO$_2$/ Fe -ZSM -5 zeolites |

Scheme 23. Synthesis of TiO$_2$/Fe-ZSM-5 zeolites.

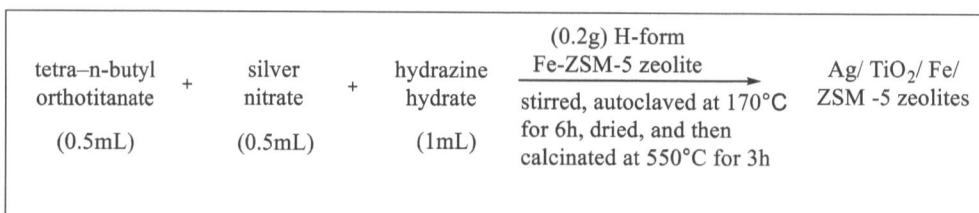

| tetra–n-butyl orthotitanate (0.5mL) | + | silver nitrate (0.5mL) | + | hydrazine hydrate (1mL) | (0.2g) H-form Fe-ZSM-5 zeolite / stirred, autoclaved at 170°C for 6h, dried, and then calcinated at 550°C for 3h → | Ag/ TiO$_2$/ Fe/ ZSM -5 zeolites |

Scheme 24. Synthesis of Ag/TiO$_2$/Fe/ZSM-5 zeolites.

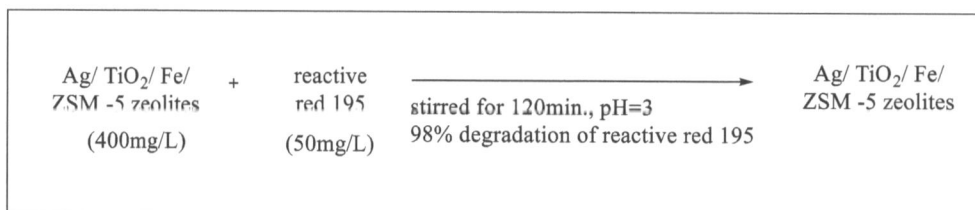

| Ag/ TiO$_2$/ Fe/ ZSM -5 zeolites (400mg/L) | + | reactive red 195 (50mg/L) | stirred for 120min., pH=3 98% degradation of reactive red 195 → | Ag/ TiO$_2$/ Fe/ ZSM -5 zeolites |

Scheme 25. Degradation of dyes.

Sol-gel and hydrothermal techniques were utilized to synthesize the zeolite nanocomposites. It was discovered that the nanocomposites had diameters ranging from 50 to 100 nm. By encouraging a longer charge separation using nanoparticles and trapping the photogenerated electrons under solar radiation, the photocatalytic activity was improved. Increases in the composite's titanium dioxide nanoparticle concentration would lead to the enhancement in the total number of active sites and, consequently, hydroxyl radical production. It was discovered that the synthesized photocatalyst was stable and highly efficient when exposed to sunlight many times.

Qi, H. et al. developed zeolite composites to degrade organic dyes from wastewater. (Schemes **26-29**) [44].

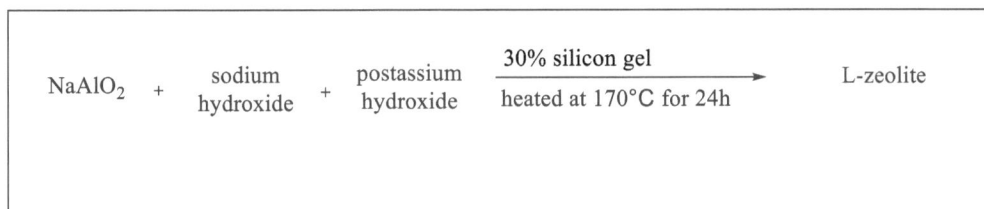

$NaAlO_2$ + sodium hydroxide + postassium hydroxide → (30% silicon gel, heated at 170°C for 24h) → L-zeolite

Scheme 26. Synthesis of L-zeolite.

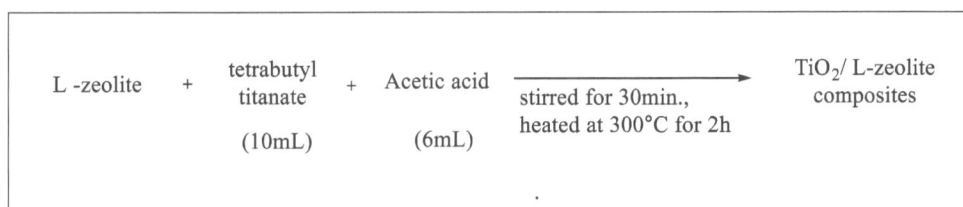

L -zeolite + tetrabutyl titanate (10mL) + Acetic acid (6mL) → (stirred for 30min., heated at 300°C for 2h) → TiO_2/ L-zeolite composites

Scheme 27. Synthesis of TiO_2/L-zeolite composites.

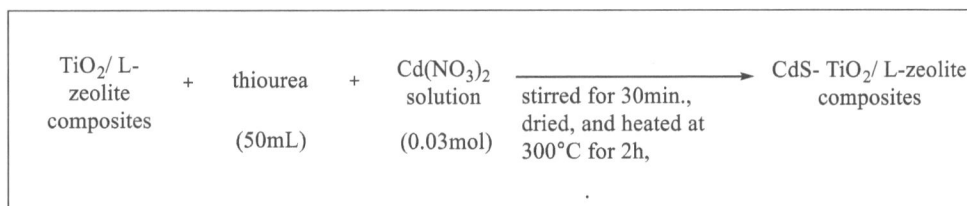

TiO_2/ L-zeolite composites + thiourea (50mL) + $Cd(NO_3)_2$ solution (0.03mol) → (stirred for 30min., dried, and heated at 300°C for 2h,) → CdS- TiO_2/ L-zeolite composites

Scheme 28. Synthesis of CdS-TiO_2/L-zeolite composites.

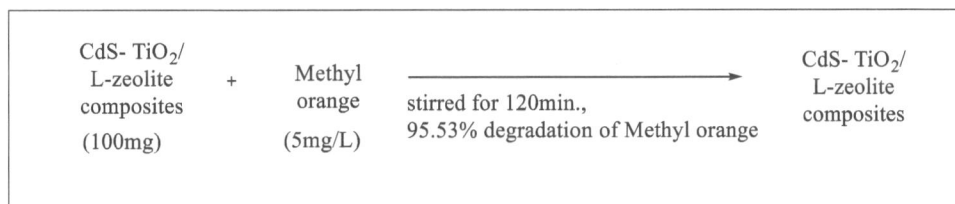

CdS- TiO_2/ L-zeolite composites (100mg) + Methyl orange (5mg/L) → (stirred for 120min., 95.53% degradation of Methyl orange) → CdS- TiO_2/ L-zeolite composites

Scheme 29. Degradation of dyes.

The zeolite composites were made using the sol-gel process. It is found that the composite photocatalyst's photocatalytic kinetics follow the first-order equation. Effective transmission of the trapped electrons to the adsorbed oxygen molecules might result in the formation of the superoxide radical, which then reacts with two protons to generate hydrogen peroxide. Hydroxyl radicals might be produced by further reacting the hydrogen peroxide created with electrons. A dopant supported by zeolite may be able to stop unwanted electron/hole recombination and encourage charge separation, which will help sustain the production of active radicals. The new composite photocatalyst has a bright future in environmental pollution degradation.

Sun, C. et al. developed zeolite composite materials to degrade organic dyes detected in wastewater. (Schemes **30, 31**) [45].

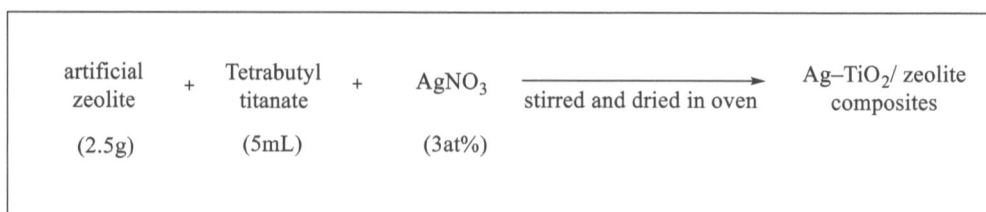

| artificial zeolite | + | Tetrabutyl titanate | + | AgNO$_3$ | stirred and dried in oven → | Ag–TiO$_2$/ zeolite composites |
| (2.5g) | | (5mL) | | (3at%) | | |

Scheme 30. Synthesis of Ag-TiO$_2$/zeolite composites.

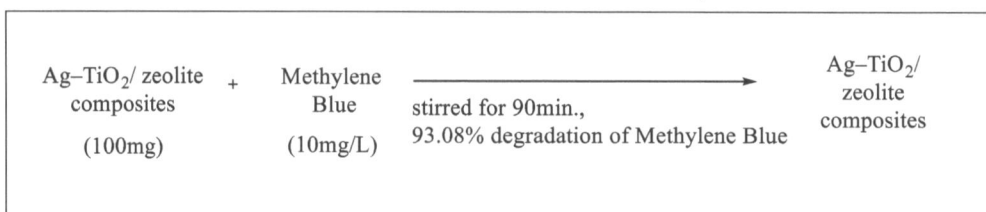

| Ag–TiO$_2$/ zeolite composites | + | Methylene Blue | stirred for 90min., 93.08% degradation of Methylene Blue → | Ag–TiO$_2$/ zeolite composites |
| (100mg) | | (10mg/L) | | |

Scheme 31. Degradation of dyes.

The sol-gel process was used to fabricate the zeolite composite. The synthetic material was discovered to have a loosely porous structure. While the holes remaining on the valence band of silver oxide nanoparticles react with water or OH⁻ to make hydroxyl radicals, the electrons are retained by the adsorbed oxygen forming •O$_2^-$. Eventually, these reactive oxygen species can interact with dye molecules, causing them to break down into smaller molecules like water and CO$_2$. The degradation of dye, whose effectiveness remained at around 89% after five degradations, was used to assess the stability of three percent zeolite composites. Zeolite composites may be considered to be the most effective photocatalysts fortreating the water cycle on a wide scale.

Du, G. et al. created immobilized zeolites to efficiently degrade organic dyes from wastewater. (Schemes **32, 33**) [46].

$Zn(NO_3)_2$ $.6H_2O$ (0.595g) + 2-methyl imidazole (0.657) + zeolite A (0.20g) → Centrifuged, dried and then calcinated at 550°C for 5h → ZnO/ zeolite composites

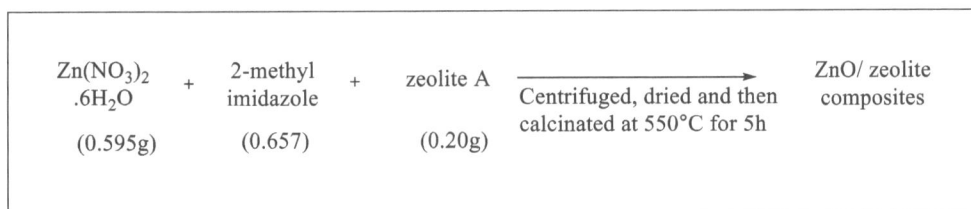

Scheme 32. Synthesis of ZnO/zeolite composites.

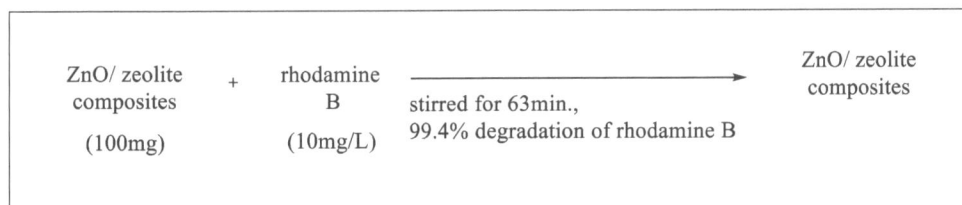

ZnO/ zeolite composites (100mg) + rhodamine B (10mg/L) → stirred for 63min., 99.4% degradation of rhodamine B → ZnO/ zeolite composites

Scheme 33. Degradation of dyes.

The sonochemical synthesis process was used to produce the zeolite-based composite materials. The composite that is calcined at 550°C has the maximum photocatalytic effectiveness of over 99%, according to the dye degradation. The samples' adsorption-desorption isotherms represent typical microporous materials, with virtually horizontal adsorption and desorption curves following the sharp uptake at low relative pressure. The specific surface measured by the BET technique decreased as the temperature rose, most likely as a result of textural porosity created by the stacking of nanoparticles. Following four iterations, the composite's photocatalytic efficiency marginally declined—it fell by only 3.9%.

Vaez, Z. et al. created zeolite nanocomposites for the removal of dyes from wastewater. (Schemes **34-36**) [47].

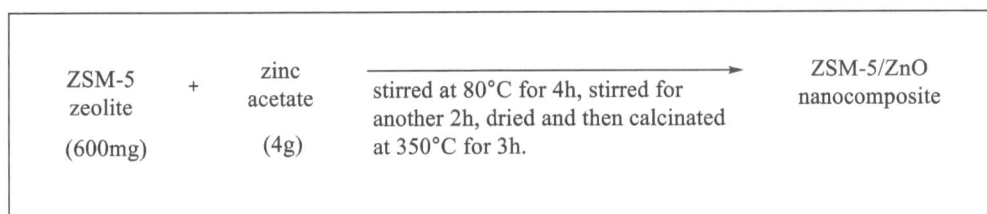

ZSM-5 zeolite (600mg) + zinc acetate (4g) → stirred at 80°C for 4h, stirred for another 2h, dried and then calcinated at 350°C for 3h. → ZSM-5/ZnO nanocomposite

Scheme 34. Synthesis of ZSM-5/ZnO nanocomposite.

Zeolite/ZnO nanocomposite	+	silver nitrate	stirred for 20 min., centrifuged, dried, and then calcinated at 350°C for 3h ⟶	ZSM-5/ZnO/Ag nanocomposites

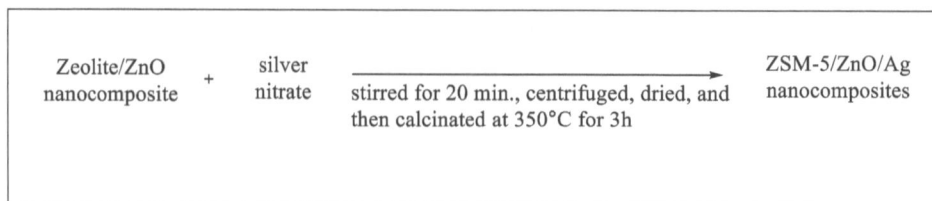

Scheme 35. Synthesis of ZSM-5/ZnO/Ag nanocomposites.

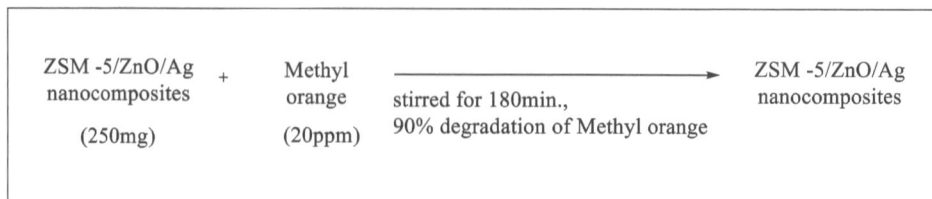

ZSM -5/ZnO/Ag nanocomposites (250mg)	+	Methyl orange (20ppm)	stirred for 180min., 90% degradation of Methyl orange ⟶	ZSM -5/ZnO/Ag nanocomposites

Scheme 36. Degradation of dyes.

The band gap studies demonstrated that adding silver nanoparticles to a synthesized zeolite nanocomposite enhances the nanocomposite's photocatalytic activity. In alkaline *vs* acidic environments, there was more dye degradation. Examining several kinetic models for dye removal revealed that the pseudo-second-order model conformed to the dye degradation process. After four successive cycles, the nano photocatalyst's regeneration and reusability show that the material's photocatalytic activity has not changed significantly.

Suligoj, A. et al. synthesized zeolite composites for the removal of organic and hazardous dyes from wastewater. (Schemes **37-39**) [48].

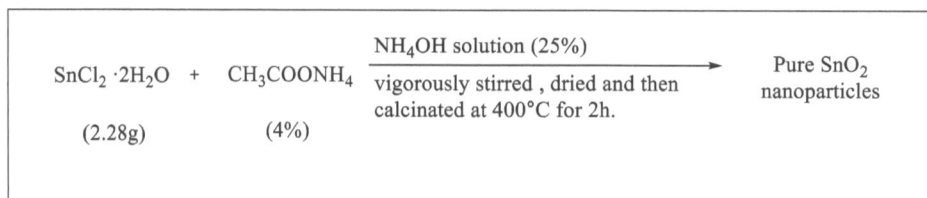

$SnCl_2 \cdot 2H_2O$ (2.28g)	+	CH_3COONH_4 (4%)	NH$_4$OH solution (25%) vigorously stirred , dried and then calcinated at 400°C for 2h. ⟶	Pure SnO_2 nanoparticles

Scheme 37. Synthesis of Pure SnO_2 nanoparticles.

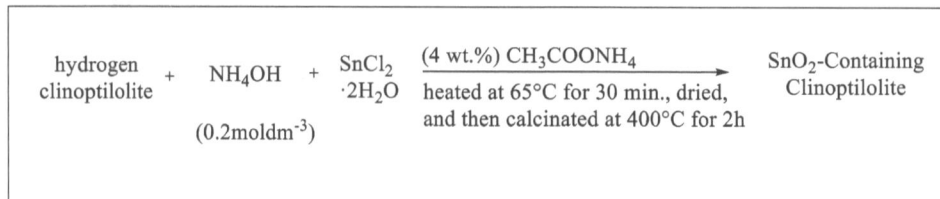

hydrogen clinoptilolite	+	NH_4OH (0.2moldm^{-3})	+	$SnCl_2$ $\cdot 2H_2O$	(4 wt.%) CH_3COONH_4 heated at 65°C for 30 min., dried, and then calcinated at 400°C for 2h ⟶	SnO_2-Containing Clinoptilolite

Scheme 38. Synthesis of SnO$_2$-Containing Clinoptilolite.

SnO$_2$-Containing   +   Methylene   SnO$_2$-Containing
Clinoptilolite   Blue   Clinoptilolite
                                                      stirred for 180min.,
(10mg)   (10mg/L)   45% degradation of Methylene Blue

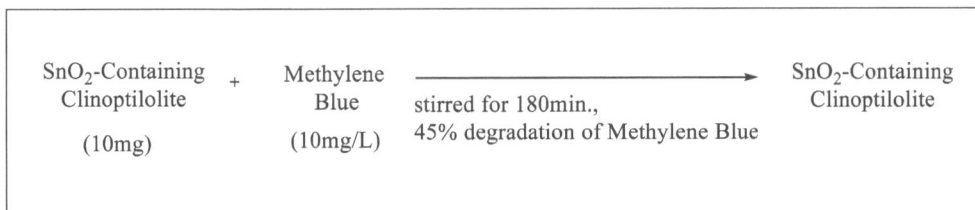

Scheme 39. Degradation of dyes.

Precipitation-deposition was the procedure used to create the zeolite composites. Because of the preferred placement of nanoparticle aggregates on the zeolite surface, there is an enhanced surface area without any disruption of the support, which might be the reason for the higher degrading efficiencies. Higher Sn loadings led to enhanced adsorption as well as increased degradation because the interaction between the zeolite and nanoparticles produced a greater negative surface potential. The composites demonstrated their ability to degrade dye even after being repeatedly recycled and treated with acid.

Abukhadra, M. R. et al. developed zeolite nanocomposites for the removal of organic dyes. (Schemes **40-42**) [49].

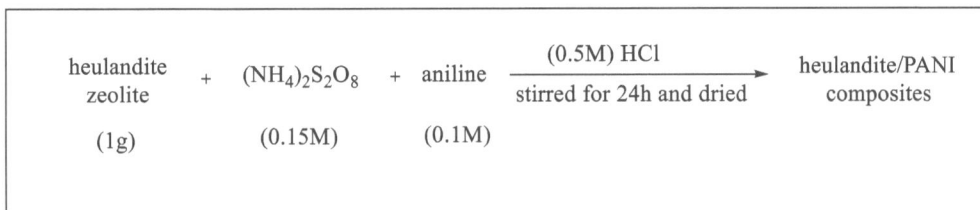

heulandite   +   (NH$_4$)$_2$S$_2$O$_8$ +   aniline $\xrightarrow{\text{(0.5M) HCl}}$ heulandite/PANI
zeolite   stirred for 24h and dried   composites

(1g)   (0.15M)   (0.1M)

Scheme 40. Synthesis of heulandite/PANI composites.

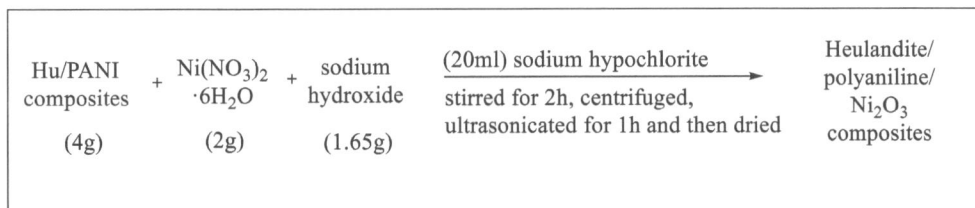

Hu/PANI   +   Ni(NO$_3$)$_2$ +   sodium   (20ml) sodium hypochlorite   Heulandite/
composites   ·6H$_2$O   hydroxide   polyaniline/
                                                                          stirred for 2h, centrifuged,   Ni$_2$O$_3$
(4g)   (2g)   (1.65g)   ultrasonicated for 1h and then dried   composites

Scheme 41. Synthesis of Heulandite/polyaniline/Ni$_2$O$_3$ composites.

Heulandite/polyaniline/   Safranin-T   ⟶   Heulandite/polyaniline/
Ni$_2$O$_3$ composites   +   dye   stirred for 30min.,   Ni$_2$O$_3$ composites

(25mg)   (5mg/L)   100% degradation of Safranin-T

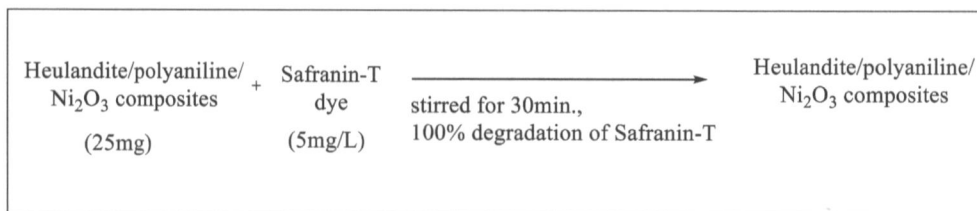

Scheme 42. Degradation of dyes.

By using greater dosages of the composite catalysts or raising the pH of the solution, the increased dye concentrations can be eliminated in a matter of minutes. The alkaline conditions yielded the highest removal percentage. Additionally, after five trials, the composites showed excellent stability, holding 84.5% of their initial photocatalytic effectiveness. Furthermore, various kinds of dyes and combined dyes may be successfully removed from the composite in the same amount of time. Amazing photocatalytic degradation capability was therefore produced by putting nickel oxide onto a hybrid composite as a catalyst support.

Zhang, G. et al. created zeolite nanocomposites to remove dyes present in wastewater. (Schemes **43, 44**) [50].

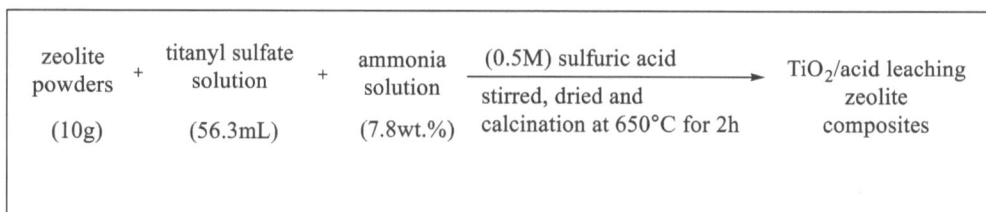

zeolite   titanyl sulfate   ammonia   (0.5M) sulfuric acid   TiO$_2$/acid leaching
powders   +   solution   +   solution   ────────────────   zeolite
                                                            stirred, dried and
(10g)   (56.3mL)   (7.8wt.%)   calcination at 650°C for 2h   composites

Scheme 43. Synthesis of TiO$_2$/acid leaching zeolite composites.

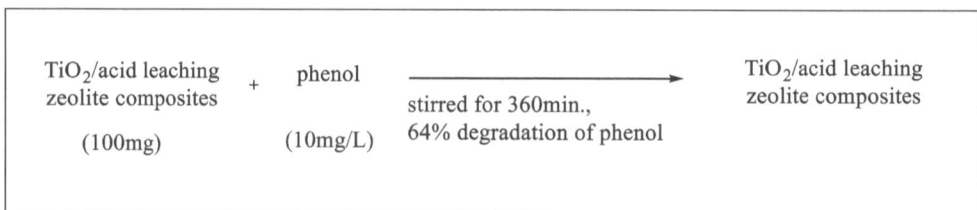

TiO$_2$/acid leaching   phenol   ⟶   TiO$_2$/acid leaching
zeolite composites   +   stirred for 360min.,   zeolite composites

(100mg)   (10mg/L)   64% degradation of phenol

Scheme 44. Degradation of dyes.

The hydrolysis deposition method, in conjunction with the calcination crystallization procedure, was used to create the composites. The larger surface area and number of hydroxyl groups present on the surface are responsible for increased photoactivity. When electron-hole pairs react with oxygen or water, reactive species like $\cdot O_2^-$ and $\cdot OH$ are created. Ultimately, the contaminants break down into CO_2, water, and other small molecules. Composites are a potential photocatalyst in pollutant degradation, as evidenced by their good activity.

Asadollahi, A. et al. produced zeolite nanoparticles to effectively degrade organic dyes. (Schemes **45-49**) [51].

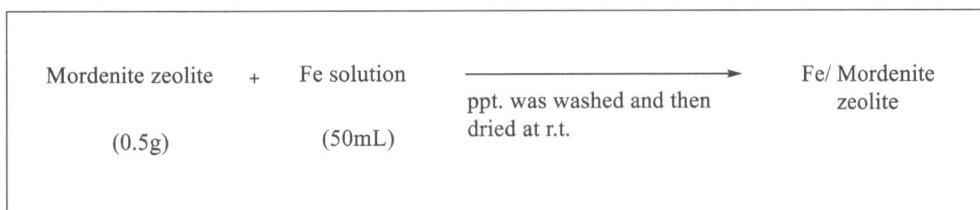

Mordenite zeolite   +   Fe solution $\longrightarrow$ Fe/ Mordenite zeolite

(0.5g)   (50mL)   ppt. was washed and then dried at r.t.

Scheme 45. Synthesis of Fe/MOR as a matrix.

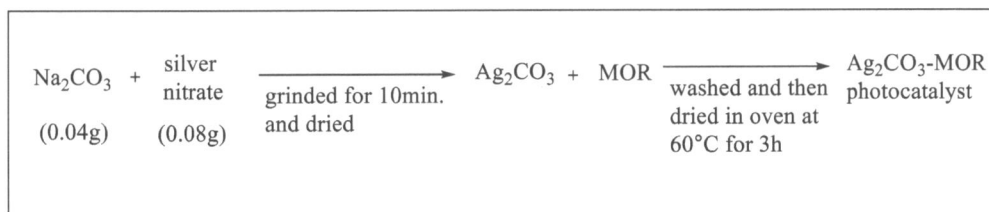

Na_2CO_3 + silver nitrate $\longrightarrow$ Ag_2CO_3 + MOR $\longrightarrow$ Ag_2CO_3-MOR photocatalyst

(0.04g)   (0.08g)   grinded for 10min. and dried   washed and then dried in oven at 60°C for 3h

Scheme 46. Synthesis of Ag_2CO_3-MOR photocatalyst.

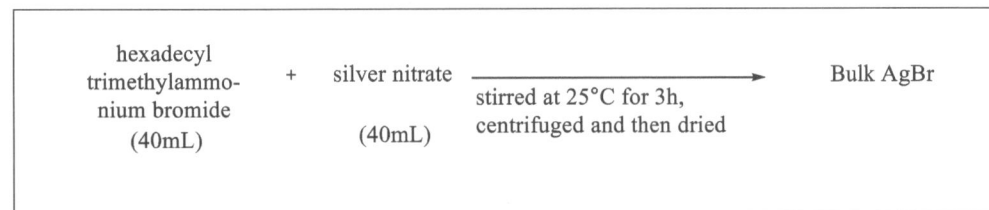

hexadecyl trimethylammonium bromide (40mL)   +   silver nitrate (40mL) $\longrightarrow$ Bulk AgBr

stirred at 25°C for 3h, centrifuged and then dried

Scheme 47. Synthesis of Bulk AgBr.

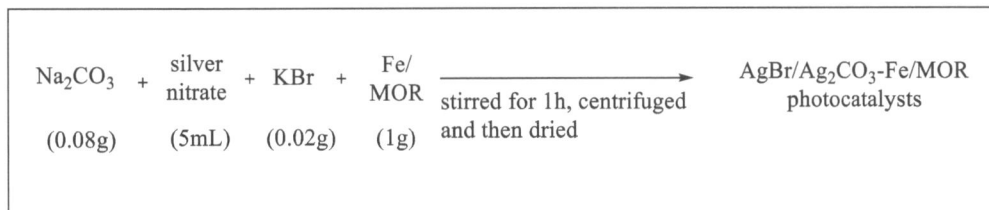

Na_2CO_3 + silver nitrate + KBr + Fe/ MOR $\longrightarrow$ AgBr/Ag_2CO_3-Fe/MOR photocatalysts

(0.08g)   (5mL)   (0.02g)   (1g)   stirred for 1h, centrifuged and then dried

Scheme 48. Synthesis of AgBr/Ag_2CO_3-Fe/MOR photocatalysts.

AgBr/Ag$_2$CO$_3$-Fe/MOR + Methylene ⟶ AgBr/Ag$_2$CO$_3$-Fe/MOR
photocatalysts Blue stirred for 75min., photocatalysts
(70mg) (3.2ppm) 90% degradation of Methylene
Blue

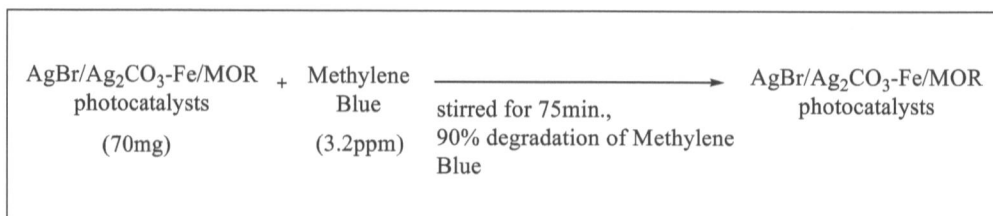

Scheme 49. Degradation of dyes.

The easy precipitation approach was utilized to synthesize the zeolite nanoparticles. The cycle studies conducted on the nanocomposite showed that the cocatalyst increases the photoinduced stability of photosensitive silver compounds in all cycles with regard to MOR and improves photocatalytic activity. The process of photocatalytic oxidation generates a lot of reactive species, including •OH, h$^+$, •O$_2^-$, and hydrogen peroxide. The degradation value dramatically decreases when EDTA is added, suggesting that two main reactive species, h$^+$ and •OH, are invloved, which isencourage the use of other photosensitive silver-based materials with improved photostability and photocatalytic activity.

Phan, T. T. N. et al. produced zeolites composite materials to eliminate organic dyes from wastewater. (Schemes **50-52**) [52].

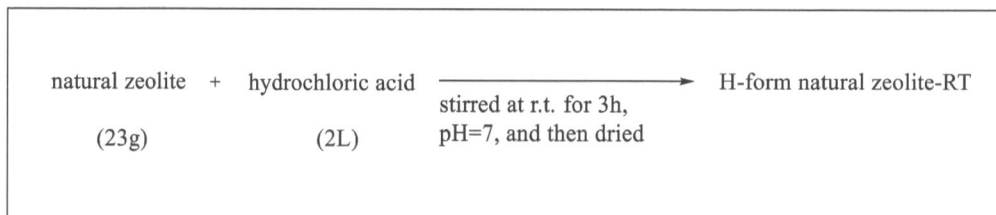

natural zeolite + hydrochloric acid ⟶ H-form natural zeolite-RT
stirred at r.t. for 3h,
(23g) (2L) pH=7, and then dried

Scheme 50. Synthesis of H-form natural zeolite-RT.

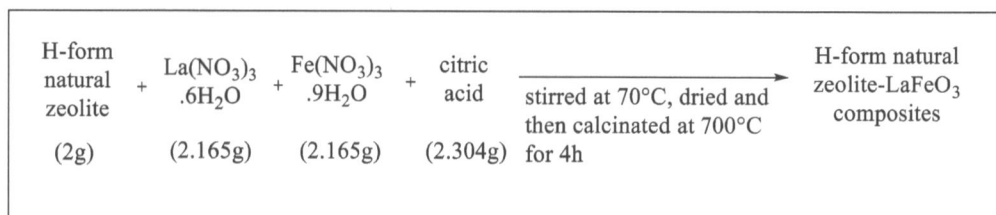

H-form La(NO$_3$)$_3$ Fe(NO$_3$)$_3$ citric ⟶ H-form natural
natural + .6H$_2$O + .9H$_2$O + acid stirred at 70°C, dried and zeolite-LaFeO$_3$
zeolite then calcinated at 700°C composites
(2g) (2.165g) (2.165g) (2.304g) for 4h

Scheme 51. Synthesis of H-form natural zeolite-LaFeO$_3$ composites.

H-form natural zeolite-   + rhodamine   ——————————————————→   H-form natural zeolite-
   LaFeO$_3$ B   stirred for 90min.,   LaFeO$_3$
   (80mg)   (10mg/L)   98.3% degradation of
                                                  rhodamine B

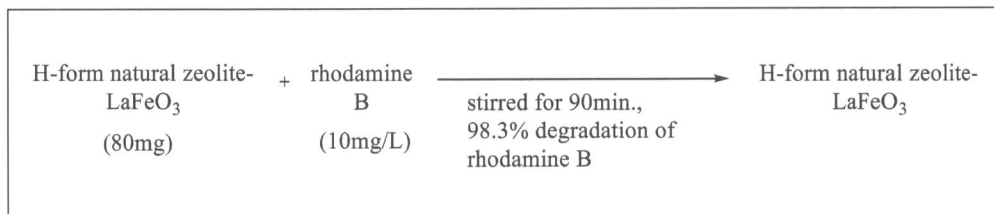

Scheme 52. Degradation of dyes.

The impregnation-calcination method was employed to produce the zeolite composites. This was ascribed to the combined action of the modified zeolite host's strong adsorption capability and the many active sites that perovskite oxides supplied for the photo-Fenton reaction. In the presence of synthetic composites, the elevation in the concentration of hydrogen peroxide boosted the photo-Fenton process, whereas an increase in the concentration of dye inhibited the degradation. The results suggest that the composites may work well as a visible-light-driven Fenton-like catalyst for wastewater treatment.

Aanchal et al. created zeolite nanocomposites to degrade organic pollutants from industrial wastewater. (Schemes **53, 54**) [53].

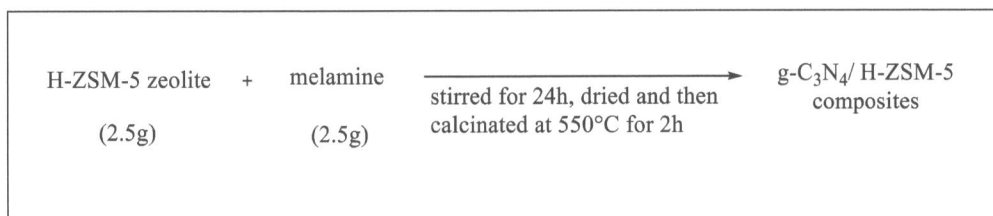

H-ZSM-5 zeolite   +   melamine   ——————————————————→   g-C$_3$N$_4$/ H-ZSM-5
                                              stirred for 24h, dried and then   composites
   (2.5g)   (2.5g)   calcinated at 550°C for 2h

Scheme 53. Synthesis of g-C$_3$N$_4$/ H-ZSM-5 composites.

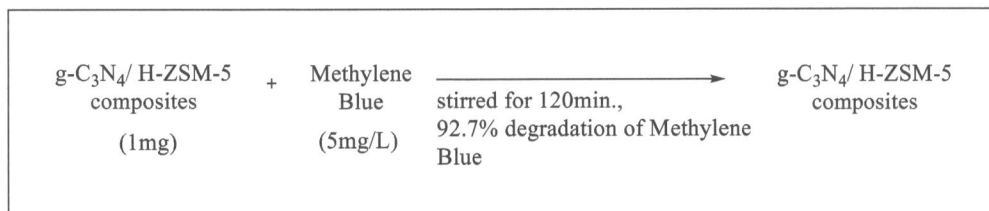

g-C$_3$N$_4$/ H-ZSM-5   +   Methylene   ——————————————————→   g-C$_3$N$_4$/ H-ZSM-5
   composites   Blue   stirred for 120min.,   composites
   (1mg)   (5mg/L)   92.7% degradation of Methylene
                                                  Blue

Scheme 54. Degradation of dyes.

The large surface area and mesopores and micropores of the catalyst were confirmed by the BET N$_2$ adsorption-desorption study. The primary reactive species responsible for dye degradation were determined to be O$_2$•$^-$ and hydroxyl radicals based on the scavenger analysis. These investigations demonstrate the catalyst's visible activity, ease of preparation, and effectiveness as a photocatalyst for the destruction of harmful pollutants. After four cycles, the photocatalyst's activity was determined to be around 70%, indicating that it is quite stable and may be utilized repeatedly.

CONCLUSION

Zeolites-based nanocomposites exhibit significant degradation capability by effectively targeting and eliminating organic dyes and heavy metals from wastewater. The present work highlights the efficacy of 20 distinct zeolite-based nanocomposite materials to effectively remove nine different heavy metals and six organic dyes, including rhodamine B, methyl orange, reactive red 195, methylene blue, safranin T, and phenol. It is evident from the analysis that the majority of these zeolite-based nanocomposites exhibit substantial adsorption capabilities for heavy metals, ranging from 17.3 mg/g to 649 mg/g for different metals. Moreover, the nanocomposites effectively degrade over 90% of the organic dyes, with most of them achieving near-complete removal of ~98-100% over time. These results highlight the possibility of zeolite-based nanocomposites as effective and efficient tools for addressing water pollution challenges, paving the way for their broader adoption in wastewater treatment practices.

LIST OF ABBREVIATIONS

BET Technique	Brunauer–Emmett–Teller Technique
EDTA	Ethylene Diamine Tetra Acetic acid
Hu	Heulandite
MORs	Mordenites
RT	Room Temperature
ZSM-5	Zeolite Socony Mobil–5

ACKNOWLEDGEMENTS

The authors would like to express their sincere thanks to the management of SGT University, Gurugram, Haryana, India, and Kurukshetra University, Kurukshetra, Haryana, India, for providing the facilities to write and submit the book chapter for publication.

REFERENCES

[1]   Isawi, H. Using Zeolite/Polyvinyl alcohol/sodium alginate nanocomposite beads for removal of some heavy metals from wastewater. *Arab. J. Chem.,* **2020**, *13*(6), 5691-5716.
[http://dx.doi.org/10.1016/j.arabjc.2020.04.009]

[2]   Mekki, A.; Mokhtar, A.; Hachemaoui, M.; Beldjilali, M.; Meliani, M.; Zahmani, H.H.; Hacini, S.; Boukoussa, B. Fe and Ni nanoparticles-loaded zeolites as effective catalysts for catalytic reduction of organic pollutants. *Microporous Mesoporous Mater.,* **2021**, *310*, 110597.
[http://dx.doi.org/10.1016/j.micromeso.2020.110597]

[3]   Medeiros-Costa, I. C.; Dib, E.; Nesterenko, N.; Dath, J.-P.; Gilson, J.-P.; Mintova, S. Silanol defects engineering and healing in zeolites: Opportunities to fine-tune their properties and performances. **2021**.
[http://dx.doi.org/10.1039/D1CS00395Ji]

[4]   Fan, J.; Wu, H.; Liu, R.; Meng, L.; Fang, Z.; Liu, F.; Xu, Y. Non-thermal plasma combined with zeolites to remove ammonia nitrogen from wastewater. *J. Hazard. Mater.*, **2021**, *401*, 123627.
[http://dx.doi.org/10.1016/j.jhazmat.2020.123627] [PMID: 33113719]

[5]   Król, M. Natural *vs.* Synthetic Zeolites. *Crystals. MDPI AG,* **2020**, 1-8.
[http://dx.doi.org/10.3390/cryst10070622]

[6]   Ali, M. E.; Hoque, M. E.; Safdar Hossain, S. K.; Biswas, M. C. Nanoadsorbents for wastewater treatment: Next generation biotechnological solution. *International Journal of Environmental Science and Technology. Springer,* **2020**, 4095-4132.
[http://dx.doi.org/10.1007/s13762-020-02755-4]

[7]   Jain, K.; Patel, A.S.; Pardhi, V.P.; Flora, S.J.S. *Nanotechnology in Wastewater Management: A New Paradigm towards Wastewater Treatment. Molecules*; MDPI AG, **2021**.
[http://dx.doi.org/10.3390/molecules26061797]

[8]   Khandaker, S.; Toyohara, Y.; Saha, G.C.; Awual, M.R.; Kuba, T. Development of synthetic zeolites from bio-slag for cesium adsorption: Kinetic, isotherm and thermodynamic studies. *J. Water Process Eng.,* **2020**, *33*, 101055.
[http://dx.doi.org/10.1016/j.jwpe.2019.101055]

[9]   Al-Sheikh, F.; Moralejo, C.; Pritzker, M.; Anderson, W.A.; Elkamel, A. Batch adsorption study of ammonia removal from synthetic/real wastewater using ion exchange resins and zeolites. *Sep. Sci. Technol.,* **2021**, *56*(3), 462-473.
[http://dx.doi.org/10.1080/01496395.2020.1718706]

[10]   Smiljanić, D.; de Gennaro, B.; Izzo, F.; Langella, A.; Daković, A.; Germinario, C.; Rottinghaus, G.E.; Spasojević, M.; Mercurio, M. Removal of emerging contaminants from water by zeolite-rich composites: A first approach aiming at diclofenac and ketoprofen. *Microporous Mesoporous Mater.,* **2020**, *298*, 110057.
[http://dx.doi.org/10.1016/j.micromeso.2020.110057]

[11]   Angaru, G.K.R.; Choi, Y.L.; Lingamdinne, L.P.; Choi, J.S.; Kim, D.S.; Koduru, J.R.; Yang, J.K.; Chang, Y.Y. Facile synthesis of economical feasible fly ash–based zeolite–supported nano zerovalent iron and nickel bimetallic composite for the potential removal of heavy metals from industrial effluents. *Chemosphere,* **2021**, *267*, 128889.
[http://dx.doi.org/10.1016/j.chemosphere.2020.128889] [PMID: 33187656]

[12]   Alabbad, E.A. Efficacy assessment of natural zeolite containing wastewater on the adsorption behaviour of Direct Yellow 50 from; equilibrium, kinetics and thermodynamic studies. *Arab. J. Chem.,* **2021**, *14*(4), 103041.
[http://dx.doi.org/10.1016/j.arabjc.2021.103041]

[13]   López-Delgado, A.; Robla, J. I. Zero-Waste Process for the Transformation of a Hazardous Aluminum Waste into a Raw Material to Obtain Zeolites. **2020**.
[http://dx.doi.org/10.1016/j.jclepro.2020.120178]

[14]   Jiang, N.; Shang, R.; Heijman, S.G.J.; Rietveld, L.C. Adsorption of triclosan, trichlorophenol and phenol by high-silica zeolites: Adsorption efficiencies and mechanisms. *Separ. Purif. Tech.,* **2020**, *235*, 116152.
[http://dx.doi.org/10.1016/j.seppur.2019.116152]

[15]   Irannajad, M.; Kamran Haghighi, H. Removal of heavy metals from polluted solutions by zeolitic adsorbents: A review. In: *Environmental Processes*; Springer Science and Business Media Deutschland GmbH, **2021**; pp. 7-35.
[http://dx.doi.org/10.1007/s40710-020-00476-x]

[16]   Kumari, S.; Chowdhry, J.; Kumar, M.; Chandra Garg, M. Zeolites in wastewater treatment: A comprehensive review on scientometric analysis, adsorption mechanisms, and future prospects. *Environ. Res.,* **2024,** *260,* 119782.
[http://dx.doi.org/10.1016/j.envres.2024.119782] [PMID: 39142462]

[17]   Prada-Vásquez, M.A.; Simarro-Gimeno, C.; Vidal-Barreiro, I.; Cardona-Gallo, S.A.; Pitarch, E.; Hernández, F.; Torres-Palma, R.A.; Chica, A.; Navarro-Laboulais, J. Application of catalytic ozonation using Y zeolite in the elimination of pharmaceuticals in effluents from municipal wastewater treatment plants. *Sci. Total Environ.,* **2024,** *925,* 171625.
[http://dx.doi.org/10.1016/j.scitotenv.2024.171625] [PMID: 38467258]

[18]   Tufail, M.K.; Ifrahim, M.; Rashid, M.; Ul Haq, I.; Asghar, R.; Uthappa, U.T.; Selvaraj, M.; Kurkuri, M. Chemistry of zeolites and zeolite based composite membranes as a cutting-edge candidate for removal of organic dyes & heavy metal ions: Progress and future directions. *Separ. Purif. Tech.,* **2025,** *354,* 128739.
[http://dx.doi.org/10.1016/j.seppur.2024.128739]

[19]   Tsotetsi, N.; Nomngongo, P.; Mekuto, L. Synthesis, modification and characterization of nano-zeolites from coal fly ash for the removal of sulfates in wastewater. *Nano-Structures & Nano-Objects,* **2024,** *37,* 101088.
[http://dx.doi.org/10.1016/j.nanoso.2023.101088]

[20]   Abbas, S.M.; Al-Jubouri, S.M. High performance and antifouling zeolite@polyethersulfone/cellulose acetate asymmetric membrane for efficient separation of oily wastewater. *J. Environ. Chem. Eng.,* **2024,** *12*(3), 112775.
[http://dx.doi.org/10.1016/j.jece.2024.112775]

[21]   Machado, R.C.; Valle, S.F.; Sena, T.B.M.; Perrony, P.E.P.; Bettiol, W.; Ribeiro, C. Aluminosilicate and zeolitic materials synthesis using alum sludge from water treatment plants: Challenges and perspectives. *Waste Manag.,* **2024,** *186,* 94-108.
[http://dx.doi.org/10.1016/j.wasman.2024.05.046] [PMID: 38870604]

[22]   Yang, L.; Zhao, Y.; Liu, Y.; Zhang, W. Sol–gel synthesis of B-TiO2(20%)/HZSM-5 composite photocatalyst for azophloxine degradation. *J. Sol-Gel Sci. Technol.,* **2020,** *93*(2), 371-379.
[http://dx.doi.org/10.1007/s10971-019-05153-6]

[23]   Rastegar Koohi, S.; Allahyari, S.; Kahforooshan, D.; Rahemi, N.; Tasbihi, M. Natural minerals as support of silicotungstic acid for photocatalytic degradation of methylene blue in wastewater. *J. Inorg. Organomet. Polym. Mater.,* **2019,** *29*(2), 365-377.
[http://dx.doi.org/10.1007/s10904-018-1007-4]

[24]   Moosavifar, M.; Bagheri, S. Photocatalytic performance of $H_6 P_2 W_{18} O_{62}/TiO_2$ nanocomposite encapsulated into beta zeolite under uv irradiation in the degradation of methyl orange. *Photochem. Photobiol.,* **2019,** *95*(2), 532-542.
[http://dx.doi.org/10.1111/php.13015] [PMID: 30225987]

[25]   Dzinun, H.; Othman, M.H.D.; Ismail, A.F. Photocatalytic performance of TiO_2/Clinoptilolite: Comparison study in suspension and hybrid photocatalytic membrane reactor. *Chemosphere,* **2019,** *228,* 241-248.
[http://dx.doi.org/10.1016/j.chemosphere.2019.04.118] [PMID: 31035161]

[26]   Kumar, A.; Samanta, S.; Srivastava, R. Systematic investigation for the photocatalytic applications of carbon nitride/porous zeolite heterojunction. *ACS Omega,* **2018,** *3*(12), 17261-17275.
[http://dx.doi.org/10.1021/acsomega.8b01545]

[27]   Gong, P.; Li, B.; Kong, X.; Shakeel, M.; Liu, J.; Zuo, S. Hybriding hierarchical zeolite with Pt nanoparticles and graphene: Ternary nanocomposites for efficient visible-light photocatalytic degradation of methylene blue. *Microporous Mesoporous Mater.,* **2018,** *260,* 180-189.
[http://dx.doi.org/10.1016/j.micromeso.2017.10.029]

[28]   Rajabi, S.K.; Sohrabnezhad, S. Synthesis and characterization of magnetic core with two shells:

Mordenite zeolite and CuO to form Fe 3 O 4 @MOR@CuO core-shell: As a visible light driven photocatalyst. *Microporous Mesoporous Mater.,* **2017**, *242*, 136-143.
[http://dx.doi.org/10.1016/j.micromeso.2017.01.024]

[29]   Djebli, K.; Tebani, H.; Abdessemed, A.; Keghouche, N. Structural, optical and photocatalytic properties of ZnS/zeolite Y nanoparticles synthesized by γ-ray irradiation. *Mater. Sci. Semicond. Process.,* **2019**, *103*, 104599.
[http://dx.doi.org/10.1016/j.mssp.2019.104599]

[30]   Liaquat, I.; Munir, R.; Abbasi, N.A.; Sadia, B.; Muneer, A.; Younas, F.; Sardar, M.F.; Zahid, M.; Noreen, S. Exploring zeolite-based composites in adsorption and photocatalysis for toxic wastewater treatment: Preparation, mechanisms, and future perspectives. *Environ. Pollut.,* **2024**, *349*, 123922.
[http://dx.doi.org/10.1016/j.envpol.2024.123922] [PMID: 38580064]

[31]   Alberti, S.; Caratto, V.; Peddis, D.; Belviso, C.; Ferretti, M. Synthesis and characterization of a new photocatalyst based on TiO2 nanoparticles supported on a magnetic zeolite obtained from iron and steel industrial waste. *J. Alloys Compd.,* **2019**, *797*, 820-825.
[http://dx.doi.org/10.1016/j.jallcom.2019.05.098]

[32]   Emdadi, L.; Tran, D.T.; Zhang, J.; Wu, W.; Song, H.; Gan, Q.; Liu, D. Synthesis of titanosilicate pillared MFI zeolite as an efficient photocatalyst. *RSC Advances,* **2017**, *7*(6), 3249-3256.
[http://dx.doi.org/10.1039/C6RA23959E]

[33]   Nassar, M.Y.; Abdelrahman, E.A. Hydrothermal tuning of the morphology and crystallite size of zeolite nanostructures for simultaneous adsorption and photocatalytic degradation of methylene blue dye. *J. Mol. Liq.,* **2017**, *242*, 364-374.
[http://dx.doi.org/10.1016/j.molliq.2017.07.033]

[34]   Sivalingam, S.; Sen, S. Swift sono-hydrothermal synthesis of pure NaX nanocrystals with improved sorption capacity from industrial resources. *Appl. Surf. Sci.,* **2019**, *463*, 190-196.
[http://dx.doi.org/10.1016/j.apsusc.2018.08.019]

[35]   Ahmad, R.; Mirza, A. Adsorptive removal of heavy metals and anionic dye from aqueous solution using novel Xanthan gum-Glutathione/ Zeolite bionanocomposite. *Groundw. Sustain. Dev.,* **2018**, *7*, 305-312.
[http://dx.doi.org/10.1016/j.gsd.2018.07.002]

[36]   Bao, W.; Liu, L.; Zou, H.; Gan, S.; Xu, X.; Ji, G.; Gao, G.; Zheng, K. Removal of Cu²⁺ from aqueous solutions using Na-A zeolite from oil shale ash. *Chin. J. Chem. Eng.,* **2013**, *21*(9), 974-982.
[http://dx.doi.org/10.1016/S1004-9541(13)60529-7]

[37]   Chen, M.; Nong, S.; Zhao, Y.; Riaz, M.S.; Xiao, Y.; Molokeev, M.S.; Huang, F. Renewable P-type zeolite for superior absorption of heavy metals: Isotherms, kinetics, and mechanism. *Sci. Total Environ.,* **2020**, *726*, 138535.
[http://dx.doi.org/10.1016/j.scitotenv.2020.138535] [PMID: 32304944]

[38]   Chiang, Y.W.; Ghyselbrecht, K.; Santos, R.M.; Meesschaert, B.; Martens, J.A. Synthesis of zeolitic-type adsorbent material from municipal solid waste incinerator bottom ash and its application in heavy metal adsorption. *Catal. Today,* **2012**, *190*(1), 23-30.
[http://dx.doi.org/10.1016/j.cattod.2011.11.002]

[39]   Figueroa-Torres, G.M.; Certucha-Barragán, M.T.; Acedo-Félix, E.; Monge-Amaya, O.; Almendariz-Tapia, F.J.; Gasca-Estefanía, L.A. Kinetic studies of heavy metals biosorption by acidogenic biomass immobilized in clinoptilolite. *J. Taiwan Inst. Chem. Eng.,* **2016**, *61*, 241-246.
[http://dx.doi.org/10.1016/j.jtice.2015.12.018]

[40]   Pratti, L.M.; Reis, G.M.; dos Santos, F.S.; Gonçalves, G.R.; Freitas, J.C.C.; de Pietre, M.K. Effects of textural and chemical properties of β-zeolites on their performance as adsorbents for heavy metals removal. *Environ. Earth Sci.,* **2019**, *78*(17), 553.
[http://dx.doi.org/10.1007/s12665-019-8568-6]

[41]   Liu, M.; Ren, Y.; Wu, J.; Wang, Y.; Chen, J.; Lei, X.; Zhu, X. Effect of cations on the structure,

physico-chemical properties and photocatalytic behaviors of silver-doped zeolite Y. *Microporous Mesoporous Mater.,* **2020**, *293*, 109800.
[http://dx.doi.org/10.1016/j.micromeso.2019.109800]

[42]   Znad, H.; Abbas, K.; Hena, S.; Awual, M.R. Synthesis a novel multilamellar mesoporous TiO$_2$/ZSM-5 for photo-catalytic degradation of methyl orange dye in aqueous media. *J. Environ. Chem. Eng.,* **2018**, *6*(1), 218-227.
[http://dx.doi.org/10.1016/j.jece.2017.11.077]

[43]   Aghajari, N.; Ghasemi, Z.; Younesi, H.; Bahramifar, N. Synthesis, characterization and photocatalytic application of Ag-doped Fe-ZSM-5@TiO$_2$ nanocomposite for degradation of reactive red 195 (RR 195) in aqueous environment under sunlight irradiation. *J. Environ. Health Sci. Eng.,* **2019**, *17*(1), 219-232.
[http://dx.doi.org/10.1007/s40201-019-00342-5] [PMID: 31321045]

[44]   Qi, H.; Liu, H.; Zhang, L.; Wu, J. Photodegradation of methyl orange over CdS–TiO$_2$/L-zeolite composite photocatalyst. *J. Inorg. Organomet. Polym. Mater.,* **2019**, *29*(2), 564-571.
[http://dx.doi.org/10.1007/s10904-018-1031-4]

[45]   Sun, C.; He, P.; Pan, G.; Miao, Y.; Zhang, T.; Zhang, L. Study on preparation and visible-light activity of Ag–TiO$_2$ supported by artificial zeolite. *Res. Chem. Intermed.,* **2018**, *44*(4), 2607-2620.
[http://dx.doi.org/10.1007/s11164-017-3249-0]

[46]   Du, G.; Feng, P.; Cheng, X.; Li, J.; Luo, X. Immobilizing of ZIF-8 derived ZnO with controllable morphologies on zeolite A for efficient photocatalysis. *J. Solid State Chem.,* **2017**, *255*, 215-218.
[http://dx.doi.org/10.1016/j.jssc.2017.07.035]

[47]   Vaez, Z.; Javanbakht, V. Synthesis, characterization and photocatalytic activity of ZSM-5/ZnO nanocomposite modified by Ag nanoparticles for methyl orange degradation. *J. Photochem. Photobiol. Chem.,* **2020**, *388*, 112064.
[http://dx.doi.org/10.1016/j.jphotochem.2019.112064]

[48]   Šuligoj, A.; Pavlović, J.; Arčon, I.; Rajić, N.; Novak Tušar, N. SnO$_2$-containing clinoptilolite as a composite photocatalyst for dyes removal from wastewater under solar light. *Catalysts,* **2020**, *10*(2), 253.
[http://dx.doi.org/10.3390/catal10020253]

[49]   Abukhadra, M.R.; Shaban, M.; Abd El Samad, M.A. Enhanced photocatalytic removal of Safranin-T dye under sunlight within minute time intervals using heulandite/polyaniline@ nickel oxide composite as a novel photocatalyst. *Ecotoxicol. Environ. Saf.,* **2018**, *162*, 261-271.
[http://dx.doi.org/10.1016/j.ecoenv.2018.06.081] [PMID: 29990739]

[50]   Zhang, G.; Song, A.; Duan, Y.; Zheng, S. Enhanced photocatalytic activity of TiO$_2$/zeolite composite for abatement of pollutants. *Microporous Mesoporous Mater.,* **2018**, *255*, 61-68.
[http://dx.doi.org/10.1016/j.micromeso.2017.07.028]

[51]   Asadollahi, A.; Sohrabnezhad, S.; Ansari, R. Enhancement of photocatalytic activity and stability of Ag$_2$CO$_3$ by formation of AgBr/Ag$_2$CO$_3$ heterojunction in mordenite zeolite. *Adv. Powder Technol.,* **2017**, *28*(1), 304-313.
[http://dx.doi.org/10.1016/j.apt.2016.10.004]

[52]   Phan, T.T.N.; Nikoloski, A.N.; Bahri, P.A.; Li, D. Enhanced removal of organic using LaFeO$_3$-integrated modified natural zeolites *via* heterogeneous visible light photo-Fenton degradation. *J. Environ. Manage.,* **2019**, *233*, 471-480.
[http://dx.doi.org/10.1016/j.jenvman.2018.12.051] [PMID: 30593006]

[53]   Aanchal, ; Barman, S.; Basu, S. Complete removal of endocrine disrupting compound and toxic dye by visible light active porous g-C3N4/H-ZSM-5 nanocomposite. *Chemosphere,* **2020**, *241*, 124981.
[http://dx.doi.org/10.1016/j.chemosphere.2019.124981]

SUBJECT INDEX

A

Agglomeration 48, 50, 56,63, 72, 74, 85
Alginate-silicate beads 58
Amination 100
Analytical tools 84, 85
Anisotropic 86
Antibacterial activity 71, 72, 76
Asymmetric membranes 214
Attapulgite 61, 62, 128
Azo dye 34, 35, 38, 39, 40, 41, 42, 43, 44, 46, 81, 83, 188
Auxochromes 138

B

Brilliant green 141, 142
Bioaccumulation 47, 199
Biodiversity 47, 181
Biofouling 191, 224, 226
Bio-molecular solids 224
Biosorbents 17, 137, 156, 181, 191
Biotechnology 1, 95, 223

C

Carbonaceous nanoparticle 4, 14, 137
Carbon 3, 12, 13, 16, 17, 40, 48, 87, 93, 94, 95, 97, 98, 99, 100, 101, 102, 103, 104, 105, 106, 107, 108, 110, 111, 137, 138, 150, 191, 196, 198, 199, 217, 218
  nanotubes 3, 12, 13, 16, 17, 40, 48, 87, 93, 94, 95, 97, 98, 99, 100, 101, 102, 103, 104, 105, 106, 107, 108, 110, 111, 191, 196, 198, 199, 217, 218
  quantum dots 137, 138, 150

Carbonization 57
Carboxymethyl cellulose 53
Catalytic degradation 71, 81, 93, 141, 143, 144, 165, 166, 235, 250
Chemical fouling 226
Chromophores 138
Coagulation 4, 7, 8, 9, 48, 183, 184, 213, 224, 225, 226
Coagulants 4, 5, 183, 184
  traditional aluminous 5
  chemical 183
Colloidal 7, 106, 225
  particles 225
  residues 7
  suspension 106
Composite formulations 93
Congo red 14, 39, 43, 45, 78, 79, 85, 86, 157
Co-precipitation 51, 75, 106, 156, 159
Crystalline 21, 39, 42, 49, 53, 71, 74, 75, 77, 78, 79, 81, 82, 86, 119, 141, 146, 234, 236
  morphology 86
  phases 53
  structure 49, 81
  nature 71, 74, 75

D

Desalination, 3, 87, 210, 211, 215, 217, 218, 225
Divalent ions 210, 211, 212
Dye degradation 34, 35, 36, 42, 63, 70, 84, 86, 117, 118, 137, 139, 142, 143, 145, 146, 148, 163, 168, 172, 190, 247, 248, 253

www.ingramcontent.com/pod-product-compliance
Lightning Source LLC
Chambersburg PA
CBHW050818220326
41598CB00006B/247